The ELEMENTS *of* POWER

The
ELEMENTS
of POWER

A STORY OF WAR, TECHNOLOGY, AND THE DIRTIEST SUPPLY CHAIN ON EARTH

NICOLAS NIARCHOS

PENGUIN PRESS   NEW YORK   2026

PENGUIN PRESS
An imprint of Penguin Random House LLC
1745 Broadway, New York, NY 10019
penguinrandomhouse.com

Brief portions of this work originally appeared, in different form,
in articles published in *The New Yorker* from 2018 to 2025.

Image credits appear on page 445.

Book design and map illustration by Daniel Lagin

LIBRARY OF CONGRESS CONTROL NUMBER: 2025009076
ISBN 9780593492017 (hardcover)
ISBN 9780593492024 (ebook)
ISBN 9798217061808 (international edition)

Printed in the United States of America
1st Printing

The authorized representative in the EU for product safety and compliance is
Penguin Random House Ireland, Morrison Chambers, 32 Nassau Street,
Dublin D02 YH68, Ireland, https://eu-contact.penguin.ie.

To the people of Congo, who deserve better

Within the techno-scientific set-up that seizes the world, there is no longer the sovereign good, nor even the happiness of humanity, its well-being and comfort. Humans have and will have to live and die without comprehending what is happening, why it is happening, how it is happening.

—KOSTAS AXELOS, *THE GAME OF THE WORLD*

CONTENTS

Part 2: TRADE AND WAR

Part 3: BATTERY BOOM

Part 4: CONSOLIDATION

Part 5: WAKING UP

CAST OF PRINCIPAL CHARACTERS

Robert R. Aronson: U.S. entrepreneur who pioneered electric vehicles in the 1960s–70s.

Clément Brasseur: Belgian lieutenant known as "Nkulukulu" for his brutal tactics.

Harry M. Caudill: U.S. Appalachian anti–strip mining activist.

Chen Xuehua: Founder of Zhejiang Huayou Cobalt.

Jules Cornet: Belgian geologist who in 1891 described Katanga as a "geological scandal" due to its mineral wealth.

George Forrest: Congolese businessman and miner.

Dan Gertler: Israeli mining entrepreneur.

John B. Goodenough: Nobel Prize–winning U.S. scientist, coinventor of the lithium-ion battery.

Aminatou Haidar: Sahrawi human rights advocate.

Alex Hayssam Hamze: Congolese mining entrepreneur.

Hassan II: King of Morocco, 1961–99.

Kalala Ilunga: King of the Baluba, circa sixteenth century.

Joseph Kabila Kabange: President of the Democratic Republic of the Congo, 2001–19.

Laurent-Désiré Kabila: Leader of the AFDL rebels. President of the Democratic Republic of the Congo, 1997–2001.

Odilon Kajumba Kilanga: Artisanal miner from Lubumbashi.

Joseph Kufi Kilanga: Government official under Mobutu's regime.

King Leopold II: King of Belgium, 1865–1909; established the Congo Free State in 1885.

Patrice Lumumba: Congo's first prime minister after independence in 1960.

Fifi Masuka Saini: Governor of Lualaba Province after 2023.

Mobutu Sese Seko: President of Congo / Zaire, 1965–97.

Mohammed VI: King of Morocco, 1999–present.

Msiri: King of the BaYeke who built an empire in Katanga in the mid-1800s.

Elon Musk: CEO of Tesla.

Richard Muyej Mangez Mans: Governor of Lualaba Province, 2016–23.

Jason Sendwe: Leader of the Balubakat during Congo's postindependence conflicts.

Suharto: President of Indonesia, 1967–98.

Félix Antoine Tshisekedi Tshilombo: President of the Democratic Republic of the Congo, 2019–present.

Moïse Tshombe: Katangese politician who led the mineral-rich province's secession from Congo from 1960 to 1963.

Wang Chuanfu: Founder and CEO of BYD Company.

M. Stanley Whittingham: Nobel Prize–winning British chemist who developed the first lithium-ion battery while working at Exxon in the 1970s.

Xiang Guangda: Chairman of Tshingshan Holding Group.

Akira Yoshino: Nobel Prize–winning Japanese chemist. Coinventor of the lithium-ion battery.

Robin Zeng Yuqun: Founder and chairman of CATL.

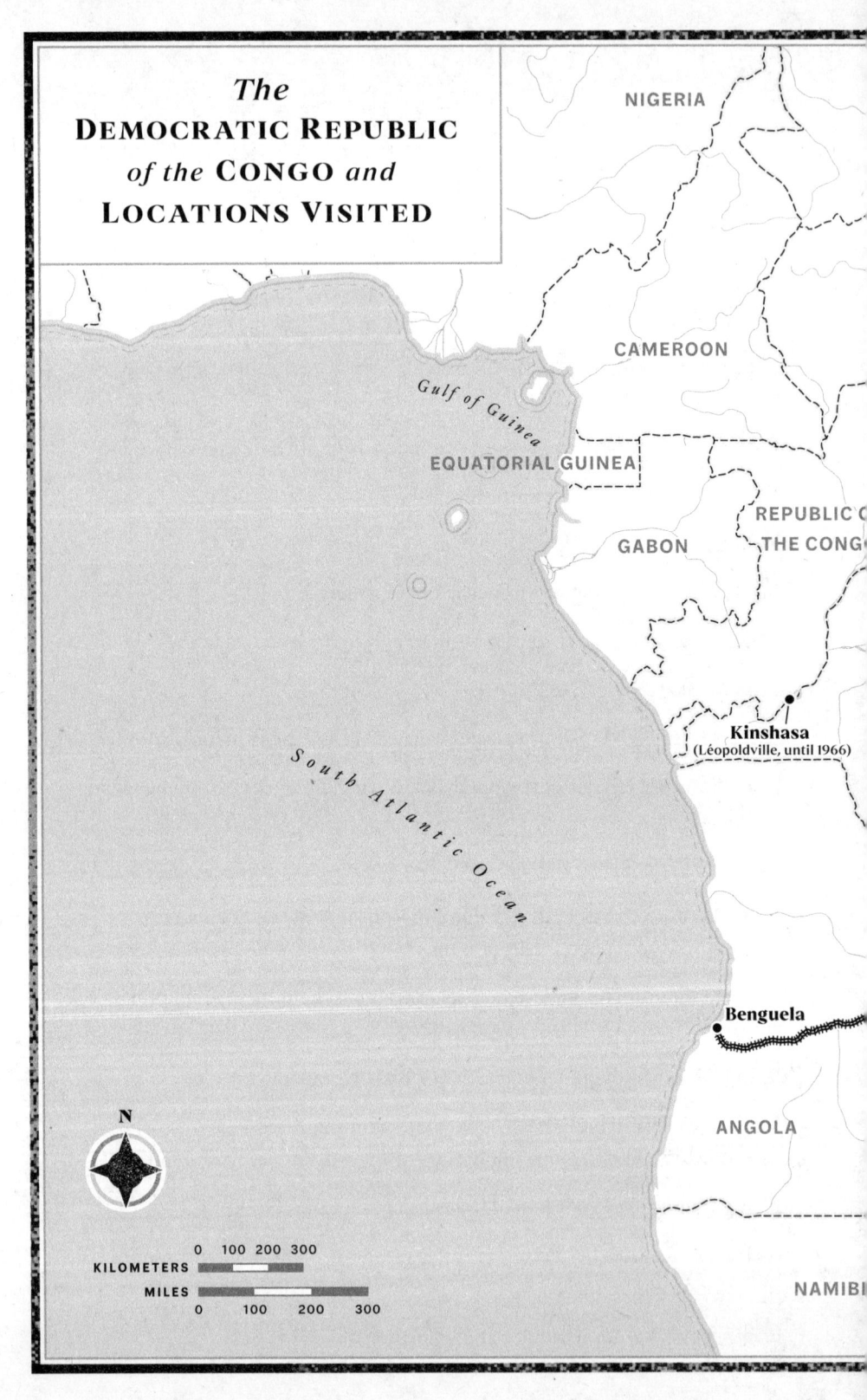

The
DEMOCRATIC REPUBLIC
of the CONGO and
LOCATIONS VISITED

NIGERIA

CAMEROON

Gulf of Guinea

EQUATORIAL GUINEA

REPUBLIC O
GABON   THE CONGO

Kinshasa
(Léopoldville, until 1966)

South Atlantic Ocean

Benguela

ANGOLA

N

NAMIBI

0 100 200 300
KILOMETERS
MILES
0   100   200   300

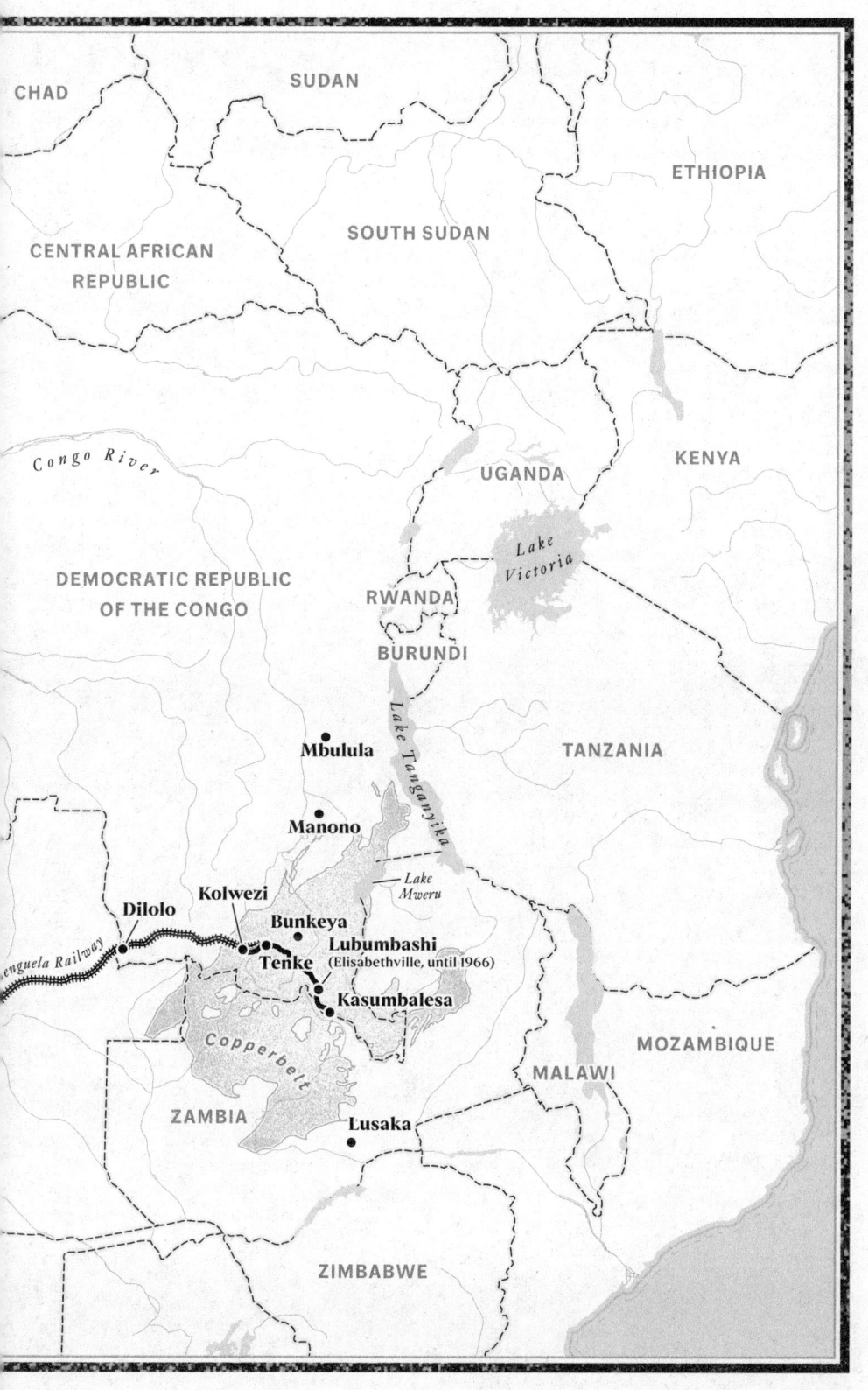

THE NEW POWER

Lithium-ion batteries make the modern world possible. First commercialized in 1991, they put powerful tech at our fingertips, tools to get around, ways to be more and more connected.

Take the iPhone 16. The top model, Apple said, can run for thirty-three hours while playing video. It has a forty-eight-megapixel camera and a built-in AI system to scan objects. You can use the phone to call your mother, to trade in decentralized cryptocurrency, and, if you had lost something under your bed, you could use it as a flashlight. None of this would be possible without a substantial source of charge from a lithium-ion battery.

Electric vehicles like Teslas and BYDs use similar batteries, just on a larger scale, to hum soundlessly and smokelessly around our streets. And as technology develops and has to meet new standards, it will need better batteries to keep its tech humming: For the iPhone 17 Air, for example, engineers worked on developing a "high density" battery to reduce its size without compromising on power.

But humanity has also made a Mephistophelian bargain with batteries. The raw materials used to build them come from every corner of the globe: metals like lithium, cobalt, and nickel, as well as materials like graphite, silicon, and phosphate. To access them, tech profiteers, politicians, and battery makers have made a trade-off: cleaner power at home for pollution and suffering elsewhere. We have, explicitly or implicitly, accepted some of the pernicious consequences of the power revolution as inevitable, including the risk of human-rights abuses and labor exploitation, without much thought about how to mitigate these externalities.

The metals and materials that have gone into these batteries are often rare. They come from complicated parts of the world. It was not for nothing that people who knew metals would say, in the first decades of the twenty-first century, that batteries and green energy had created the greatest dislocation of demand and supply in their lifetimes.

The twentieth century was powered by oil, but by the second decade of the twenty-first century, we have developed myriad ways to store power without using fossil fuels. Among these methods, lithium-ion batteries have come to dominate. Batteries are globalized products—they are built from materials mined in one place, refined in another, assembled somewhere else, and eventually sold in yet another, crisscrossing a multitude of borders in the process—and without globalization, it would be impossible to build them or the computers, telephones, and cars they power. Politicians may claim otherwise, but global trade and supply chains are the only way that such products, especially at the scale modern society demands, can exist. Understanding these batteries and how they are made is key to understanding how a new form of power is being created, one that is measurable in dollars, strategic influence, and volts.

Lithium-ion batteries are produced using supply chains that are vulnerable to geopolitical rivalries, most notably the rivalry between China and the United States and Europe. Through strategic investment, Beijing has taken control of large parts of the extraction and refining of battery metals, and by 2024 one report estimated that 70 to 90 percent of lithium-ion batteries were being made in China. Critical metals have been as valuable in times of war as in times of peace. History is rife with examples of nations that failed to secure access to raw materials and suffered defeat in conflict or serious disadvantages in peacetime.

Like all revolutions, the battery revolution has had its winners and its losers. Lithium-ion batteries have helped create great fortunes. Elon Musk's share of the electric car company Tesla was the single greatest contributor to him becoming, in 2021, the world's wealthiest man, and Apple and Microsoft, two companies whose portable devices rely on the batteries, vie for the title of the world's richest. Meanwhile, at the other end of the spectrum, people have toiled in squalid conditions, in poverty and human bondage, to extract the minerals needed to make these batteries.

And, as with oil, battery power has increasingly become political power. Musk arguably became one of the most influential people in U.S. politics after putting an outrageous $288 million behind Trump's 2024 presidential bid and buying himself a seat at the table of governance, only to flame out, mid-2025, partly over disagreements over electric vehicle policy. But even his grasp of the complex issues behind where these batteries come from does not seem solid: In a 2023 shareholder meeting, Musk answered a question about child labor by saying he would put webcams in cobalt mines to check whether underage miners were descending into unsafe pits. His answer belied ignorance of how and where this mining actually happened.

In the end, Tesla provided only a monthly satellite image of its main cobalt source, the Kamoto Copper Company mine in the Democratic Republic of the Congo.

THE CRITICAL METALS CONUNDRUM HAS BEGUN TO REGISTER IN THE HALLS of power. The rush for critical minerals is fueling geostrategic shifts and shaping conflicts around the world. Soon after his inauguration in January 2025, President Donald J. Trump's team began scouring the globe for minerals, from Greenland to Ukraine, even if the people who lived over them were less than willing to countenance signing them over to the U.S.

Then, on a Wednesday in mid-March 2025, Bret Baier, the host of Fox's prime-time *Special Report*, turned his attention to another mineral-rich nation, the Democratic Republic of the Congo. Many of the minerals used in lithium-ion batteries exist in gigantic quantities in Congo,*

---

*For ease here, and to avoid littering my prose with acronyms, I'll use *Congo* from now on (except in citations) to refer to the Democratic Republic of the Congo. During King Leopold II's but they are hard to access, partly because the country has been racked by near-continuous conflict since 1997 and suffers from deep-seated corruption. *Special Report* was one of the most popular cable news shows on television, and Trump was known to be a regular viewer. That Wednesday, a special guest appeared on videolink: the president of Congo, Félix Antoine Tshisekedi Tshilombo.

Dressed in a midnight-blue tunic, Tshisekedi at first looked a little uncomfortable. It was his first time on Fox. But the president soon found his flow as he began to talk about the subterranean wealth of his country and a potential deal with the new U.S. presidential administration. "We have established partnerships with many other countries, and we think the United States of America, given its role and influence around the world, is an important partner to have," Tshisekedi told Baier. He was offering Congo's ores for metals like lithium and cobalt, as well as for tantalum and copper, which are key to other parts of electric devices. China had long controlled his country's natural resource trade, he said, and the U.S. was "waning" in Africa.

Tshisekedi had a problem: A rebel group, supported by Congo's pugnacious neighbor, Rwanda, had seized large areas of his country's East, and his corrupt army and government were ineffective at defending his people. While the sources of the conflict were many, it was partly fueled by a competition for valuable resources. He was turning to Trump to fund his fight using revenues from mining. If the president helped him, Tshisekedi implied, the U.S. would be allowed privileged access to battery minerals. "Whatever deal is made, Tshisekedi wants to make sure that it benefits the DRC," Karl Von Batten, the chairman of the Development Committee of the District of Columbia's Republican Party and one of Tshisekedi's lobbyists in Washington, told me. Tshisekedi wanted, as Von Batten put it, a deal that benefited his population, people who would "not just extract like the Chinese," but mostly he wanted money, weapons, and military might to help fight the invaders. In response, he was promising

---

time, the country was called the Congo Free State. During the Mobutu epoch, from 1971 to 1997, it was known as the Republic of Zaire, and I'll refer to that country, accordingly, as Zaire. The country to the north of the Congo River is called the Republic of the Congo, or Congo-Brazzaville.

access to the rudiments of power, and a chance to beat China at its own game.

Trump was looking at foreign policy in transactional terms, a senior Republican official involved in shaping Africa policy told me on a visit to Washington that spring. The official was cautiously optimistic about a deal, but worried that the Congolese, who had a history of backsliding, might be preparing to stab them in the back. The prize was worth investigating, but he was worried that the U.S.'s prime foreign policy adversary, China, was also maneuvering to remain in control of vast swaths of the world's critical minerals. "Let's recognize that Tshisekedi could be just using this to get leverage in another deal with the Chinese," he said.

Trump's administration would end up penning a deal that would increase U.S. investment into Congo. Firms like KoBold, a mining firm backed by the likes of Jeff Bezos and Bill Gates—men whose fortunes had been made in tech and were almost as large as Musk's—began eyeing potential deals in Congo for critical metals like lithium. "Under Tshisekedi, Chinese interests have continued to expand in Congo," the official told me. "As much as he purports to be Western-orientated, a question mark remains. It's always in the back of my mind: Are we being played?"

BATTERY WARS HAD ALREADY CAUSED UPHEAVAL AROUND THE WORLD, BUT people who use lithium-ion gizmos have also begun to focus on the detrimental effects at the bottom of the supply chain. When the iPhone 16 was released on September 20, 2024, for example, demonstrators gathered outside Apple Stores in over a dozen cities. Many of them waved Congo's colors—the vivid blue, red, and yellow of the country's flag—in the early-autumn air. They were there to decry what they called a "silent genocide" in Congo, and they accused Apple, a company worth some $2.6 trillion at the time, of profiting off the Congolese, some of the poorest people in the world. Apple was fueling a series of conflicts and condoning child labor, they said. The protesters exhorted would-be Apple buyers: "Don't scroll with bloody hands."

Lawyers for the government of Congo had filed suit against Apple earlier that year in France and Belgium. (The French case was later dismissed.) "Apple has sold technology," they wrote, "made with minerals sourced from a region whose population is being devastated by grave violations of human rights."

It was not just left-wing activists who were critical. The company had based its supply chain in China, which enraged human rights advocates and hawks alike. The former argued that pollution and abusive labor standards were tolerated by Chinese companies; the latter said that Apple was farming out its supply chain to the U.S.'s chief adversary.

Apple's representatives had long insisted that its batteries were produced virtuously. The company, after all, audited its supply chain to make sure that minerals connected to human-rights abuses did not wind up in its devices. It was true, moreover, that there were many different kinds of lithium-ion batteries, some with more negative externalities than others. And Apple insisted that its cobalt—a metal whose supply chain was one of the dirtiest among all those that went into making the phone—was recycled. A press release touted how much was reused: 100 percent of the cobalt in the iPhone 16 battery and 95 percent of the lithium.

But Apple was not telling the whole truth. It had built a vastly profitable supply chain on the back of cheap extraction from poor countries like Congo and cheap manufacturing, especially in China. Its supply-chain audits only disclosed a limited range of information. The company said it was not buying hand-mined cobalt. But its very own approved smelter list, released that year, included a company, Zhejiang Huayou Cobalt, that was involved in some of the worst human-rights abuses—like child labor and wage slavery—in Congo.

A few months after critical metals lawsuits were filed, Apple quietly issued a statement that it would completely halt buying minerals like tin and tantalum from Congo and Rwanda, parties to the conflict in the East. (Other essential minerals like cobalt, which overwhelmingly comes from mines in southern Congo, were not included in this announcement.) "We believe that this statement is a very significant development in the international consensus and the international world of supply chain," Peter Sahlas, one of Congo's lawyers, told me. Still, he said that Apple's announcement was only the beginning. "It doesn't mean that we're going to go away." Other companies needed to be scrutinized, he said. "Apple didn't become a $3 trillion company without using conflict minerals, and we're going to hold them to account for that."

FOR HALF A DECADE, I HAVE SEEN THE BATTERY TRANSITION UP CLOSE, IN some of the places where it has had its greatest effects. In Congo, in Indonesia, and in the Western Sahara, I saw where battery minerals are mined and spoke to people whose lives have been changed by the new energy map. I traveled to France and Germany to see how cars are being produced using lithium-ion batteries, and to Belgium, where the original colonial supply chain was forged. In Japan, I saw where the lithium-ion industry first arose and later collapsed, and in China, I saw where a new one was being built. In London and Switzerland, I spoke to traders of the minerals that make the batteries. In the United States, I learned about how the country has been trying to revamp its mining and battery industries.

I made my first trip to Congo in May 2019. There, I saw firsthand how some of the poorest people toiled in the country's copper-and-cobalt mines, employing techniques that had been abandoned as far back as the 1700s in many other countries. I returned home to news that Elon Musk had raised $2.7 billion for Tesla. Later that year, I visited Congo again. I showed my phone to Ziki, a fifteen-year-old former child miner, and said that the latest smartphone models, which contained some of the metals whose ores he used to dig up, cost upward of $1,000. "I have sadness in my heart when I think of people who buy the minerals," Ziki said. "They make so much money, and we have to stay like this."

On these trips and four additional visits to Congo, I saw mine sites, spoke to child laborers who dug out the minerals, and met with executives who ensured that these lucrative extracts flowed into factories and ultimately into the hands of consumers. I chatted with the politicians and the spies shaping the geopolitics of this new power. As an expert witness, I gave testimony to the United States Congress about just how dangerous this supply chain could be, geopolitically and environmentally.

I was lucky. Some of my friends and colleagues who tried to speak the truth, in Congo especially, were locked up and tortured.

I was detained, twice, as I tried to investigate aspects of the battery supply chain. A reporting assignment in 2017 occasioned my ejection from the Western Sahara by Moroccan authorities. They were concerned about their control over the supply of phosphate, which had started to be used in some lithium-ion batteries. While working with the Congolese journalist Jeeftour "Jeef" Kazadi Kamwanga in 2022, in the city of Lubumbashi, I was detained on spurious charges by the Congolese secret police. Congo's copper, cobalt, and lithium resources were their priority. I was released after six days and banned from the country. Kazadi was detained for almost three weeks.

Over and again, in less dramatic ways, I was stymied, frozen out, and told not to report this story. Emails went unanswered and calls were cut short. People would often only meet me in private locations and on the condition of anonymity. Governments even refused me entry into their countries. In 2018, as I was first conceiving this book, I traveled to China on a broad informational trip, where I met some of the players in that country's international business scene. When I tried to go back, in 2023, the consulate in New York would not issue me a visa.

This was a book on batteries, I thought, a topic that might cause eyes to glaze over during dinner talk. But as I delved further into the world of lithium-ion, I realized that it was not simply a story about sockets and charges—it was also a story about control and immense power.

IT MIGHT HAVE SEEMED THAT POWER HAD BECOME MORE DISPERSED IN THE first decades of the twenty-first century, placed through digital wizardry into the hands of the multitudes, but the inverse had also happened: Power had become more and more centralized, not just by Musk and executives at tech firms but by governments. States used technology, and especially technology powered by lithium-ion batteries, to extend their reach into every aspect of life.

In a single fortnight in 2024, for instance, the Canadian government announced 100 percent tariffs on Chinese electric vehicles, which were threatening to flood the market and push out cars made by traditional manufacturers, and thousands of lithium-ion-powered pagers had been detonated in Lebanon, killing and maiming civilians and militants alike. The pagers, which had been bought by Hezbollah, a secretive terror-group-cum-political-party, were widely thought to have been detonated by the group's enemies in Israel's intelligence service. Lithium-ion batteries were everywhere—even in some of the most secretive hideouts on the planet.

These batteries and the materials used to make them were themselves protagonists in the wars—hot and cold and commercial—that have come to scar the history of the twenty-first century. The people who controlled these resources wielded an extraordinary amount of political power, the power to shape world affairs.

China had taken an early lead in battery making, but by the early 2020s, governments in the U.S. and in Europe were cottoning on to the conundrums raised by batteries and their manufacturing. President Joe Biden's Executive Order 14017 occasioned a strategic review of critical mineral supply chains conducted by the U.S. Department of Defense in 2021 and 2022 that recommended the Pentagon focus on U.S. production of critical minerals, engage with allies and partners, and mitigate foreign control of the supply chains. As one former Defense Department expert told me in 2024, this meant, for example, that the U.S. military stopped using Chinese-produced cobalt to build military equipment.

It wasn't just technologists who had become filthy rich off lithium-ion batteries. Thanks to the preponderance of metals and minerals that go into them, mining firms and commodities traders were also getting flush. Lithium, cobalt, nickel, phosphate, graphite, and other materials were feeding what came to be called "the revenge of the miners."

The miners were passed over by the finance and media worlds, and by the world at large, in the 2000s, as tech companies promised a shiny digital future and conjured dollars out of thin air. But they instinctively understood a basic truth about human society: Everything we use, everything we eat, everything we work and play with—it's all either grown or dug out of the ground. Mines and mining companies had long been considered dirty, a pollutive anachronism, but, all of a sudden, everyone was realizing that they needed people who carved open the earth in search of its riches.

And somewhere in this mix was hope. Hope that the people who controlled this new power might have answers to the complex questions around humankind's impact on the climate and the planet. After all, multiple studies have shown that a lithium-ion battery, if used correctly and long enough, can help reduce the carbon footprint of our society, which was only racing faster and faster.

———

BY THE MID-2020S, ALMOST EVERYONE ON THE PLANET (EXCEPT, PERHAPS, members of uncontacted tribes) had some kind of exposure to lithium-ion batteries. They powered cell phones, breathed life into laptops and other portable devices, allowed for smokeless cars and motorbikes.

A tourist in London, say, might be knocked off her electric bike by an electric bus, all while looking at her lithium-ion-powered iPhone. Lithium-ion-propelled drones had changed the face of modern warfare and of videography. And it wasn't just the phones and cars and drones: Garden tools, cattle prods, gas detectors, vapes, air enrichers, industrial robots, and children's toys all used lithium-ion technology. Batteries were being touted as the solution to the climate crisis, since they produced no emissions (at least not when they were used, that is—production emissions were a very different story). Militaries relied on lithium-ion when wars were waged; doctors used lithium-ion-powered medical devices to save lives—pacemakers, drug pumps, defibrillators. In fact, if you're reading this book on a screen, you're probably using lithium-ion technology right now.

How did we get to this point, and how might we make things better? This book is the story of the genius that produced the batteries, the men—and it has mainly been men—and women who turned them to their profit, and the people who have suffered on account of our lust for concentrating ever more power in ever smaller devices. It is also the story of how the geology and colonization of Congo were key factors in shaping the supply chain we use today. Without Congo, the battery revolution would have been much slower.

This story is one of tremendous innovation, of political intrigue, of the rise and fall of empires. But it is also a tragic saga whose protagonists are some of the poorest people in the world. And it is the tragedy of Congo. Why is a country so rich in minerals still so poor? How could a U.S. mining executive who lived in Congo and loved the country tell me, in 2023, "It's always going to be the land of maybe, and it's never going to be the land of dreams"? Threaded through this narrative is the history of the country, which sits at the bottom of the supply chain for many of the metals that power our devices, and exemplifies the consequences of the renewable energy revolution. Congo, after all, is the place that people say will power the green, fossil-fuel-free future.

Part 1 tells the tale of the conditions that wrought the current rush for minerals and batteries, during the colonial days in Congo, and in the United States, during the 1960s and '70s.

Part 2 examines how batteries were perfected, how Cold War–era geopolitical rivalries and political competition allowed for a predatory system to emerge, and how globalization set the stage for China's primacy in lithium-ion battery production.

Part 3 focuses on how batteries came to be in cars, and how new and more powerful batteries were created to power them; how China understood it could control and profit from that industry; and how the mining system that feeds this industry pollutes and exploits.

Part 4 describes the system that is currently in place, how Congolese children still mine for cobalt and other metals in abysmal conditions, and how industrial mining functions. It shows how seas are poisoned in Indonesia in the search for metals like nickel and how China consolidated control of the supply chain. Seen from this vantage point, the green transition looks more like a displacement of pollution from wealthy cities to poor, rural communities.

Part 5 explores how remarkable new technologies have been created to address issues up and down the supply chain, as well as how, in Congo and in the Western Sahara, there still exist stumbling blocks when it comes to technologies like batteries based on iron and phosphates. I also tell the tale of my own detention to show just how the status quo is enforced. Lastly, I look at the U.S. and Europe's rude awakening to China's dominance of the supply chain, and their attempts to wrest back control.

AN INCREASING NUMBER OF PEOPLE WORLDWIDE BEGAN TRADING IN THEIR combustion-engine cars for electric vehicles at the end of the second decade of the twenty-first century. By the end of 2024, there were forty million electric vehicles in use globally. In the States, they accounted for 10 percent of the market. And they were set to grow in popularity: That summer, researchers at Bloomberg predicted that more than thirty million electric vehicles would be sold in 2027 (33 percent of all vehicles globally) and that by 2070, this number would reach seventy-three million (73 percent of global car sales).

By late 2024, even Trump, a longtime EV skeptic, was getting on board, despite announcing that he would cut tax credits for electric vehicles. Trump wasn't exactly consistent, though: During his first administration, Apple's CEO, Tim Cook, was involved in a successful lobbying effort to exempt consumer electronics from tariffs on Chinese goods; during Trump's second administration, the company looked vulnerable to tariffs, especially since many of its lithium-ion-powered devices are manufactured by BYD, a huge electric vehicle manufacturer and Tesla competitor.

Then, in March 2025, after a precipitous fall in Tesla's share price, partly thanks to Musk's political activities but also thanks to a slowing pace of new vehicle deliveries, Trump used the South Lawn of the White House to promote the electric car company. The company's Cybertruck, he said, had the "coolest" design, and he tweeted that "Radical Left Lunatics" were trying to destroy Tesla.

But people were starting to question whether the electric vision of the future is quite as green as its advocates have liked to trumpet. Some of the questions they are asking revolve around how ethical electric vehicles are and how battery minerals are produced. "Last night Teslas burned again in Berlin," a German anarchist website announced after a night of car destruction in June 2024. "In Congo, children work themselves to death for cobalt, and new toxic lithium mines are being opened all over the world to satisfy the hunger of the car industry."

By 2025, such attacks had spread to the U.S., with Tesla cars being vandalized across the country, often in response to Musk's political involvement with Trump. A disgruntled veteran blew himself up in a Tesla Cybertruck in Las Vegas on New Year's Day. In Oregon, someone fired bullets into a Tesla dealership. Tesla superchargers from Massachusetts to Italy were set on fire.

The powerful are violently lashing out in response, not just at people who are committing vandalism, but at people who dare ask questions about Musk and Tesla. Pam Bondi, Trump's attorney general, said that Tesla vandals would be prosecuted as domestic terrorists, and Elon Musk told Bret Baier on Fox that the "real villains" are not the vandals but "the people pushing the propaganda." On-screen, he issued a threat: "The president's made it clear: We're going to go after them." When Musk fell out with Trump, the proclamations stopped but the prosecutions continued.

THIS IS NOT A BOOK THAT ADVOCATES FOR SIMPLISTIC SOLUTIONS. I DON'T believe violence and vandalism are the answer, and I don't think we should turn away from mining, at least mining ethically, that is, and the power of human endeavor.

What's more, I do not believe that a return to fossil fuels would solve any of the problems that have been unleashed by the battery revolution. I have seen firsthand how global warming is drying out the African Sahel, how it is melting ice caps in Antarctica, and how it has prolonged drought conditions in El Salvador. And, in any case, cobalt, one of the metals I focus on most, is used in some types of fossil-fuel production.

Don't get me wrong—batteries, and the cobalt, lithium, and other metals that are used to create them, are not intrinsically bad. In fact, we need more solutions to tackle a warming planet, and lithium-ion batteries should be among them. But electric vehicles as they are currently built are not always a great solution. If you count scope one, two, and three emissions—that is, indirect emissions and those that arise from production as well as from the tailpipe—several studies have shown electric vehicles are more polluting than hybrid vehicles in most places. If you take a Tesla Model 3 and compare it to a Toyota Prius, the Model 3 is more polluting than the Prius in China, the U.S., and Germany over both the short and the long term. Only in countries that use a majority of clean energy—countries like Congo, incidentally, although the grid there is so unreliable that an electric vehicle would be a risky prospect for any owner—are Teslas less polluting. This is because the manufacturing of electric vehicles and their batteries is hugely emissions intensive, and since people tend to upgrade more quickly in the electric realm, they end up adding more pollution to the environment. That's to say, if you buy a new Tesla every 3.6 years, like the average U.S. customer, you're actually doing more harm than good.

I don't have grand answers to the questions that arise from this book. I have approached them as a journalist, and tried to understand them in their complexity. I believe that before we jump into a better future, we must know something of the past and present. This book is an attempt to delineate crises of the status quo and look at the forces shaping where we are going.

We know how to make batteries. We know that they can help us cut down on emissions. And we know how to sustainably mine the metals that make them. But at the moment, with a few exceptions, some of which I explore in this book, we are not doing these things. Instead, we are desperately scrambling into the abyss. We are, by and large, cloaking a lack of real action in hollow virtue signaling while ignoring solutions that are sitting right in front of us, solutions that mean that local communities can share in the world's mineral wealth without destroying their lives and homes, and at the same time turning a blind eye to those whose only motive is squeezing out profit from a dying world.

Part 1

FUNDAMENTS

"When Katanga is hurt, money screams,
and money has powerful lungs."

—CONOR CRUISE O'BRIEN, *TO KATANGA AND BACK*

CHAPTER 1

A BEND IN THE LUFILIAN ARC

W hen Odilon Kajumba Kilanga got home, usually late at night, it was hard for him to get any rest. He lived with his brother, Amos, and another man in a tiny one-bed room. It was the mid-2010s, and he was in his late twenties, working as what Congolese refer to as a *creuseur*, or digger, and what the Anglophone corporate world calls, somewhat deceptively, an "artisanal miner." What this meant was that he made his living searching for copper and cobalt in the mineral-rich soil of the Democratic Republic of the Congo using only the most rudimentary of tools.

Kajumba's* life was shaped by riches in the soil beneath his feet, for he lived in an area of southern Congo called Katanga, in the city of Kolwezi. To certain foreign ears, "Katanga" meant valuable metals that could be traded on international markets, and fortunes were being made as Katanga's earth was traded, refined down into metals, and made into batteries and electrical wire. Kolwezi had particularly rich mineral deposits:

---

*In the early 1970s, reforms made by the dictator Mobutu Sese Seko called for people in the Democratic Republic of the Congo to have "authentic" African names. Today, Congolese usually have three names: a first name, a surname, and a *postnom*, or postname, which comes after the surname. Hence, Odilon's surname is Kajumba; his *postnom* was given to him in honor of his paternal uncle.

After the rain, puddles would sometimes shimmer green with oxidized copper.

Kajumba was a laconic man, rail-thin, with the type of largish head that suggested nature had intended him for corpulence, even as circumstance had conspired to keep him slender. The air in his cramped room was thick, stuffy with the stench of three sweating men. Coming back late at night, Kajumba would call beforehand to let his roommates know it was him and not one of many strangers who might rob them in the early hours. Then he would knock at the door, and one of the slumbering men would groggily open it up for him. Trying to make as little noise as possible, Kajumba would crawl into bed, crowded with hot limbs, and attempt to sleep.

The buzz of the city, with its thousand megaphones and beery canteens, would crackle around him; he would sweat; the mosquitoes—not as many as in some other parts of the city but annoying all the same—would flit about him, filling his ears with a thin whine. As he drifted off, his brain would play back a series of images from his time in the deep pits, where he wrenched minerals from the soil with his bare hands and a metal bar.

These were places where he had seen his friends crushed and suffocated, where he had seen lives given over to immense toil simply as a means of escaping the poverty that weighed down on them every day. "The worst souvenir of my time working as a miner is the nightmares," he said. He described the horror of hearing a mine shudder and collapse, the sadness he felt upon seeing metal-poisoned children with swollen, malformed heads and pinprick eyes peering out of hovels. "I would never accept for my children to become miners." Lives like his, he reasoned, were being sacrificed for another generation, one that he hoped would never have to descend into the sweltering underground of Congo's copper-and-cobalt mines.

When I met Kajumba in 2019, Amos had left, and Kajumba was sharing his room in Kolwezi with Trésor Mputu, a friend and colleague from Likasi, a town on the road to Lubumbashi. They were two-thirds of a three-man team of *creuseurs*. The third member was Trésor's brother Yannick, who lived nearby. Other people worked as *porteurs* (mineral carriers), *laveurs* (mineral washers), and *ramasseurs* (groups, consisting mainly of children, who pick up slag that has been discarded near the mines).

Trésor had, like Kajumba, come to the city to dig riches from the soil, only to discover a daily cycle of miserable hours, backbreaking work, and anemic pay. (Kajumba's brother was now looking for work in a town called Kasaji, maybe a day's drive to the west when the roads hadn't been turned to soup by rains.) Kajumba's small room was in a poor neighborhood, but he didn't mind; everyone who lived there knew one another and, Kajumba was sure, would keep their neighbors safe.

The mining economy that Kajumba was part of in Kolwezi was brutal, and it was key to powering the global economy. On six trips to Congo between 2019 and 2022, I visited the city four times. There I saw children and mothers on the roadside, sorting through minerals. Former child miners explained to me how they had learned to pick out the purest ore from rock slabs. Soon enough, they were lugging ore for adult *creuseurs*. Then, as teenage boys, they worked perilous shifts that saw them navigating rickety shafts. Near large mines, the prostitution of women and young girls was pervasive. Women washed raw mining materials, which are often full of toxic metals and, in some cases, mildly radioactive.

Researchers have estimated that thousands of children work in Kolwezi's mining industry alone. It is hard to ascertain how many children actually work in the mines; figures range from thousands into the hundreds of thousands, although the higher numbers are often extrapolations based on small sample sizes. In the summer of 2024, UNICEF's representative in Congo suggested that 361,000 children might be laboring in mines in southern Congo, though this number seems implausibly high and drew quick opprobrium from Congolese NGOs that work on the issue. The truth is that there are no reliable census data, a problem compounded by the rhythm of artisanal mining: While some children work consistently at specific sites, others enter the mines during particularly hard times for their families. "I don't think the government has any capacity to monitor children's involvement in this," said Mark Canavera, who has spent time in Kolwezi and worked as the codirector of the Care and Protection of Children Learning Network at Columbia University. "Even if it did," he told me, "it doesn't have a framework for thinking about what is child labor and what isn't."

The health effects that attend the extraction of this precious haul from the soil are not limited to just radioactivity. In March 2022, I spent a day with Dr. Billy Mukong, a local physician in Kolwezi. As he made his rounds, he introduced me to women whose children were born with defects. Some had distorted heads. Some were developmentally disabled.

The abnormalities, Dr. Mukong said, were due to exposure to dust, either directly, down in the mines, or secondarily, by breathing in what was blown off the backs of trucks carrying raw and processed ore around Kolwezi. "The illness rate is also very high because they work with no protective equipment," said Charles Carron Brown, a consulting mining engineer who spent much of the 2010s working in Katanga. "They have no protection against dust. They're breathing in silica dust and other metals as well. From a health standpoint, it is very bad for them. But they have no real choice, because it's the only way they can make a living." The respiratory effects of such minerals have long been known: In 1965, a biochemist hired by the region's largest mining company noted that around a third of deaths among workers at the company had been caused by respiratory illness, even as confirmed cases of tuberculosis were dropping, and that such ailments were the primary cause of sickness in workers' children.

Dr. Mukong's observations were also backed up by more contemporary science: If, for example, a pregnant woman works with heavy metals such as cobalt, or other metals that are by-products of the mining process, her chances of having a stillbirth or a child with birth defects increase. According to a 2020 study in *The Lancet*, women in Lubumbashi "had metal concentrations that are among the highest ever reported for pregnant women." The study also found a strong link between fathers who worked in mineral extraction and fetal anomalies in their children, noting that "paternal occupational mining exposure was the factor most strongly associated with birth defects."

———

THE POLITICAL AND ECONOMIC GEOGRAPHY OF THE COPPERBELT, THE MINING zone that snakes under Katanga and northern Zambia, has been fashioned by external forces and outside greed since at least the time of Belgian colonization. In the 1880s, Belgium's King Leopold II annexed the entire Congo Basin, more than nine hundred thousand square miles of Central African jungle, highlands, savanna, and forest, Katanga included. (Belgium wasn't even twelve thousand square miles.) The regime that Leopold put in place was responsible for the deaths of up to thirteen million people, as his agents frantically plundered ivory and rubber in what the novelist Joseph Conrad famously capsulized as "the vilest scramble for loot that ever disfigured the history of human conscience." Katanga, with its copper mines, escaped the worst of the violence. It first began to be exploited for its metals by the Belgians who came after Leopold. By the 1930s, the Union Minière du Haut-Katanga, the Anglo-Belgian concern that ran the mines of the southern Congo, was the largest copper-producing company in the world. Over two decades into the twenty-first century, the Central African Copperbelt is still globally important—a little more than a tenth of copper mined worldwide in 2020 came from Zambia and Congo alone, and 70 percent of the world's cobalt came from Katanga.

Not that Kajumba or many of his fellow miners saw more than a trickle of the massive profits produced by those minerals. In Congo, vast wealth, much if not most of it accrued from selling off the country's abundance of natural resources, coincides with crippling poverty. Kajumba was one of seventy-four million Congolese who lived below a poverty line of $2.15 a day. Forty-three percent of children in the country were malnourished.

And yet, if you knew where to look, there was a tremendous amount of wealth circulating in Congo. Patrick Masengo Kalasa, a Katangese politician who worked as a tax adviser until 2017, told me that the large companies he worked with had to pay their tax bills at a government office in Kinshasa, Congo's capital. Masengo worked for a large gold-mining firm but his colleagues worked with Tenke Fungurume Mining, the largest copper-cobalt mine in the country. "You would go to the window at the government office, and they would say, 'Wait for a special account number.' You would wait, and then sometimes at five o'clock in the afternoon, you would hear that taxes had to be paid into a special account," he explained. "Most of those accounts were in foreign countries. Sometimes they would be private accounts at private banks. This would happen at the level of the central bank, but we would never be paying into government accounts." I asked him to whom the accounts belonged, but Masengo said he was never told. The answer became partly clear when a series of leaks known as the Congo Hold-Up showed that Joseph Kabila Kabange, the former president, and his entourage had used a bank called BGFI to siphon at least $138 million from the country.

Such practices, Masengo heard from people he worked with, were "still very current" in 2024. Large companies were to blame too: "Accountants fake earnings, so there is a work of corruption, and taxes are not paid at their real value." The result? "There is no money circulating in Katanga. Can you imagine? The governor's office is in arrears—it's supposed to be at the head of the province, and people haven't been paid for eight months." On the same day that the iPhone 16 was released, September 20, 2024, Masengo was detained by the Congolese government for publishing a charter demanding rights for the Katangese people and calling for the mineral wealth of the province to flow back to them.

Only a small number of Congolese have profited from their country's riches. When I visited Congo for the first time, in 2019, I sat on the terrace of a high-end mall in Kinshasa with Onasis Kavul, a thirtysomething political functionary whose loyalties lay with the party of Kabila, then the outgoing president. Kabila had privatized Congo's mining sector and, in doing so, made fortunes for his family and the people around him. Kavul crowed about how quickly he could drive a Porsche Cayenne from Lubumbashi to Kolwezi. The car's price point in the U.S. started somewhere around $80,000—more than 170 years' salary for many Congolese at that time—and prices were always higher in Congo, thanks to steep import fees and tariffs. Outside, the children of the elite, many of them people who had made money from mining, revved their late-model sports cars in the humid Kinshasa night. They couldn't go far, though: Beyond a few stretches of road in the upscale Gombe neighborhood, the roads were too rutted and chewed up for the fleet of low-riding Ferraris and Aston Martins to handle.

Under Congo's next president, Félix Antoine Tshisekedi Tshilombo, the cast of characters would change, but the principle would remain the same. And, as bad as all that looting was, the wealth that remained in Congo constituted only a small percentage of what was being made on markets globally, and from the technology those minerals powered.

There is a cruel dichotomy that separates the haves—the rest of the world—from the Congolese and other people who inhabit the zones where the minerals that will power the future are mined.

The minerals Kajumba and hundreds of thousands of impoverished Congolese were digging up allowed the rest of the world to enjoy the benefits of electricity, but the benefits to the people of Congo were woefully limited. Almost all cell-phone batteries contain cobalt, and although more and more people in Katanga had cell phones, few of them could pay to use them regularly. And away from Congo's town and city centers, there was little or no reliable electricity.

COBALT WAS FIRST REFINED INTO A METAL IN 1735. AT THAT TIME, IT WAS seen as junk, an impurity in nickel, silver, or copper—useful in some forms as a blue pigment perhaps, but not particularly valuable. Its name derived from the German *Kobold*, the term for an impish household spirit that miners blamed for making their silver impure. Cobalt is almost always a byproduct of mining another metal: In Congo, it comes with copper, but elsewhere, like in Indonesia and Australia, it is bound to nickel ore.

As the energy transition ramped up, the demand for such metals exploded. Many electric vehicles also use cobalt in their batteries. A 2021 analysis by Jon Lynch of the Chicago Mercantile Exchange showed that electric vehicles required more than double the amount of copper used by cars with traditional internal combustion engines. The infrastructure needed to build them also gobbles up massive amounts of metals and energy. "Copper is the metal of electrification, and electrification is much of what the energy transition is all about," Daniel Yergin, the vice-chairman of the financial analytics firm S&P Global, told CNBC in mid-2022. Much the same could be said of cobalt.

Solar power, wind power, and the storage of that electricity also demanded these and other metals, such as lithium, which governments in the U.S. and in Europe describe as "critical." But Kajumba could barely afford his twenty-five-dollar-a-month room in Kolwezi, let alone support his wife and two children, whom he had left behind in his hometown of Lubumbashi, some 180 miles to the southeast.

According to the ASM Database, a web-based archive of artisanal mining documents, somewhere between 1.8 million and 2.5 million people in Congo work as artisanal miners. Some search for diamonds, gold, tantalum, and tin ore in other parts of the country. Many live in conditions similar to—or even worse than—those endured by Kajumba. When I met him in Kolwezi in 2019, Richard Muyej Mangez Mans, the governor of Lualaba Province, told me that around 170,000 artisanal miners like Kajumba worked to extract cobalt and copper from the mines around Kolwezi.

———

THE CONGOLESE PROVINCES OF HAUT-KATANGA AND LUALABA SIT ATOP A bend in what is known by geologists as the Lufilian Arc, where geological formations holding veins of mineral riches curve under the region's beige hills. The arc was formed by a folding of the earth's crust that pushed up the hills upon which the Central African Copperbelt sits. Congo and Zambia share copper deposits, but the Copperbelt's cobalt is found almost entirely in Congo, above the border line set out between the colonial powers as they carved up Africa. Congo sprawls above something like 6 million metric tons of the metal, almost half of the world's supply, and accounts for nearly three-quarters of the global production of the world's cobalt. This is why Congolese politicians these days like to boast, as a governor of one mining province told me, that Congo is the "Saudi Arabia of Cobalt." In fact, in terms of pure market share, Saudi Arabia is a somewhat modest comparison. Congo controls a far greater proportion of the world's cobalt supply than Saudi Arabia does of oil. Even at its height, in 1973, the Organization of the Petroleum Exporting Countries (OPEC) controlled, as a whole, about 55 percent of the world's oil supply.

The rocks containing the metal ores were laid down on the bed of a shallow ancient sea around eight hundred million years ago. "They were deposited in an environment that would look like the Persian Gulf today," explained Murray Hitzman, a geologist who worked for several years in Katanga and now teaches at University College Dublin. As the sedimentary rocks and salts filled the sea over millions of years, the rocks containing the ores were buried. "It progressively got buried by more and more sediments," Hitzman told me. "At some point—we don't know exactly when—they were mineralized" into the deposits that are found there today.

Around three hundred million years ago, through a process known as orogeny, those ore-containing rocks were pushed upward when a southern part of the African continent collided into the landmass that would later lie beneath Congo. "The ore bodies are blocks that got carried up in the salt, several kilometers from where they were formed," Hitzman said. "And they're now higgledy-piggledy, folded, broken upside down, back-ass-wards, every geometry you can think of. And predicting where the next one is in the bush—because some are mineralized and some aren't—is almost impossible."

In some locations—in the suburb of Kasulo, for example, or in the vast Mutoshi superpit, or around the towns of Tenke and Fungurume—rocks containing cobalt were pushed to the surface, forming what some people called cobalt "super-caps," mushroomings of rock where concentrations of the metal were particularly high.

As a result of this geology, the Copperbelt contains some of the world's richest reserves of copper and cobalt. "We have found our largest mineral deposits in the places that are the confluences of all of these factors that are, in fact, geological flukes," said David Evans, a Yale geologist whose field of study includes the formation of continents. "If they weren't flukes, then we wouldn't be in this situation—you could go out to your backyard and scoop up some cobalt." In fact, in the towns and cities of the Congolese Copperbelt, that's exactly what people would come to do.

CHAPTER 2

THE GREAT ROCK THAT
SPREADS ALL OVER THE LANDS

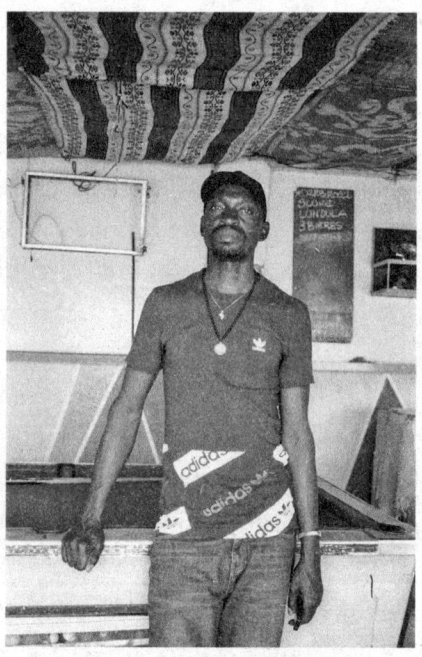

Odilon Kajumba Kilanga in Kolwezi's
Gécamines neighborhood in 2022

B y the time I met him, in September 2019, Odilon Kajumba Kilanga
was thirty-two. He had lost his right front tooth in a brawl four
years before. ("I was in a group of people who clashed with another group of people," he said. "We didn't get along.") His eyes were
sunken and hollowed, and his level gaze suggested he had experienced
much more than a man should in several lifetimes.

Kajumba knew that some people had made it big in Kolwezi, but their number was vanishingly small, and once they'd made their money, they had often gotten as far from the pits as possible, moving on to brighter, airier places in South Africa or even as far away as Malta. The name of the island in the middle of the Mediterranean didn't mean much to someone like Kajumba, though—it was just an impossible distance. Rather, he dreamed of making enough money to buy a restaurant, a small place where he could serve food and build a stable life, one in which he could afford to send his four children to school. He said that although many people he knew in Kolwezi wasted all their earnings on partying, alcohol, and even narcotics, he tried to avoid such temptations. Whenever I met up with him, he made a point of drinking cola.

For the moment, though, Kajumba was stuck in that suffocating little room, where the smells of manioc root mingled with human sweat and rushed the nostrils. The room was in a small stand-alone cinder-block structure on the edge of one of Kolwezi's teeming *cités populaires*—slums where wastewater would run in rivulets down dirt hills into patches of garbage. Two walls of the room had been painted green in an effort to lighten the mood, but the color had become caked with grime. The windows were fitted with sheets of metal rather than glass panes. On one of the other walls, which were painted a liverish red, there was a picture of his brother. In the image, Amos, bathed in light, was depicted as an acolyte of the church that Kajumba and the two Mputus—Trésor and Yannick—attended. The church was the "thirtieth Pentecostal community in Congo," a fading sign painted on the building's facade proclaimed. It was better to have faith if you were poor and lived in Katanga.

IT WAS NOT ALWAYS SO. THE PEOPLE OF KATANGA KNEW THEIR UNDULAT-ing country was rich long before the Europeans conducted their soundings and their surveys. Katanga's wealth wasn't just in red metal; it lay in its land, hills, and high-altitude savanna, scattered with lakes and trees. If you had stood atop one of the region's many hills in AD 1600, say, you would have seen swaths of woodland pocked with termite mounds. Perhaps, in a *dembo* or a *dilungu*—an area of low-lying ground where the drainage was poor—there might have been areas of high grass where an-telope grazed. "The plains pullulate with strange animals whose equals don't exist anywhere else," an early European traveler to the region wrote, "and it's an endless pleasure to contemplate the myriads of antelope mov-ing from place to place." By the great lakes of the Upemba Depression, which drained into the Congo River, there were strips of marshland where hippos bathed. Fishermen had lived in villages around the lakes for at least a thousand years.

The people living in Katanga's millions of acres of clear forest had learned to live among the trees. In the *miombo*, as the forest was known, there were trees for heat and cooking (*musamba*, which makes a good char-coal), trees for building (the sturdy-branched *muputu*, or zebrawood), trees for healing (*kafissi*, whose roots had medicinal powers), and trees for harming (the bushman's poison, whose sap was used to coat deadly ar-rows). There were, too, trees for eating, trees of myriad shapes, sizes, and colors that bore nourishment: the orange fruit of the *mubambangoma*, the single green thorn, the pulp of the yellow-flowered *kabalala*, and the wild golden custard apples from the tree known as *mulolo*. The land was lush and verdant after the rains, but during the dry season, it became cold and parched. The soil turned to dust and stained the trees ocher.

You also would have noticed villages—perhaps sending smoke into the sky, perhaps vibrating with drums and dances to please ancestors whose spirits were all-important. And mines had always been an impor-tant part of Katanga's lived landscape. The most profitable of them were ruled over by powerful rulers. As the historian Eugenia W. Herbert has noted, the ability to work metal may have conferred regality or mag-ical powers upon kings. In Katanga, the secrets of master smiths were passed from generation to generation, and sorcerers invoked ancestral spirits before mines were dug into the ground, chanting, "You who have preceded us, it is you who have opened for your children the entrails of the mountain. Grant that we may find treasure." By the fifteenth century, cross-shaped copper ingots smelted in Katanga had become currency in regions across Central Africa, and the people of Katanga began to band together.

The kingdoms grew and the peoples of Katanga created armies and fought one another for control of resources, human and otherwise. Slaves were captured and traded between rulers and chiefs, sometimes over long distances.

Looking about the region sometime around 1600, you might have also seen a conflict between two kings, one known as Red, the other as Black. The date is only approximate because the peoples of Katanga did not use writing, although some used *lukasa*—"memory boards" studded with beads—to help them remember their history, which was passed down orally.

According to the most common version of the story, Nkongolo Mwamba, the Red King, was the descendant of peoples from east of the upper Congo River. As a boy, he had watched a colony of driver ants destroy a more numerous colony of termites, and he had resolved to dominate other men. Kalala Ilunga, the Black King, was a hunter who grew up in Nkongolo's capital and helped him subdue some of the copper-rich lands to the south. One day, the Black King beat the Red King at a ceremonial game played with a rubber ball, causing the latter's mother to burst into a fit of laughter. The Red King was so upset that he buried his mother alive and planned to kill the Black King. The younger man escaped, however, and fled across the upper Congo River. He ultimately returned with an army to defeat the Red King.

The dynasty that the Red King founded, that of the Luba people, would last until Belgian colonists arrived in the region. Praise phrases—short mnemonic poems that are still passed down in Katangese villages—composed for Kalala Ilunga reflect the expansive understanding of his kingship: *Ami ne dibwe dya kyalantanda; kekudipo ntanda ya shile* ("I am the great rock that spreads all over the lands; there is no land that it does not reach") and *Ami nkidopo mukalo na muntu* ("I have no boundaries with any man"). The Luba king might well have been talking about Katanga today, or at least its minerals, which have spread, through technology, to every part of the globe.

Over the next two hundred years or so, the Luba Empire grew and split into other kingdoms. By the time the first written records of Katanga began to appear, the country was divided, broadly, into three kingdoms, or empires—the Luba, Lunda, and Yeke. (Other groups of people, including the Sanga, existed at the peripheries of these realms.) The alliances formed, and the wars waged, in the days before the colonialization of Katanga would continue to profoundly affect Congo into the twenty-first century. In his work on mining in Katanga, the social scientist Claude Iguma Wakenge points out how "politico-ethnic relationships," many of which can be traced to the separation of the early Katangese kingdoms, continue to create informal governance structures and corruption in the Congolese mining industry. As one local administrator put it when speaking to Iguma in 2018, "The governance of the Katangese extractive sector is shaped with politics and ethnicity."

IN 1806, TWO ENTERPRISING MIXED-RACE PORTUGUESE TRADERS, OR *pombeiros*, arrived in the area and described for the first time to the outside world a hilly country governed by powerful master smiths. One, Pedro João Batista, wrote in his diary that "green stones (malachite) are found in the ground, called 'catanga.'" This was probably the first written instance of the name Katanga. Batista's words were later translated into English as an exploration mania gripped the colonial European powers; to a certain type of Victorian Brit, Frenchman, or Belgian, the mere mention of the journey of the *pombeiros* would have conjured some magic. The same green stones that Batista wrote about are the ore from which cobalt and copper are extracted today.

When the German explorers Richard Böhm and Paul Reichard traveled to Katanga in 1880, they reported back that the area was ruled by a king named Msiri,* who, they claimed, kept a large collection of human skulls hanging "like hats on pegs" near his abode. Msiri's empire was known as Garenganze, and his people were the Yeke, named after the Sumbwa word for a guild of elephant hunters. Other missions ensued, and, in 1894, Jules Cornet, a young Belgian geologist in the service of King Leopold II, declared that the land was a "geological scandal" because it was so rich in minerals. By that point, Leopold had claimed the

---

*Also called Msidi or Mushidi by contemporary writers.

Congo Basin as his personal territory, saying that he wanted to bring trade and civilization to the heart of Africa. Missions like Cornet's suggested that extraction for profit, not humanitarianism, lay at the heart of Leopold's intentions from the very first days of colonialism.

The king urged Cornet to keep his discoveries of the country's rich minerals secret. It was rumored that Katanga's rivers and streams were full of gold. The British also had designs on the region and its fabled wealth, and the competition between Britain and Belgium became known as the "scramble for Katanga."

Cornet began his work on the quiet, drawing up the detailed maps of minerals and ore bodies, replete with charts and cross sections of the mineral-rich earth beneath. They are still consulted by miners looking to make a fortune. In a drafty annex of the Archives of the Royal Palace, in Brussels, the librarian looked at me askance when I asked to see boxes of archival material from the Union Minière, the Belgian company that came to mine southern Congo. It was 2022. "Most people who come here to look at those boxes," the librarian said. "They're looking to find gold."

MSIRI WAS NOT A LOCAL—HE WAS SOMETHING OF A COLONIST HIMSELF. From his homeland to the east, in what would become Tanzania, he had followed his father, a successful copper trader, to Katanga in the 1830s. Lured by the province's riches, Msiri began to set up a state in Katanga around 1856. To build his empire, he used strategic marriages and brute force and benefited from rich mines of the Copperbelt. A decade and a half later, he had seized a chunk of territory roughly the size of Alabama. Msiri brooked no opposition and used his superior army to capture slaves on a massive scale. He sold these slaves to traders and made them into concubines. By 1891, the Sanga people, to the south, began to fight back against Yeke intrusion into their ancestral lands.

Msiri needed foreign trade not only to acquire modern weapons but also to fight against his enemies. Arab slavers were cutting into his bottom line as they seized more and more men from Central Africa to work as slaves on clove plantations in Zanzibar, on Africa's east coast. Although

European nations had forbidden slavery early in the nineteenth century and used antislavery campaigns as pretexts for creating colonies in Africa, they often treated the Africans they had supposedly liberated from bondage as little more than slaves.

Global markets were having a growing effect on Katanga: Cloves were a commodity buffeted by the winds of international commerce. The king, who had an almost preternatural understanding of such forces, opened new trade routes with the west coast of Africa. At first, Msiri welcomed European visitors to Bunkeya, and they were impressed by the Yeke capital: Frederick Stanley Arnot, a missionary who chose to live in a series of huts near the king's palace, wrote in a letter that "life and property, I have no hesitation in saying, are safer here than in much-favored England." But king or no king, African territory not occupied by a European power was considered *terra nullius*, "no-one's-land," ripe for the taking.

Leopold was determined to have Katanga for himself. His men rolled into Bunkeya in late 1891. A Belgian officer, Omer Bodson, shot and killed Msiri after he refused to accept Leopold's rule. The Belgians hoisted the flag of the Congo Free State, a yellow star on a deep-blue background, over Bunkeya, and then they began to plunder the territory. They needed ivory—tons and tons of elephant tusks—to satisfy the bottom line for Leopold.

By 1893, the Belgians had created a regency in Katanga that included Msiri's descendants. They used imported labor too. Long caravans of migrant workers, often from the Kasai region, were brought to Katanga on foot from hundreds of miles away. The Sanga people, known for their elephant-hunting prowess, intensified their guerrilla warfare, which they now focused on the new Yeke-Belgian order. Clément Brasseur, the Belgian lieutenant who had been assigned to run part of Katanga, decided to employ brutal tactics to suppress the rebellion, allowing the Yeke to raid and take slaves. According to a later inquest into violence in the Congo Free State, Brasseur was known by the locals as Nkulukulu, after a bird "whose inner wings are bloody red." As a magistrate later explained, "the natives say Mr. Brasseur was only happy when he had blood up to his armpits." By early 1899, just over eight years after Msiri's death, the Sanga were defeated and Katanga had been forced into submission.

MUCH HAS BEEN WRITTEN ABOUT LEOPOLD'S MURDER OF MILLIONS OF CON-
golese in his crusade for rubber, in Congo's North, but little about what
happened in the South. Kajumba's living conditions in 2019 harked back
to the stage of colonization that came next. In fact, his room seemed to
have been lifted directly from *Vocabulaire de ville de Elisabethville*,* one of
the first collections of Congolese testimonies of life in southern Congo
after Leopold had been censured and the Belgian state took control. The
Vocabulaire describes the lot of "Domestics" who worked under Belgian
colonialists in the early part of the twentieth century. The Belgians
"brought a bad sort of slavery to us, the Congolese," the unnamed narra-
tor bitterly recalls. "In that respect the Whites had a very bad spirit indeed. Because they thought [it good] to build for us black people just a
one-room house." The Belgian masters would have large homes with
extra rooms, but they would often keep animals in them rather than give
them to their servants.

In the twenty-first century, the only difference was that Katanga's big
houses were inhabited by Congolese politicians and businessmen who
had figured out how to hawk their country's wealth and keep most of the
profit. A few foreign businessmen—mostly Chinese, Lebanese, Belgian,
and Israeli—joined their ranks. The lives of this select elite were spent
in palatial mansions decked out in acres of marble. They frequently chartered private planes to South Africa, Zambia, the Middle East, and Europe, flying out of Kolwezi's tiny airport. For fun, they rode Jet Skis on
nearby Lake Nzilo.

Around Kajumba and Trésor's home, a few of the trio's meager pos-
sessions lay scattered about. At first, they were embarrassed to show me
around. An old cathode-ray television sat in one corner, and two shoe-
boxes (MADE IN P.R.C.—the People's Republic of China) filled with combs
rested on the shelf beneath it. Atop the TV and on a nearby ledge had

---

*The Belgians founded the city of Lubumbashi and called it Elisabethville after Queen
Elisabeth of Bavaria, wife to King Albert I, the third king of the Belgians. The name was
changed to Lubumbashi in 1966.

been placed some shampoo bottles, tubes of toothpaste, and a single toothbrush; in a bucket, some cassava and a long knife. Behind the bed, a hanger that was suspended from a nail in the wall held neatly pressed suits, shirts, and jackets, many with natty checks and stripes. It was essential for Kajumba and his roommates to keep up appearances. "No matter how you live, you still have to look good," Trésor told me. And then there was the bed itself, barely big enough to fit one person, let alone three grown men with long, lanky limbs.

CHAPTER 3

THE BEGINNINGS OF A BATTERY

The ink was barely dry on M. Stanley Whittingham's chemistry doctorate. The twenty-seven-year-old chemist was keen to get away from Britain's postimperial grayness. It was February 1968, and he wanted, as he remembered, "to go someplace where the sun shines." Whittingham, who sported a sweep of black hair and thick spectacles, began to look for jobs in California.

Nobody could have known it at the time, but the decision of the young Englishman to relocate in search of warmer weather would trigger a chain of events that ultimately led to ubiquitous smartphone-computers tucked into billions of pockets and noiseless electric vehicles gliding down city streets. It would also kindle a potential energy revolution; offer one solution to climate change; and spark a competition for resources unseen since the days of the Cold War, with states pitted against one another in a mad rush to mass-produce batteries and thus eliminate carbon fuel. Whittingham's work eventually garnered him the highest scientific recognition: In 2019, he won the Nobel Prize in Chemistry—along with two other scientists, John B. Goodenough and Akira Yoshino—for his contribution to creating the first lithium-ion battery.

For his master's degree at Oxford, Whittingham had studied alloys made out of tungsten metal. "We were looking at how fast the ions moved inside them," Whittingham said when we spoke on a winter's day in early 2020. Ions are atoms or clusters of atoms that hold an electric charge; when they move, they create an electrical current. The Cold War loomed large. "Part of that was tied to what was then the beginning of the space race," he told me, "and part of my research at Oxford was in fact paid for by the U.S. Air Force."

After completing his master's, Whittingham continued to study the tungsten bronzes, undertaking a PhD with a fellowship from Britain's Gas Council. "One month before I started, Britain hit natural gas in the North Sea," he said. "So I was told, 'You can work on anything you like. You've got your money. Don't bother us with reports or anything else.'"

The freedom the Gas Council gave Whittingham facilitated his work on the tungsten bronzes. He showed that some ions moved rapidly within them. "That was it," he said. "The beginning of a change, because at that point, I was going to be working on catalysis to turn coal gas into the equivalent of natural gas. Lucky changes, if you like." The transport of ions would be his occupational focus for much of the next decade and beyond—their movement would be key to the creation of lithium-ion batteries. After his Gas Council fellowship, Whittingham decided to move to the U.S. for a postdoctoral fellowship at Stanford.

Whittingham quickly fell in love with America, conducting his research during the week and hiking in California and Oregon's national parks on the weekends. On a trip to the San Francisco Opera, in August 1968, he met his wife, Georgina, who was at that time studying for her master's, with a concentration on Latin American literature and poetry in Spanish. Within seven months, they were married. Their four grandchildren are all native Californians.

The U.S. was intellectually stimulating for Whittingham. At Stanford, he continued to work on fast-ion transport, focusing on a compound called beta-alumina that Ford had been developing since 1967 for use in its own research into electric cars. He began to grasp the implications of his work—how it might be used to create a new, powerful battery—and the need for such a battery to power vehicles. Ford's battery, however, was impractical, as it needed to be kept at three hundred degrees Celsius— far too hot for the inside of a car.

The next year, 1972, Whittingham traveled to northern Italy to attend a conference at Belgirate, a shrine-studded village clung to the shores of Lake Maggiore. The meeting was sponsored by NATO. Among Belgirate's faded waterfront villas, Whittingham presented a paper on tungsten bronzes. The Belgirate Conference, as it came to be known, had a seismic impact on the new study of ion transport, and it was crucial to the discovery of lithium-ion batteries. "The original Belgirate Meeting made for the first time visible the technological potential related to the phenomenon of the fast ionic transport in solids," Bruno Scrosati, an Italian electrochemist, later wrote.

The stage had been set for a new scientific field, known thereafter as "solid state ionics." Studies in this field have led to a plethora of innovations we rely on today, such as electrochromic windows and mirrors, chemotronics (the development of optical and chemical sensors), fuel cells, nanotechnology, and water electrolysis. At Belgirate, another idea had come into sharper focus, and the outline for a new kind of battery had implanted itself in Whittingham's head. The conference would end up having its greatest impact on a technology that would make its way into billions of pockets around the world—batteries.

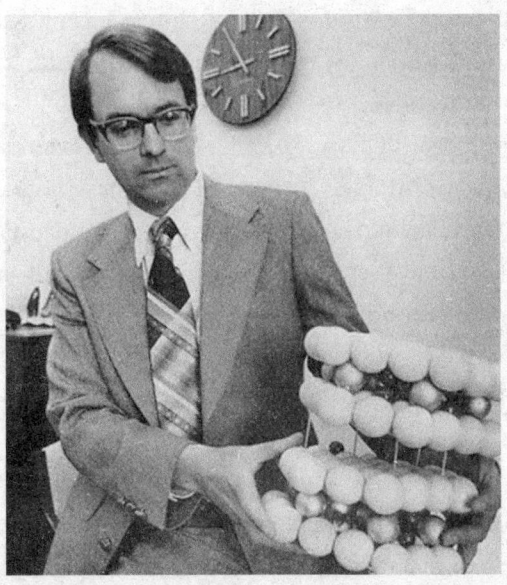

M. Stanley Whittingham in his Exxon lab

CHAPTER 4

THE LAND OF THE THREE KINGS

Growing up in 1990s Congo, Odilon Kajumba Kilanga did not look to copper and cobalt miners as his role models. He and his family, who lived on the outskirts of the city of Lubumbashi, were poor, but they had a sense that there was a larger world beyond the suburb they called home. After all, his father, Joseph, a tire salesman, could claim official connections. Joseph's uncle Joseph Kufi Kilanga was a government minister, and when he had a child Odilon's father gave his son the postname Kilanga.

Kufi exemplified success to Kajumba and to his extended family, even if they barely knew him. He was born in 1935, in Mbulula, a small town not far from a tributary of the Congo River in the northernmost reaches of Katanga. At that time, Congo still belonged to Belgium, and the Belgians had warped the concept of ethnicity there through conquest and games of colonial favoritism—divide-and-conquer strategies with far-reaching consequences for the way profits from battery minerals are distributed in the twenty-first century.

"My family are Hemba," Kajumba told me. The Hemba have a reputation among other Congolese, deserved or not, for being quick of temper. "The Hemba are a disciplined people. In Katanga, we are known as an excessively brutal people, a bit like the cowboys of Texas—you know, a warrior has lots of discipline," Gaylord Kilanga, Kufi's son and Kajumba's cousin, told me. "We raise our children in a very disciplined way, as if we were in the army." Kilanga told me that many Hemba, including himself, think of themselves as descendants of a "Nilotic" people from Ethiopia, though scholars also trace their roots to a nineteenth-century migration from Tanzania and to a northward progression of Lunda groups from the Copperbelt. Many Congolese consider them to be originally from Katanga, distant cousins and neighbors of the far more populous Baluba-Katanga, or Luba, people, who can trace their origins to the fight between the Red and Black Kings.

The Hemba are perhaps most famous internationally for their art and sculptural figures, the most well-known of which are known as *singitis*. These sculptures, which occasionally have two faces, emphasize different attributes of kingship. *Singitis* were held inside special huts at the centers of their towns, only meant to be accessed by designated caretakers. During Belgian rule, however, the sculptures were looted by colonialists, who prized them for their careful craft; in the 1970s, many were sold to unscrupulous dealers. The catalog entry for one Hemba statuette at New York's Metropolitan Museum of Art explains how such a large-eyed, big-bellied sculpture emphasized two important factors in Hemba society: the eyes, for education and experience, and "the umbilicus," which signified an ancestral link, or "the point of connection with one's extended lineage."

Kufi, Kajumba's great uncle, was born, according to his son, "into a family of very weak fortune." By the age of five, he had lost both his mother and his father, and he was entrusted to a Belgian Catholic priest for his upbringing. A bright boy, he was sent to study at a seminary in the nearby town of Kongolo. In the mid-1950s, when he was in his early twenties, the priest who had cared for Kufi sponsored his study in Katanga's capital, more than six hundred miles to the south. The education of Black Africans in Congo was a novelty: The colony did not open its first university until 1954. "*Pas d'élite, pas d'ennuis*," the Belgians would say. "No elite, no trouble."

If things were getting a little better for Congolese, they were getting much better for the colonialists, especially those who were involved in the mining sector. The companies were some of the most profitable global stocks: the Apples or Nvidias of their time. This is a fact that is barely acknowleged these days: More than sixty years after independence, some Belgian journalists still cite a study from 1957 that claims that Belgium put more into Congo than it took from it over the fifty-two years of colonization. The money that Belgium supposedly "lost" in Congo stands at more than $1.5 billion in 2025 U.S. dollars. But this ignores the specific nature of Belgium's colonial project, which was for the state to support private enterprise through investment and cloak private gain in corporate structures, a practice that was passed down to the rulers of an independent Congo and their foreign backers. And, incidentally, a practice that is still alive and well today.

In fact, the levers of business and commerce had made Congo into one of the most purely extractive colonies in world history. Huge fortunes were made in the private sector, especially by way of mining. Shareholders, including the Belgian state and Belgian and international investors, were raking it in. Through the mining of copper and cobalt, the Union Minière, a single company, produced net profits of over thirty-one billion Belgian francs (more than $6.8 billion in 2025 U.S. dollars) in the years between 1950 and 1959 alone. During the 1950s, the company was one of the largest suppliers of copper to a swiftly electrifying world. "The colonial economy was a goldmine for Belgian investors and the yields of Congolese mining companies were especially impressive," the political scientist Emizet François Kisangani wrote in a 2023 reassessment of Congo's colonial economics. "They ranked among the world's best financial performers."

Congo did not exist in a vacuum. Its corporatist structure meant that global market trends determined the colony's economic fate. In the mid-1950s, after Kufi arrived in Elisabethville, the world economy slipped into a recession, and with its outsize reliance on exports, Congo suffered.

In 1958, the head of a Belgian tin-mining concern's economic studies unit summed up the situation: "Unemployment has appeared in the Congo, whereas previously the country had always had a shortage of labor." Elisabethville found itself in the middle of a political foment aggravated by a lack of jobs for Congolese, who had increasingly begun traveling to cities to find work. Long mistreated by their colonial masters, the Congolese found a voice in the calls for liberation that were starting to be heard across the African continent. People began to agitate for a Congo run by the Congolese, not by the Belgians.

AS KUFI CONTINUED HIS STUDIES IN ELISABETHVILLE DURING THE SECOND half of the 1950s, a new category of ethnic identity was becoming central to the politics of the era: Which people a Congolese belonged to, what language they spoke, and what region they came from suddenly became dividing lines. Katanga was hurtling toward decolonization, but a new class of politicians was asking: *Decolonization for whom?*

By the late 1950s, an unsuccessful Katangese trader named Moïse Kapend Tshombe was one of those politicians. He was local royalty by virtue of his marriage to the daughter of the Mwaant Yav, a traditional king of the Lunda people whose title meant "lord of the vipers." Since colonizing Congo, the Belgians had used so-called *chefs coutumiers*, or customary chiefs, often the relatives of kings and lords they had vanquished, to reinforce their control. Under European rule, the privileges of these chiefs were mainly symbolic, but as the move toward decolonization gained momentum, they became rallying posts for political factions.

Tshombe used his royal connections to become one of the most important political voices in Katanga. The Lunda were especially worried that a postcolonial Congolese government headquartered outside Katanga might strip the South of its mineral wealth. This, they pointed out, would simply mean replacing their rulers in Brussels with another faraway elite who had little to do with them. This dynamic continues to be key to the politics of Katanga during the current era. Katangese point out that the province does not receive its fair share of the wealth generated by its copper-and-cobalt mines. As the sociologist Claude Iguma Wakenge has noted, when a politician who was considered "not a Katangese" was appointed to a ministerial position overseeing mining for the Haut-Katanga provincial government in 2017, controversy erupted, and the politician was shifted out of the mining role.

The seeds of an ethnic dilemma that continues to plague southern

Congo were planted in the 1920s, when the Belgian colonial government embarked upon a policy of "labor stabilization." At the time, scores of African workers were dying in their mines. The heads of the Union Minière and the colonial administration decided to create workers' compounds with health care and schools for workers who had transferred to Katanga. Foremost among them were a people called the Luba-Kasai, who quickly rose in the Belgians' estimation. In 1951, a Belgian scholar described the Luba-Kasai as "intelligent, robust, hard-working and entrepreneurial; they have a decided willingness to cooperate with the occupying whites to adapt and elevate themselves in the new state of things which has come to throw their primitive situation into disarray." By 1956, the Union Minière employed a staff that was 53 percent Luba-Kasai.

The Lunda—along with the Yeke, Sanga, and Luba—called themselves "autochthonous." They saw themselves as the true inheritors of the wealthy land that the Belgians were about to leave. To them, the Luba-Kasai were "foreigners" who were interested in stealing minerals.

In October 1958, Tshombe founded the Confederation of Tribal Associations of Katanga, better known by its French acronym as the Conakat. Kufi did not agree with its policies, which he thought of as "tribalist," but many Southerners began to side with Tshombe as he promoted a Katangese ethnostate.

In many ways, Tshombe's ideology rested on a foundation that the Belgians had laid. The party's leaders cherry-picked among colonial-era studies by European anthropologists to create a vision of a Katangese ethnic group. Godefroid Munongo Mwenda M'Siri, another of the Conakat's cofounders and a direct descendant of King Msiri, adeptly baked these into an ideology. "When the first white explorers discovered the part of Africa called Katanga," he wrote, "they found three monarchies which were not only bound by family, economic and social links but—and this is by far the most important—their historic destiny had been linked for centuries."

NOT ALL CONGOLESE THOUGHT A POSTCOLONIAL STATE SHOULD BE SPLIT along ethnic lines. People who saw a future for a united Congo drew inspiration from around Africa. The late 1950s were heady years across the continent. In Ghana, Kwame Nkrumah, a teacher and organizer who was inspired by the Indian independence movement of Mahatma Gandhi, used nonviolent protest, or "positive action," to eject the British in 1957. The Congolese, and especially a young post office clerk named Patrice Émery Lumumba, looked up to Nkrumah and other African nationalists. Lumumba was ten years Kufi's senior, but his education had taken a similar form: He had studied at seminaries and worked his way through the colonial hierarchy. In October 1958, the same month Conakat was founded, he created the Congolese National Movement (MNC).

The MNC was a party that espoused nationalism and stood in opposition to other Congolese parties that had been founded along ethnic or tribal lines. That year, Nkrumah, who had become Ghana's prime minister, invited Lumumba and African activists from twenty-seven other African countries and territories to Accra for an All-African Peoples' Conference. "Hands off Africa!" the event's slogan shouted. "The wind of freedom currently blowing across all of Africa has not left the Congolese people indifferent," Lumumba declaimed during a speech that would launch him into the world's consciousness. He inveighed against "colonialism,

Patrice Émery Lumumba visits Brussels in 1960.

imperialism, tribalism, and religious separatism, all of which constitute a serious obstacle to the blooming of a harmonious and fraternal African society."

For Kufi, the Unitarian politics of Lumumba were more to his liking. "He was a student but Lumumbist in his way of being," Kilanga told me of his father. By 1959, many Luba, too, despite their position as "autochthonous Katangese" in the ethnic hierarchy that the Conakat had drawn up, began to support the ideas of Lumumba. The Association of the Luba People of Katanga (Balubakat) allied itself with the MNC. Kufi looked up to the Balubakat, particularly its charismatic leader, Jason Sendwe, and during breaks in his studies, he began to organize for the movement to keep Congo whole.

After years of repression under the Belgians, Congo was now barreling toward independence, but the country had been barely developed by its colonizers. Roads led to mines and coffee plantations, but away from the axes of profitability, and many Congolese were completely cut off from the world. There were, among them, few of the functionaries, lawyers, and doctors who would be needed to staff a modern state. And there were no senior soldiers to run the country's army and prevent mutinies. When the competing Congolese political parties met in Brussels at the beginning of 1960, Tshombe and the Conakat wanted to retain control of the province's mineral wealth and maintain an oversight role for the Belgian king. The other political groups, led by Lumumba, overruled Tshombe. The stage had been set for a rupture that would tear Congo to shreds.

CHAPTER 5

THE PRIME MINISTER'S TOOTH

About thirty miles into the drive eastward from Lubumbashi, the road forks. Take a left and you will eventually end up in Kolwezi, the mining city where Kajumba lived, and where most of Congo's copper and cobalt is mined. On the right, near the little village of Shilatembo, stands a painted concrete gateway topped by a cheery greeting: BIENVENUE AU SITE TOURISTIQUE LUMUMBA.

Congo quickly fragmented after it won its independence on June 30, 1960. The army revolted and bands of young people, or *jeunesse*, as they became known, began forming into militias. Parts of the country began to separate from the new government's rule. Just eleven days after Congo gained independence, on July 11, Moïse Kapend Tshombe and the Conakat proclaimed Katanga's own. In northern Katanga, Jason Sendwe and his Balubakat party, still committed to a unified Congo, began to fight Tshombe. In August, the Luba of Kasai declared their own independence as a diamond-fueled "mining state." Their leader, Albert Kalonji Ditunga, took the traditional title *Mulopwe*, an honorific referring to a sort of king who drew his power from the spirits of ancestors. "The basic idea was to obtain a province for the Baluba [Kasai] because they were persecuted everywhere," Kalonji would later write. South Kasai would be an Israel of sorts for the Luba-Kasai, who were sometimes called "the Jews of Africa." The country, which had so recently radiated with hope and proclamations of liberty, was now a series of fragmented fiefdoms, each at war with the other.

One of the most famous, and earliest, victims of the scramble to control Congo was Kufi's ideological hero from his student years, Lumumba, who had become Congo's first prime minister, the ruler of a state deeply riven with divisions and outside suspicion. He quickly fell victim to Cold War shadowboxing and was branded a Communist by many in the United States, including the administration of President Dwight D. Eisenhower. The country was regarded through a Cold War lens. In fact, the way Congo was seen was not too dissimilar to the way it would be seen in the first decades of the twenty-first century, a prize in a game of great powers, all jockeying for resources. And, just as in the 2010s and 2020s, the people of Congo would be completely ignored in the struggle for power.

Within a few weeks of Lumumba taking power, foreign actors were scheming to have him deposed. Larry Devlin, the CIA station chief in Léopoldville, began to pay Congolese agitators to stage protests against Lumumba's government. In August 1960, Eisenhower became the first U.S. president to order the murder of an overseas head of state when he suggested to a meeting of the National Security Council that Lumumba be killed. Lumumba's enemies at home were fiercer; chief among them was his former confidant, Joseph-Désiré Mobutu, who overthrew the government and placed Lumumba under arrest. A Belgian adviser suggested sending him to near-certain death in Katanga. Devlin did not protest, sealing Lumumba's fate.

On January 17, 1961, Lumumba, as well as Maurice Mpolo and Joseph Okito, two politicians who had tried to flee with him, were bundled onto a four-engine DC-4. The plane was crewed by two Belgians, a Frenchman, and an Australian. A group of Luba-Kasai soldiers was on board. They hated Lumumba for a massacre that the Congolese military had carried out in Kasai. The plane's radio operator vomited after witnessing how savagely Lumumba and his companions were beaten. At the airfield, Tshombe's officials, supported by Belgian advisers, took Lumumba and the two politicians from the plane, tortured them in a bungalow, shot them in the field at Shilatembo, buried their bodies, and then exhumed their corpses to dissolve them in acid.

———

I VISITED THE SITE IN 2022, SIXTY-ONE YEARS AFTER LUMUMBA'S DEATH. A sign at the entrance welcomed Félix Antoine Tshisekedi Tshilombo, Congo's president, and King Philippe of Belgium. A few weeks beforehand, a gold-capped tooth in a coffin had been paraded there in front of the president and the king. The tooth was the only known remnant of Lumumba. It had been yanked from his bullet-riddled corpse as a trophy by two Belgian brothers who helped the Katangese secessionists dissolve the bodies in acid. In 2016, the tooth was retrieved by Belgian authorities. Now they were returning it to Congo. It was a grisly way to show contrition, but it was one with which the Congolese would have to make do: Even by the mid-2020s, Belgium would refuse to make a full apology for the crimes of colonialism and its aftermath.

The site was empty. Willy Nkuwimba, a semi-reformed rebel, joined me on my visit. Nkuwimba had belonged to a militia that had operated in northern Katanga through the country's long civil wars in the late 1990s and early 2000s. The group he belonged to had morphed into a separatist enterprise that used Tshombe's flag. Nkuwimba, whom I had picked up in the town of Likasi, wore a pink three-piece suit and made a point of chatting into two cell phones at the same time. He liked the idea of stopping at the Lumumba site. He said he hadn't visited before, and he listened intently as a caretaker told us about the landmark and Lumumba's struggle to unify Congo.

Nkuwimba told me that the visit to Shilatembo had affected him deeply. "I am ready for a fate like Lumumba's," he remarked solemnly, despite the difference between both of their political ideologies. "I want to be a great leader." I wondered what would have happened to the country had Lumumba managed to hold on to power. The site was empty, save for a few people sweeping dust and dozing beneath trees in the midday sun. Perhaps he wouldn't have managed to keep Congo together and Katanga would have become its own state; maybe Congo would have transitioned to democracy; or maybe the conditions for the wars that would plague the country for more than half a century had already been baked in under colonialism. In any case, the Congolese had been robbed of the chance to decide their future.

At the site, two workers desperately tried to reaffix the severed head of a statue. It kept rolling off.

CHAPTER 6

A PATRIOT WITH A CAUSE

Mining had provided a full 20 percent of the entirety of Congo's gross domestic product at independence in 1960. Robbed of it, the rest of Congo was swiftly going bankrupt. In Katanga, Moïse Kapend Tshombe funneled mining revenue into building his state, buying weapons, and hiring mercenaries. He also purged the Luba-Kasai, herding them into internment camps through policies the academic Thomas Bakajika Banjikila has called "ethnic cleansing."

Despite its wealth, Katanga was now completely divided between Unitarians and secessionists. A UN envoy decried the situation, which was said to be approaching civil war. Kufi, who was studying in Elisabethville when Katangese independence erupted, supported the Unitarians. "He was not a rebel; he was a civil activist, one who believed in the unity of the republic," Gaylord Kilanga, his son, told me.

Throughout the Katangese secession, the Union Minière and its Belgian and European shareholders continued to make money from Congo's minerals. Internal memoranda show that the firm even exported *more* metals during these years than it had during the 1950s. In 1960, the company made commercial agreements that would last for ten years after decolonization. They were banking on little changing. "Africa is still administered in great part by Europeans," summed up the author of an internal memo circulated at the Union Minière in 1960. "Even if Europeans have become advisers, their needs and their numbers shall not diminish before many years have passed." In fact, the report suggested that the European presence would double over the next twenty years. In a copy of the note held at the Union Minière's archives, the word *double* is underlined, as is a section contending that foreign enterprise could set advantageous terms, given the "risks" inherent in working with newly independent states. The implication was clear: The Union Minière could turn the chaos in Congo to its own advantage.

IN JANUARY 1963, THE UNITED NATIONS, FED UP WITH WHAT IT SAW AS BELgian meddling, crushed the fledgling Katangese state. Tshombe made a last stand at Kolwezi. Protected by several thousand of his gendarmes, he announced that he had rigged the giant Union Minière factory there with explosives. He threatened to destroy it, along with Congo's economy. After thirty-six hours of negotiations through a pair of Belgian intermediaries, however, Tshombe capitulated in return for amnesty for his secessionists. Tshombe's gendarmes melted westward and began establishing bases in Angola.

As the secession ended, Kufi graduated from college. He took a job as a lecturer, teaching economic science. Katanga enjoyed an uneasy peace, but the rest of Congo was aflame. The central Congolese government used the former leaders of the Katanga secession to consolidate power, and then eliminated them. In 1964, less than a year after his capitulation at Kolwezi, Tshombe was brought back to serve as prime minister; he used his gendarmes, as well as Western mercenaries, to fight against Chinese-backed Simba rebels in the East.

China's role in the crucible years of the Congo crisis was a considerable one, and it foreshadowed the influence Beijing now wields in the country. Antoine Gizenga, a minister who now led a faction of the rebels, had first been in touch with Beijing in September 1960, when he requested food, finances, and weapons for Lumumba's government. But the Chinese had vacillated, seemingly unsure as to where they fit between the machinations of the U.S. and the Soviet Union.

Beijing claimed to be spreading Communism through its support of rebel groups like Gizenga's in the country's heavily forested East. A secret Chinese intelligence memo from January 1961 judged that Congo was surrounded by enemies whose interests weren't particularly aligned: the "new American colonialism," hoping to install a "puppet state" in Kinshasa; Belgium, bankrolling the Katangese separatists; and Britain, stationing troops along the Congolese border. "In their concerted action to enslave and slaughter the Congolese people they cannot conceal the acute struggle among themselves," the report's writer averred. "The road of struggle of the Congolese people is winding, but the future is bright. The key to achieve final victory is the further awakening of the Congolese people, the political ripening of the Congolese people," the report continued. "The situation is favorable, but the leadership is weak."

In the early 1960s, Beijing generally contended that it took an ideological stance when approaching assistance; another Chinese intelligence analysis from 1961 railed against the "American aggressive plot" in Congo. But Beijing's agents were also fanning the flames of a brutal war. China supported rebel movements such as the Simbas, a group led by Pierre Mulele, another ex-Lumumbist minister. Mulele fled Congo in 1961, trained in China, and returned as the head of a quasi-religious cult of rebels in 1964. He elevated the struggle to spiritual heights, telling his men, who had trained with Chinese help elsewhere in Africa, that if they were baptized in water that he had blessed, they would be impervious to bullets, a practice that would last into the twenty-first century.

In mid-1964, the *People's Daily*, China's state newspaper, judged that there was an "excellent revolutionary situation" in Congo, and it compared the rebellion to guerrilla action in another Cold War battlefield—South Vietnam. Shortly after that, in October, the Simbas took nearly a thousand U.S. and European citizens in Stanleyville hostage. Belgium and the United States sent troops and planes to crush the Chinese-backed insurgency.

When the government took Stanleyville, the rebel capital, it began its own brutal purge of the city. Foreign mercenaries hired by Mobutu's government went house to house, shooting and beating people to death. As one observer to the slaughter put it, "Anything black was killed indiscriminately, blindly." The government troops rounded up the remnants of the population, then led them into a packed stadium one by one. If they were applauded, they were set free; if they were booed, they were taken off to be machine-gunned.

The mass executions caused outrage in the West, and when some of the mercenaries were later interviewed on television about what they had done in Congo, the anger over their overt racism intensified. One young man in sunglasses and a dark hat told a BBC interviewer, "A Black man is like an animal to me." But the furor did little to stop the U.S. from supporting the Congolese government, which it now had more or less under its control through Mobutu and his circle. Devlin, the CIA station chief, had become one of the leader's closest confidants.

IN 1964, ZHOU ENLAI, THE FIRST PREMIER OF THE PEOPLE'S REPUBLIC OF China, delineated the Chinese framework for aid most robustly as he toured Africa. His speeches show how China saw its aid to African governments, and how many Chinese continue to see their presence on the continent in the twenty-first century. In 2018, when I visited the Beijing headquarters of the National Development and Reform Commission, an arm of China's overseas economic policy efforts, the commission's chairman referred to a similar set of principles as the basis for China's development policy overseas.

In a 1964 speech, Zhou outlined eight guidelines for the provision of economic aid and technical assistance "on the principle of equality and mutual benefit." Beijing would respect the sovereignty of the countries it was helping and place no conditions on the aid it provided. Low or interest-free loans and "the best-quality equipment and material," as well as training, would be provided so that countries China was helping could start off on the road toward "self-reliance." The point, Zhou insisted, was not to make countries reliant on China but to make a world full of healthy— and presumably socialist—states for the betterment of humanity.

But the Chinese were no less enthralled to strongmen than were their rivals in Moscow and Washington. Beijing quickly came to abrogate aid to favored rebel groups, even as it extolled their virtues. In Congo, as the Mulele revolt failed and rebels retreated to the mountains around Lake

Tanganyika, China's leaders adopted what can most generously be described as a spirit of pragmatism. Across Africa, China rapidly began abandoning rebel groups that it had once supported in order to gain acceptance with established governments, even those of former enemies like Mobutu.

A CIA report from 1972 explains that China's "flexible approach to foreign policy" in Africa involved coupling assistance with ideology. "Peking uses its foreign aid as more than just a lure for recognition. Provision of aid helps project an image of a dynamic, expanding, and modern Chinese economy," the report states. China was competing with the United States not only in Africa but also in the Soviet Union, which it worried was "colluding" with the U.S. to divide the world. "Peking loses no opportunity to establish the image of China as a champion of third world—hence, African—interests against those of the two superpowers."

MOBUTU, RIDING ON THE SUCCESS OF THE ANTI-SIMBA CAMPAIGN, FORmally seized power through a bloodless coup in November 1965. He was thirty-five years old and close to the CIA's Devlin. "The Congo was utterly lacking in leaders with international experience, and Devlin had a close personal relationship with a dynamic, popular *pro-West* young sergeant who trusted him," a U.S. intelligence officer of the era once explained to me. Mobutu would rule Congo more or less absolutely for the next three decades, and Devlin would be haunted by his descent into greed.

Mobutu consolidated power quickly, and Katanga was not exempt from his machinations. "Mobutu recruited several Katangese, and my father became a parliamentary deputy in 1966," Kilanga told me of his father, Kufi. The Union Minière, which under Tshombe had slowly been training Congolese staff to fill roles occupied by Belgians, was first in Mobutu's sights. The government passed a series of laws that increased taxes, but high copper prices allowed the company to weather the levies. New laws were then passed that allowed Mobutu, at the start of 1967, to nationalize the company, which was valued at around forty billion Belgian francs (almost $10.2 billion in 2025 U.S. dollars).

A month later, some among the company's European workforce tried to send their wives and children home. Mobutu's men stopped them from boarding planes out of Congo: The dictator still needed foreigners to make his machine run. The company was renamed Gécomin, then Gécomines, and then, in 1971, Gécamines, or the *Générale des Carrières et des Mines.*

In 1967, Kufi was promoted once more by Mobutu. "He was one of the first Black mayors of Elisabethville," Kilanga told me, proudly. Administrators like Kufi were tasked with yoking Katanga to Mobutu. "He was recruited as a nationalist; he was an ultra-Mobutist," Kilanga told me of his father. "His job was to eliminate pockets of the Katangese secession that remained." The Hemba, he explained, were liked by Mobutu despite their support for the secession—he thought they made good soldiers. The same year that Kufi became mayor, Mobutu finally decided to be done with Tshombe once and for all. The former prime minister had fled, so the general sentenced him to death in absentia, and more of the gendarmes began to filter into Angola. Tshombe died under murky circumstances in Algeria in 1969.

Mobutu also dealt with other enemies. Mulele was lured back from exile in 1968 with a promise of amnesty. Once back in Congo, the former rebel leader was arrested and publicly castrated. His eyes were gouged from their sockets and his limbs sliced from his body.

In the wake of Mulele's death, Beijing's qualms with Mobutu were quickly forgotten. Mobutu decided to extend full diplomatic recognition to China in 1972. Six weeks later, less than a decade after quashing the Simba rebellion, the leader visited Beijing. There he met Mao Zedong, who confided in him that China had "lost much money and arms" supporting the Simba rebels. Mobutu agreed to let bygones be bygones and sought investment. Money was forthcoming—in the form of a $100 million interest-free loan. Returning home, he outwardly adopted Maoist vestiges and decreed that everyone would call each other *citoyen*, or citizens.

After Mobutu's visit to Beijing, the Chinese government was seemingly unperturbed that Mobutu had a close relationship with the United States, or that he depended on U.S. funding and weaponry, or even that he had ordered the death of Mulele and other socialists with abandon. They closed their eyes, too, when he deployed troops around Africa to quell other left-wing uprisings and when he happily agreed to provide minerals like cobalt and uranium to the U.S. for its defense industry. Ideology had taken a back seat in China's policy and Beijing's pragmatic policies that led to the nation's supremacy in the metals markets of the twenty-first century, and, indeed, in the globalized battery industry, had begun to take shape.

IT WASN'T JUST THE CHINESE WHO WOULD SWITCH TACK. THE BELGIANS had also become comfortable with the new reality in Congo. In 1969, Mobutu's government finally came to an agreement with the remnants of the Union Minière on profit sharing. In exchange for technical assistance, a new Belgian company would market Congo's minerals for fifteen years and receive 6 percent of the profits from Congo's copper, cobalt, and other metals mined by Gécamines. The next year, King Baudouin of Belgium traveled to Kinshasa to sign a treaty of friendship with Mobutu.

Congo's leader was a close reader of Machiavelli, and from the medieval Italian philosopher he absorbed lessons about how kings must treat their nobles. One way in which Mobutu maintained power over his subjects was by regularly rotating positions in his government so that no one could establish a power base and challenge his rule. As Kilanga put it, "Mobutu liked to send his people left and right." (He would also make a point of sleeping with his ministers' wives.) He would often move his administrators and his apparatchiks, his generals and his functionaries, weakening any chances of on-the-ground opposition. The writer V. S. Naipaul, who was in Congo during the aftermath of such a rotation to cover the rule of what he described as a "new king," explained how Mobutu's inner circle was "abruptly dismissed, packed off to unfamiliar parts of the bush."

In one such shake-up, Kufi left his job as mayor in 1969, and he was sent by Mobutu far to the north, to administer Bandundu, a swampy port on the east bank of the Kwango River. Kufi is still remembered in Bandundu "as the progenitor author of the town" because of his efforts to build a centralized settlement around the crumbling colonial post that the Belgians called Banningville. "It was he who created the city of Bandundu," Kilanga told me. "It hadn't existed beforehand."

Then, in 1972, after impressing Mobutu with his work in Bandundu, Kufi was called to Kinshasa, Mobutu's capital, renamed a few years beforehand. Under a program that Mobutu called *Authenticité*, many things were renamed. "Authentic" music was promoted nationwide; people were encouraged to take a third "authentic" Congolese name; the country, its river, and its currency were all "authentically" renamed Zaire. Katanga became Shaba, Elisabethville became Lubumbashi, and Jadotville became Likasi. Even Mobutu changed his name, becoming Mobutu Sese Seko Nkuku Ngbendu Wa Za Banga, "the all-powerful warrior who, because of his endurance and inflexible will to win, will go from conquest to conquest, leaving fire in his wake."

Kufi, eager to please Mobutu, headed to Kinshasa. He was named as director of secondary education planning. "The evidence is legion to show that this patriot had a single cause," the Bandundu journalist Edouard Nyindu wrote in a hagiographic memorial article for Kufi in 2019, "to develop Congo and to make it so that the Congolese lives as a civilized man."

MOBUTU REMAINED CLOSE TO CHINA FOR THE REST OF HIS RULE. ACCORDing to a declassified CIA report from 1982, "since 1974, Beijing—at no cost to Zaire—has trained more than 230 Zairean military personnel in China, furnished a variety of weapons, including 16,000 small arms and some heavier ones, over 80 tanks, several small naval vessels, and numerous vehicles." In 1979, the Chinese completed the Palais du Peuple, or the People's Palace, an enormous gray convention hall in Kinshasa that Mobutu had ordered when he made his 1973 visit. The building is still in use: It now serves as the seat of Congo's National Assembly and Senate.

The edifice's nomenclature suggested pluralism, but the Zaire that Kufi was working for was a decidedly authoritarian kleptocracy. These were the fat years in Zaire, remembered fondly by elites in the capital and in a few of Congo's cities. Those close to Mobutu became known as the *Gros Légumes*, or Big Vegetables, and they profited enormously, both from the country's natural resources and from foreign aid, whose liberal flow was ensured by the leader's Cold War balancing act.

Everyone else, however, became poorer. Those who critiqued Mobutu or his Popular Movement of the Revolution (MPR) party were wont to end up in prison and in the torture chambers of the National Documentation Agency, the country's secret police. Children were taught to chant "MPR—*discipline!*" and songs composed by Mobutists urged people to *songi songi*—beware of spies and subversives in their ranks. "We became like parrots," Jeef Kazadi Kamwanga, the investigative journalist who accompanied me in Katanga, remembered as we were driving through Congo's South. "We didn't even know what we were saying by the end."

No amount of chanting or singing, however, could stave off the fact that Congo's wealth was being stolen once again. Where had the money gone? The economics of Zaire only made any sense if you counted the profligacy of the country's officials, Mobutu chief among them with his villas in Belgium and the South of France, his castle in Spain, and his lavish ski holidays in the Swiss resort town of Interlaken. Officials in the West (and in China) turned a blind eye.

Even more money was pumped into the dictator's vast patronage system. "Generals had become businessmen over being soldiers, which provoked a terrible disorder in the ranks and as a result an unprecedented weakening of the army," George Arthur Forrest, a Katanga-born businessman, would later remember. Corruption was an open secret—institutionalized, even. Mobutu neatly acknowledged it in a 1976 speech: "If you want to steal, steal a little cleverly, in a nice way. Only if you steal so much as to become rich overnight, you will be caught." The problem was that Mobutu didn't take his own advice.

CHAPTER 7

LIBERTY IN A WASTELAND

As Stanley Whittingham was moving to California and Mobutu Sese Seko was consolidating his grip on Congo, people in the United States were becoming more and more concerned with their negative effect on the earth. Environmentalism as an intellectual movement would come to push the concepts of the electrification of vehicles and the storage of renewable energy (along with the lithium-ion batteries that would be used to store this energy) to the forefront of public policy discourse. In the 1960s, an increasingly vocal environmental lobby began calling for the end of fossil fuels, starting with efforts to reduce smog in Los Angeles. Polling data showed that in the U.S., concern for the environment jumped from 25 percent in 1965 to 75 percent in 1969.

During the 1960s, the destructive consequences of mining also began to weigh on the national consciousness. The movement against strip mining started in Appalachia, where residents opposed the decimation of the natural landscape to harvest coal. In 1960, Raymond Rash, a retired coal miner from Kentucky whose thirty-eight-acre farm was being threatened by the expansion of a mine, wrote a petition to the governor of his state, urging legislation against the mines. "Strip mining in our steep mountains," he wrote, "destroys the natural beauty which God has so lavishly placed in our region."

Five years later, at what would later be called the Battle of Clear Creek, armed protesters confronted strip miners in Knott County, Kentucky. "We have seen industrial devastation spread like a gargantuan cancer across hundreds of thousands of acres and marveled that a democratic society could care so little about its future as to calmly destroy the land its descendants will inherit and inhabit," Harry M. Caudill, chairman of the Congress for Appalachian Development, a pressure group formed to highlight issues affecting Appalachia, told a Senate committee in 1968. "Liberty in a wasteland is meaningless."

The movement that started in Appalachia would set off the slow death of many U.S. mines, especially in the West, where minerals like cobalt, lithium, and nickel were being dug out of the ground. In the 2010s, China's supremacy in mining those metals would prompt soul-searching and plans to reopen pits from the hills of Idaho to the deserts of Nevada.

———

IN 1969, JOHN C. WHITAKER, A GEOLOGIST TURNED NIXON WHITE HOUSE staffer, believed that the coming of the environmental movement from the Republican side was a pivotal moment for the new president, Richard M. Nixon. "There is still only one word, *hysteria*, to describe the Washington mood on the environment in the fall of 1969," he later wrote. "The words *pollution* and *environment* were on every politician's lips."

In California, where Whittingham had landed, attention was being focused on the environmental consequences of industrialized life. Toxic smog hung over American cities from Los Angeles to New York, and both Democrats and Republicans were turning a political lens on pollution. In 1970, the Democratic senator Edmund Muskie gave a speech on the first-ever Earth Day, calling for an "Environmental Revolution," while on the right, California governor Ronald Reagan penned an article about how pollution was "bad for business," citing figures that pegged the cost of air pollution at almost $13 billion a year.

For Nixon, the focus on environmental issues was one of political expediency. In private, he told his general counsel, John Ehrlichman, an avowed environmentalist who would later serve jail time for his role in the Watergate cover-up, that "in a flat choice between smoke and jobs, we're for jobs. . . . But just keep me out of trouble on environmental is-sues." This calculus had far-reaching effects. In 1970, Nixon signed the Clean Air Act, which set standards for air pollution in the U.S., and in the same year, he created the Environmental Protection Agency. "I real-ize that the argument is often made that there is a fundamental contra-diction between economic growth and the quality of life, so that to have one we must forsake the other," Nixon said in his State of the Union ad-dress that year. "The answer is not to abandon growth, but to redirect it. For example, we should turn toward ending congestion and eliminating smog [using] the same reservoir of inventive genius that created them in the first place."

DURING THE 1970S, AT HOME AND ABROAD, THE UNITED STATES' POSITION as the world's dominant superpower was also being questioned, not least because the nation's power had been built on access to cheap petroleum. After a decade of crisis and cartel formation, it looked like the U.S. might no longer have access to Middle Eastern oil at the low prices it had en-joyed since World War II. Companies started to look at deploying the Nixonian "inventive genius" in the hope of finding a solution to the en-vironmental and geopolitical conundrum. Scientists and engineers began scrambling to develop an electric car, although strictly speaking, electric vehicles were not new at all—they predated the internal combustion engine.

The battery has existed since 1800, when Alessandro Volta first stacked pairs of copper and zinc discs, separated them with brine-soaked card-board, and then demonstrated that the contact of metals produced elec-tricity. The device's potential in automation was not lost on the scientists of the early nineteenth century, and several European innovators created small-scale electric vehicles; one of the first was created in Scotland by Robert Anderson, who in the 1830s used a disposable battery and crude oil to build an electric three-wheel carriage. After the lead-acid battery was developed by the Frenchman Gaston Planté in the second half of the nineteenth century, the technology was used to power modern devices, such as telegraphs and streetcars. In 1881, another French scientist, Gustave Trouvé, made an electric tricycle using several lead-acid batteries, which he paraded "as rapidly as a good hackney cab" down the rue de Valois in Paris.

Over the next few decades, electric cars, which were easier to start, quieter, and cleaner than fuel or steam cars, grew in popularity. In 1899, the first car to drive faster than one hundred kilometers an hour was electrically powered; at the same time, in the U.S., 38 percent of all automobiles were electric, outnumbering petroleum-powered cars (a slim majority of that period's cars were powered by steam). Even Henry Ford's wife, Clara, used one.

Battery technology, however, would also spell the electric vehicle's doom; petrol-powered cars had, since their invention by Carl Benz in 1885, needed to be started by a hand crank. These rudimentary mechanisms were dangerous, and when cars misfired upon starting, the crude contraptions would break motorists' arms, crack their wrists, and shatter their shoulders. In 1911, Charles F. Kettering would create an electric starter, which, after it was first introduced in the Cadillac Model Thirty, became wildly popular. By the 1920s, the U.S. had fallen in love with gas vehicles, and electric cars were largely forgotten all over the world. Fifty years after the demise of the original EVs, as politicians and CEOs started to weigh the "energy-environmental balance," scientists focused once again on electric vehicles.

IN THE 1960S AND '70S, PROMPTED BY OIL SHORTAGES, U.S. CAR COMPANIES—as well as French-, British-, and Japanese-funded projects—began looking for new ways to power cars, prompting a mini battery boom. Even Greece produced a pint-size electric vehicle. One Enfield 8000, as the car was known, is today still on display at the industrial museum on the island of Syros.

Perhaps the greatest evangelist of battery vehicles during this epoch was Robert R. Aronson, the president of a U.S. company with the futuristic name Electric Fuel Propulsion Inc. As a teenager in Los Angeles, Aronson had taken a job delivering telegrams for Western Union, and after being caught in cloud after cloud of diesel bus exhaust smoke, he found himself dreaming of a vehicle that would not produce emissions. Later, Aronson became a salesman for a battery company in Los Angeles, and he started to envision a "smokeless car" that ran on batteries.

In 1966, while living in Louisiana, Aronson created his first electric vehicle, which he called Mars, not so much for the associations with interplanetary travel but for the chocolates his son loved to munch on. He managed to drive it at 55 miles an hour up and down the Mississippi River, and it had a range of around 190 miles. Aronson believed that his invention had geopolitical implications. "The energy crisis has turned on the electric vehicle," Aronson told *The New York Times*. "The interest is very keen due to gas shortages. People realize there is not an unlimited supply of oil."

His vehicles used a lead-acid battery that he called "tri-polar" because of the way the positive plates were connected to one another. The battery, which had originally been designed for the Esso Standard Oil Company in 1953, was mainly used in Puerto Rico, where temperatures ran high and degraded the cells. The solution that scientists came up with was dissolving a small amount of cobalt sulfate in the electrolyte. As Aronson's firm would later put it: "The cobalt sulfate, after a few charge-discharge cycles, forms a protective layer on the surface of the positive plates, protecting the grids from oxidation." This was decades before the metal's inclusion in lithium-ion batteries.

By the late 1960s, Aronson had built manufacturing facilities in New Orleans and Puerto Rico, and he had even set up a predecessor to the fast-charging infrastructure for which Tesla is known today: six fifty-kilowatt electric charging stations, located at Holiday Inns on I-94 between Detroit and Chicago, that allowed drivers of his company's electric car to charge their vehicles in forty-five minutes while having a cup of joe and maybe a bite to eat. He called it "the World's First Electric Car Expressway."

In the end, however, Americans just didn't want to buy Aronson's cars. Only about a hundred of them were ever produced. The history of batteries is punctuated with mavericks like Aronson, people who were just a little ahead of their time. Had Elon Musk been tinkering around in the

1970s, his life trajectory in many ways might have followed Aronson's—
a life of constant striving and limited success.

———

THE HIGH-WATER MARK OF '70S BATTERY FERVOR CAME IN JULY 1976, WHEN
Congress passed an act that directed the federal government to purchase
at least seven and a half thousand experimental hybrid or electric cars by
1981 and spend some $160 million on research, development, and dem-
onstration. For a brief moment, everyone seemed to be focused on clean
energy: In 1979, President Jimmy Carter installed solar panels on the roof
of the White House. In Japan, the auto giant Toyota boasted that it had
created a battery vehicle "capable of providing a performance matching
that of gasoline engine driven automobiles."

And then, as the 1980 election results came in, the impetus stopped.
The new Republican administration, under Ronald Reagan, would decide to focus on oil. Despite his support of clean-air policies earlier in the
decade, Reagan would have little interest in curbing emissions once in the
Oval Office, and even less interest in electric vehicles.

The downturn relegated firms like Aronson's to the sidelines. His company still exists, though its name has been changed to Apollo Energy Systems Inc. It focuses on, among other things, creating batteries for small
electric vehicles and renewable energy systems for "mini-grids" (used in
remote places off the grid where mains electricity or power for utilities
isn't readily available).

Apollo also markets something that recalls Aronson's invention from
the 1970s: a "tri-polar lead cobalt battery." In mid-2022, I tried to find
Apollo at its listed address in Pompano Beach, Florida, at a warehouse in
the (aptly named) Powerline Business Park. The space had been vacated,
and a woodwork shop had moved in. Outside, a Tesla Model 3 was charging under the blazing Southeast sun.

CHAPTER 8

INTERCALATION STATION

In 1971, Theodore H. Geballe, an applied physics professor who was recruiting for Exxon, read one of Stanley Whittingham's studies on beta-alumina. Exxon had begun to plow oil dollars into developing new energy projects, and Geballe thought Whittingham would be a good person to lead a lab for a recently created subsidiary called Exxon Enterprises. "They wanted to be an energy company, so they were into batteries," Whittingham said. "They were building up their corporate lab, so they were just looking for, basically, top scientists." Whittingham soon found himself hard at work in suburban New Jersey, in a lab across the way from an oil refinery. By 1976, the company was spending $25 to $35 million (the equivalent of $144 to $206 million in 2025 U.S. dollars) on ten projects in what one executive called "long-term, high-growth areas." In addition to high-powered batteries, Exxon's subsidy had established arms that were developing everything from nuclear fuel production to text editors to a test-scoring machine.

In his research, Whittingham narrowed his focus on the movement of ions—tiny particles with net electrical charges—and on how when they traveled in a specific direction, they produced an electric current. Surely, then, they could be moved around in a battery to store electricity. But what kinds of ions would he use?

One day, at his Exxon lab, Whittingham read an article about a rechargeable carbon fluoride battery developed by Japanese scientists for night-fishing floats. The battery included lithium, the lightest of all metals. Whittingham knew that lithium, a reactive metal, will give up an electron to achieve a more stable state, resulting in a positively charged ion that easily forms compounds with other elements. (The Japanese scientists weren't the first to use lithium in batteries; it had been mooted as a battery material since the days of Thomas Edison.) Still, as he studied the metal, he realized it had drawbacks. Because of its tendency to shed an electron, lithium is known to be unstable—liable to catch fire in air and water.

Whittingham's early work at Exxon involved combining lithium with tantalum (incidentally, a metal whose ore is, like cobalt, extensively mined in Congo). But tantalum was too heavy for use in batteries, and another metal, potassium, was too unstable. The work was fast-paced, Whittingham remembered: "You were making breakthroughs almost every month."

Perhaps the most important of those breakthroughs came when Whittingham and his team decided to insert lithium ions into a metal compound called titanium disulfide to see whether it could be used to create an electrode in a battery. Whittingham didn't want to use actual lithium metal in the battery, he said, "because the lithium forms these dendrites that can then short the battery out, and then you get thermal runaway, and then potentially fires." Dendrites are microstructures that grow in vegetal, branchlike shapes as the lithium metal warps on a cell's anode after charging and recharging. They can cause fires and deplete batteries.

To introduce the lithium ions, the scientists at Exxon used a technique called *intercalation*. In common parlance, *intercalation* means "to insert something into an existing sequence," usually a calendar—for example, to slot February 29 between February 28 and March 1 in a leap year. But in chemistry, as Whittingham wrote a few years later, "it describes the reversible insertion of guest species into a lamellar host structure with maintenance of the structural features of the host." A lamellar structure is one in which very thin sheets of material are stacked one atop another. Titanium disulfide's atomic structure is lamellar, so the team at Exxon slotted lithium ions between the sheets. Clustered in the sheets at the positive electrode of the battery, the ions were attracted to the negative electrode, which would also have a structure that they could nestle in.

In the process, electricity would be released. Each time the battery was charged and discharged they would jump back and forth, endlessly bouncing between electrodes. (The word *intercalation* has filtered into the battery-and-EV vernacular: A popular Substack newsletter on batteries, founded in 2020, is called *Intercalation Station*.)

To perform the feat of intercalation, the scientists devised a chemical process that included mixing a lithium-containing solution with the compound, allowing it to dry in a box for between three days and three weeks, filtering the results, and then washing them with hexane, a compound produced by distilling crude oil.

Soon, Whittingham and his team had created a workable battery—the first lithium-ion battery—with lithiated titanium disulfide as the cathode material, a liquid electrolyte of dissolved lithium salts and lithium foil as the anode.

On September 10, 1973—almost a year to the day from the beginning of the NATO-sponsored Belgirate Conference, where fast-ion transport was first theorized—Whittingham filed a patent for a new type of battery, a "chalcogenide battery," named after the structure of the titanium compound. The move was not heavily publicized: Exxon and its scientists were locked in intense rivalry with other laboratories, especially Bell Labs, which at that time was still a behemoth to be reckoned with. In fact, the battery would not be publicly announced until 1976. Whittingham still remembers the time as the heyday of his work. "Everything was new," he told me. "There weren't many people in the field."

When Whittingham announced his discovery to Exxon, the company jumped at the opportunity to produce a lithium battery. The timing could not have been more on the mark. As the philosopher Ivan Illich observed in 1974: "It has recently become fashionable to insist on an impending energy crisis." Critics such as Illich, a Catholic priest by training, were arguing for a world with *less* energy, but companies like Exxon were fervently searching for more sources to keep up with the world's insatiable appetite for power. As the batteries that Whittingham developed have proliferated across the planet, that obsession has only deepened.

CHAPTER 9

A SPY IN PRIEST'S CLOTHING

On a Wednesday in August 1975, a mustachioed man descended from a Learjet at an airfield outside the small town of Silva Porto, now Kuito, Angola. John Stockwell was dressed in a short-sleeved black shirt, complete with a cross that dangled from a large silver chain. One might have mistaken him for a missionary. Stockwell, however, had come to Silva Porto not to preach the gospel but to evangelize on behalf of the United States.

Stanley Whittingham might have been testing newfangled batteries in New Jersey, but Africa was not yet fully decolonized. Angola, lurching toward independence, was being fought over by the West and the Soviet Union. The country's strategic position and proximity to Congo's critical minerals made it quite a prize. The jet that Stockwell had arrived on had been lent to a rebel named Jonas Savimbi by a British businessman to whom the rebels had promised control of a swath of the country's mineral riches after the war had been won.

Stockwell was in his late thirties, and he worked for the CIA. He had spent part of his early schooling at a mission in the West of the Belgian Congo; his career had taken him to Katanga, to Burundi, and to Vietnam. On that Wednesday, he was anxious and tired. His mind was spinning from dysentery, and his stomach swilled with thermos coffee that the jet's copilot had proffered. What's more, he was deep inside rebel territory and he did not yet know whether the rebels would be friendly to a U.S. agent like him.

Angola was the principal means of egress for Zaire's minerals, and Savimbi's camp lay athwart the Benguela Railway. In the early twentieth century, a Scottish entrepreneur had built the railroad to export the mineral wealth of southern Congo, and it was owned jointly by British and Belgian companies. Until the late '60s, the Benguela Railway was how Europeans arrived to work in the Congolese Copperbelt, embarking on a shaky three-day journey from the Port of Lobito in Angola. Trains going in the other direction were filled with Katanga's copper and cobalt, bound for international markets. Before 1975, this line curved through more than eleven hundred miles of savanna from the copper mines to the ocean along a route upon which slaves were driven toward the Atlantic in centuries past.

When Stockwell arrived in Silva Porto, the rebels took him to see the train tracks. "[I] stared down the visual infinity of rails and cross-ties," Stockwell wrote. "I projected my mind beyond the horizon, six hundred kilometers to Benguela itself and the crystal waters of the Lobito harbor. Turning a hundred and eighty degrees, I looked nine hundred kilometers to Lubumbashi, and remembered the huge mines which disgorged tons of copper ore to be transported to waiting ships by an endless string of open railroad cars."

In Angola, however, Stockwell saw no trains: The Katangese gendarmes, the force comprised of Tshombe's erstwhile supporters who had fled Congo during the political foment of the 1960s, had sabotaged the tracks. "They wanted Mobutu to fall," Jean Dusausoy, a mining technician for Gécamines at the time, told me. "They wanted to do that by attacking the regular operations of Gécamines."

The gendarmes' attacks in 1975 would continue to have a resounding effect on world trade well into the 2020s: After them, the Congolese began exporting their copper and cobalt via road through Zambia to ports in South Africa and Tanzania. This was a circuitous and expensive means of transportation, but it was much safer (rebels like Savimbi also began attacking the railway as Angola spiraled into civil war). Roving work gangs would travel up and down the tracks, fixing broken sections, but nothing could induce the Congolese to risk transporting their valuable minerals out through Benguela. The Angolan war ended in 2002, but even then, the trains did not begin running: It had become more profitable on the Congolese side of the border for politicians to ship minerals out by truck.

STRANGE THINGS BEGAN HAPPENING AROUND KOLWEZI IN 1976 AND 1977. Dusausoy, the technician, started to hear Africans in Kolwezi speaking Portuguese, which he attributed to the surreptitious infiltration of the city by gangs of Katangese gendarmes. They were "poor guys, lost, drugged, drunk, (often) armed and therefore dangerous because out of control," Dusausoy later remembered. The metropole began to fill with soldiers.

In October 1976, Mobutu's secret police announced a curfew that began at 6:30 p.m. and upset the easygoing rhythm of the city, which had, for expatriates, more or less reverted to the cosseted status quo of the colonial era. "No more cinema and tennis in the evening, and it wasn't even possible to go round to a friend's for a drink in the evening," Dusausoy remembered. Rumor had it that American agents were in town. One day, while going to the bank, Dusausoy saw a white man in civilian clothing with a Colt .45 in his belt, a U.S. national who was purportedly working on a high-voltage power line project. "When you knew it was completely forbidden to have a gun, and that the town was full of soldiers, well, you had to ask yourself who this guy really was," he said.

In March, the gendarmes struck. Dusausoy, who was often at the military's headquarters or at the airport, heard the attack unfold over the airwaves. In the end, the gendarmes couldn't hold the town, or choke off Mobutu's economy by seizing the copper-and-cobalt mines of Kolwezi. Moroccan troops, airlifted to Congo by King Hassan II, an ardent anti-Communist, roundly defeated them, and they fled back to Angola.

COBALT WAS BASICALLY IGNORED IN THE ECONOMIC CALCULUS OF THE COP-perbelt until 1978. Copper was the hot commodity. Cobalt was a nice-to-have, a niche metal that in Congo offset movements on the copper market;

it was bought and sold by a handful of traders who ensured that the global demand for the mineral was fed and used price moves based on scarcity and glut to turn a profit. A former commodities trader with the firm Marc Rich & Co. remembered that people who bought and sold cobalt worked with other metals because there was such little volume. The mineral was a small fry in comparison to metals like copper, iron, and even lead. The conflicts in Katanga, however, were about to thrust the blue metal into the spotlight.

That process had begun in the depths of the freezing London winter that marked the beginning of 1978, when a series of peculiar orders appeared at the city's metals trading houses, where most of the world's metals were bought and sold between trips to beery pubs and steak-and-chips lunches. Cobalt prices were about to balloon, but no one—apart from those, perhaps, who were doing the buying—quite knew why.

At the time, cobalt was only traded by around twenty dealers. The traders in this tiny community played their cards close to their chests: None of them, apparently, remarked upon or connected the dots between the steady stream of orders coming from the Eastern Bloc, from Poland, the German Democratic Republic, and, most importantly, the Soviet Union. The Soviet Bloc had been building up its stockpiles since 1976, just before the first gendarme attack, but now the effort would, in retrospect, seem more concerted.

And only about 10 percent of cobalt at the time was traded on the free market; the rest was bought in bulk through transactions between large suppliers and large consumers. It was fairly dull business. The mineral, a byproduct of copper and occasionally nickel, had a fairly fixed demand and a fairly regular supply—give or take the occasional strike at the mines producing it. Most of those mines were controlled by one company—Gécamines.

Cobalt may have been niche, but it was strategic: The main uses of the metal were military-industrial—it went into jet engines, missiles, and submarines. Seventy-one percent of the United States' cobalt was imported from Zaire; worldwide, Mobutu's nation enjoyed a 65 percent share of the global market. All of Zaire's cobalt came from Katanga, and almost all of that from within the direct vicinity of Kolwezi.

The amounts that the Soviets had requested in the winter and early spring of 1978 were large in aggregate, but each order was small enough not to rouse suspicion: around thirty to fifty tons at a time. (The overall market was "around twenty thousand tons a year, if that," a former cobalt trader told me.) The Eastern Bloc buyers also bargained hard for their cobalt, insisting on buying it for around $7.60 a pound. "They had a shrewd inkling of what was going to blow," Lef Lubett, then chairman of the Minor Metals Traders Association, told *The Washington Post* after the extent of their buying became clear. It "was pre-emptive buying in expectations of a shortage."

What "blew" in May 1978 was the second invasion of Katanga by the gendarmes. While there is no suggestion that the Soviet Union engineered the attack, the Soviets had been training the gendarmes in military intelligence, and their Cuban allies had given them artillery and other military instruction. This time, the gendarmes managed to take Kolwezi.

GEORGE FORREST WAS LIVING IN KOLWEZI WITH HIS FAMILY AT THE TIME. Forrest's father, Malta, had come to Congo and grown a transport business into a subcontracting empire for Union Minière. Malta even ran many of the firm's most important mines for four years after the Wall Street crash of 1929. George, who turned twenty the year Congo became independent, traced his roots to New Zealand and the Greek island of Rhodes. He did not feel particularly Belgian—Greeks and Jews had been relegated to a rung below the Belgians in the caste system of colonial Congo. Rather, he and his family felt Congolese, African, and especially connected to Kolwezi. George decided to remain in Congo after decolonization, unlike many Belgians, who had fled, and he took the reins of his father's business.

In May 1978, several hundred civilians were killed in the fighting, many of them by Zairean troops. Those who survived were left with scarring memories. Forrest, for one, saw his home invaded by armed rebels who threatened to shoot him. As massacres intensified in the city, French paratroops scrambled to be deployed, and Mobutu hurried to the front, piloting a plane full of journalists into Kolwezi to show his bravado.

The gendarmes were defeated by the superior firepower of the government and the paratroops, and they again headed back to Angola. Mobutu's army, so ineffective at defending Kolwezi, began to terrorize the Lunda population living around the city, and tens of thousands of civilians fled to Angola and Zambia.

This time, cobalt prices skyrocketed. By the end of May, prices for the mineral on the free market had risen by some 500 percent. Officially, the metal was worth forty-two dollars a pound. On the black market, a pound of the metal was worth fifty bucks. "It generated some of the highest prices for cobalt ever seen," the former cobalt trader told me. Engineers and Europeans left Kolwezi in droves. Forrest decided to move away with his family, only to return decades later as one of the most important people in cobalt mining. Production plummeted and prices soared. For a brief moment, it was even profitable to fly cobalt out of Kolwezi; usually, only precious metals like gold and silver warranted such white-glove treatment.

The U.S. government decided to put together a strategic stockpile of cobalt, which was key for the armaments business. The stockpile reached more than twenty thousand tons in the mid-1990s, but by the year 2000, it had been sold off, prompting soul-searching when cobalt became vital again. Questions about Soviet stockpiling were raised in Congress, and people started to look for alternatives to cobalt. Some businessmen suggested that deep-sea mining would reduce dependence on Zaire.

Reading reports from the time, I found it all a bit eerie: The exact same solutions were being proposed in the late 2010s and early 2020s, when I began reporting on cobalt and battery minerals. Even as cobalt prices fell back to normal levels after the 1978 invasion, and as the commodities market resumed its inevitable pace, something had changed. Washington was sitting up and taking notice. Cobalt had become critical.

Part 2

TRADE AND WAR

"Trade cannot be maintained without war, nor war without trade. The times now require you to manage your general commerce with your sword in your hands."

—GOVERNOR-GENERAL JAN PIETERSZOON COEN, DUTCH EAST INDIES COMPANY, 1614

CHAPTER 10

PUTTING OUT FIRES

The fire began in the midriff of a Harlem apartment building at around 2:00 p.m. on February 23, 2024. The flames spread quickly, eating through the 120-year-old frame and wooden floors. Thick black smoke began to billow from windows. By the time the New York City Fire Department arrived, the fourth and fifth stories were burning, and firefighters had to rappel down the side of the building. They made three rope rescues; one resident even had to jump from the sixth story. Seventeen people managed to escape with injuries. Fazil Khan, a twenty-seven-year-old journalist who had moved to New York from New Delhi, a person described by his friends as "the most honest person I've ever met," did not make it out.

At a press conference later that day, the fire department announced that Khan was the first person in New York City to die from a fire caused by a lithium-ion battery in 2024. The year before, 268 fires had been started in the city by batteries bursting into flame. And that wasn't even counting fires that had been started at places like recycling facilities. "It is an enormous problem in our industry, and it is probably the biggest risk that we face," Tom Outerbridge, the CEO of Balcones Recycling, told me in 2024. He pointed out that even modern greeting cards use lithium-ion batteries to play music and light up. The cards, placed into recycling bins with paper, have frequently caused fires. "It's growing exponentially," Outerbridge said. "Fires were really uncommon ten years ago, and now they are happening every week, every day sometimes." As Eric Frederickson, vice president of Call2Recycle, a battery-recycling nonprofit, put it: "Keeping batteries out of mixed recycling is the single biggest challenge that the general waste and recycling industry is dealing with."

Fires have bedeviled lithium-ion batteries since the very beginning. Exxon researchers in Stanley Whittingham's Linden, New Jersey, lab had to call the fire department a handful of times after inadvertently starting lithium fires, which would burn ferociously and only intensified if doused with water. One early version of Whittingham's lithium-ion battery had to be carefully unscrewed at the end of each day to release a gas that ignited in contact with air.

The Exxon scientists had made a risky trade-off: So much power packed into such a small space meant the batteries were liable to combust.

FOR EARLY ELECTRIC VEHICLES, LITHIUM-ION BATTERIES WERE A MOOT point: Almost all of the batteries being used in '60s and '70s electric vehicles were old technology, improvements on Gaston Planté's original lead-acid batteries. These were heavy and had short ranges, but they had the advantage of being incredibly stable. Some experimental vehicles used other forms of batteries, including sodium-based technologies, but these were largely shelved, since sodium cathodes had to be kept at molten temperatures.

The lithium-ion battery needed to be stable enough for commercial viability. A deadly battery fire early on could doom the technology; some of the first versions of the battery were used in bedside clocks and watches—imagine if they exploded. And from the earliest stage of development, the Exxon scientists were thinking of putting their batteries into cars. Executives at the company wanted to create something that would future-proof their firm against any potential exhaustion of the world's oil supply.

Despite the fires, Whittingham and others at Exxon held out hope for their new battery; the titanium disulfide cells delivered 2.4 volts, more than the most powerful lead-acid cells. The scientists decided to remove lithium from the battery and use an aluminum compound, which was more stable. In 1977, the company developed a watch that used its battery. A year later, Whittingham made a series of small lithium cells that he kept in his personal collection. When he tested them thirty-five years later, he found they had retained around 50 percent of their original capacity. A small solar-powered clock that the team had designed still kept time.

Executives at Exxon were starting to wonder about long-term profits. "They said to us with some rationality, 'We're a multibillion-dollar company. Why do we need this?'" Robert Hamlen, a scientist working with Whittingham, told *Inside Climate News* in 2016. "If I were to characterize Exxon's new ventures in two words, it would be 'unrealistic expectations.' They thought that because they were good at oil, they could handle these new ventures in other areas, too." Whittingham and his team needed more time, but as the decade came to a close, the bottom was dropping out of the excitement around electric cars and the lithium battery.

THE CATALYST FOR THE DOWNTURN WAS THE SAME THING THAT HAD PROvoked the hand-wringing about America losing out at the beginning of the decade: oil. The Iranian Revolution and the subsequent hostage crisis spiked oil prices to a historic high in April 1980. Then they started to drop as new fields around the world began to come online. In response, Saudi Arabia and other Gulf oil producers ramped up production to increase their market share, adding to an "oil glut." Prices would continue falling until 1986. Carbon energy seemed abundant, even infinite.

Exxon went into cost-cutting mode and began to question how its battery division fit into its overall business model. Under Reagan, federal grants for alternative energy dried up. "Two things happened," Whittingham said. "One, oil prices dropped. And the second thing is the market—they decided the market wasn't big enough. They wanted a market, if I recollect correctly, of about one hundred million dollars a year."

Over the first few years of the 1980s, Exxon sold off its Enterprises businesses as it "refocused on its core" and doubled down on oil and gas. After all, Enterprises had always been a blip on the balance sheet: In 1976,

The New York Times noted that Enterprises had made investments total-ing between $40 million and $50 million. The firm's oil sales totaled around $49 billion. Whittingham left the company in 1984. Exxon later decided to license the technology its scientists had created to three firms in Japan, Europe, and the U.S.

A cycle had begun that would lead directly to China's dominance of the battery supply chain in the 2020s, a cycle in which a U.S.-developed technology was to be perfected and commercialized abroad. The battery had left the country, and its fate would now be decided elsewhere.

CHAPTER 11

A COBALT CATHODE
AND A CARBON ANODE

obalt may have become a critical metal in the 1970s, but it was not until the next decade that it found its most critical use—in batteries. April 1980, the month that oil prices peaked, was also the month that a paper landed on the desk of the editor of a scientific journal called the *Materials Research Bulletin*. The paper was sponsored by the U.S. Air Force and by the European Energy Commission. It would go on to become the most cited article the journal published that year, and it would eventually win its lead author a Nobel Prize. But at first, nobody, not even battery companies, took much interest.

The paper's main author was John B. Goodenough, a tall, genial man who had taken the inverse of M. Stanley Whittingham's transatlantic journey. Goodenough's father was a scholar of religion at Yale whose major field of study was how Hellenistic culture had influenced Judaism. John decided to focus on science: He studied physics at Yale and the University of Chicago, and then went on to Oxford to work on energy storage. Like Whittingham, Goodenough had spent time contributing to U.S. government projects while at MIT's Lincoln Laboratory, an institution dedicated to the research side of national defense.

At the dawn of the 1970s, Goodenough began to focus on the problem of energy storage. His reasons chimed with those of Exxon Enterprises and the various global initiatives on electric vehicles that sprang up around the oil crisis, though his were shot through with an altogether more spiritual strain, a legacy of his father's interests. In his memoir, *Witness to Grace*, Goodenough writes:

> It was obvious already in 1970 that our dependence on foreign oil was making the country as vulnerable as the threat of ballistic missiles from Russia. Solar energy was an obvious renewable source to be harnessed; our profligate use of energy made conservation an obvious target also. Since solar energy is variable in time and location, it was also obvious that we needed to find a way to store the solar energy that is converted into electricity.

In 1976, after turning down work at a solar-power institute that the shah of Iran was trying to establish, Goodenough accepted a position leading an inorganic chemistry class at Oxford. There he began to read the papers Whittingham had produced doing his research for Exxon. Goodenough wondered whether he could make a more stable cathode. In 1978, an undergraduate thesis spurred him to ponder whether a metal oxide cathode would be more stable. Metal oxides, especially those formed from transition metals like cobalt, tended to collapse if ions were taken out of them. But Goodenough knew that if he could find a way to *order* those ions and not take too many of them out, he might stand a shot at creating a hardy enough structure to serve as a cathode.

Koichi Mizushima, a young postdoctoral student of experimental physics from the University of Tokyo, was visiting the inorganic chemistry lab at Oxford at the time, and Goodenough put him to work on the problem, alongside Philip Wiseman, a postdoctoral research associate in chemistry. Goodenough directed them to work on inserting lithium ions into, and removing them from, different metal oxides. "We found that over half of the lithium could be removed reversibly with cobalt or nickel," Goodenough later wrote. The cathodes of both these oxides, moreover, produced a far greater electrical potential than Exxon's disulfide; with cobalt or nickel, the battery cell had an output of four volts, more than one and a half times greater than the cells Whittingham had made.

The paper Goodenough submitted to the *Materials Research Bulletin* represented a breakthrough in batteries, but it confounded conventional (and commercial) logic. Cathodes were normally built charged, but the Oxford team was proposing creating one that was discharged. Initially, European and U.S. companies had little interest in the work, because selling charged batteries seemed unwieldy, a logistical and safety headache. In Mizushima's home country of Japan, however, the idea would gain traction.

ON NEW YEAR'S EVE IN 1982, DR. AKIRA YOSHINO DECIDED TO CLEAN HIS OF-fice. The workspace that he occupied at Tokyo's Asahi Kasei Corporation was littered with reams of paper stapled into bundles and spread out haphazardly across a small desk. Across Japan, people were piling into narrow overheated bars to indulge in *bonenkai*, that traditional Japanese ritual of drinking with colleagues and friends "to forget the previous year." But Yoshino certainly didn't want to forget the year he had spent researching high-tech battery materials at Asahi Kasei.

Indeed, he could not, for he had a problem. The batteries he and his team dreamed of building were just too unstable when it came to the real world. They would catch fire; they would blow up. And so, this New Year's Eve, Yoshino was trawling headlong through mountains of scientific research.

In the 1980s, Japan had embarked upon a quest for technological excellence. The government's Ministry of International Trade and Industry had fostered domestic competition by distributing research widely, and Japanese companies had bought boatloads of licensing agreements to import foreign technology.

The resulting progress had catapulted Japan to the foremost ranks of high-tech producers, and worried policymakers in the United States predicted that Tokyo might vanquish Washington in a battle over technology. The concerns voiced in the 1980s over a putative Japanese technological primacy (a "hi-tech Pearl Harbor," as Robert Reich famously described it in *The New Republic*) would be echoed decades later in the 2010s and '20s by commenters worried about China besting the U.S. through Beijing's current dominance of tech. The cast of characters even has some overlap:

President Donald Trump, whose administration initiated a U.S.-China trade war through levying tariffs on Chinese imports, formed many of his views in the 1980s, arguing that America was being "ripped off" by Japan.

In those years of discovery and growth, however, it wasn't just U.S. critics who depicted the competition between Washington and Tokyo in bellicose terms. In the summer of 1982, a few months before Yoshino decided to clean his desk, *The Yomiuri Shimbun*, Japan's most widely circulated newspaper, called the competition for tech primacy the "*Nichi-Bei gijutsu sensō*," or "Japan–United States technology war." (These days, headlines read HOW THE U.S.-CHINA TECHNOLOGY WAR IS CHANGING THE WORLD.) Yoshino had positioned himself in one of the advance trenches of that war. His laboratory at Asahi Kasei was at the front line of the rush to discover a type of rechargeable battery that would better power the gadgets and futuristic devices that, thanks to Japanese companies, had begun to flood global markets. "I just sort of sniffed out the direction that trends were moving," he once said. "You could say I had a good sense of smell."

The most famous of these Japanese devices was perhaps the Sony Walkman, a personal tape player. Released in 1979, it almost immediately revolutionized the way people listened to music. In the first two months, Sony sold ten times more units of the Walkman than the company had expected. But such devices were powered by mercury-containing batteries that had, by the early 1980s, sparked environmental concerns. "The battery problem is especially serious in Japan because there are few sites to dispose of wastes," *The New York Times* noted in 1984. Yoshino was on a mission to make a cleaner battery.

———

STACKS OF YET-TO-BE-READ PAPERS HID YOSHINO'S DESK. "IT WAS COMpletely disorganized," he recalled forty years later, laughing at himself. "Papers were everywhere." For almost two years, Yoshino had been working on various materials for the anode of his battery, specifically polyacetylene, a silvery-gray substance that has been called a "plastic that conducts electricity."

Yoshino had hit a roadblock. He had found the anode material he

wanted to experiment on, but what about the cathode? "I needed a positive electrode material that contained a lithium ion," he told me. "I was
having trouble finding the right material." At the time, a whole host of
metals and elements in combination with others had been proposed, but
to Yoshino, they all presented fundamental problems. Those that contained actual lithium metal were likely to explode or catch fire; those that
did not contain lithium had no lithium ions to exchange with the anode
during the charging and discharging progress; and Whittingham's titanium disulfide was expensive to produce and released foul smells when
exposed to air.

As the chemist began to sift through the mess that had accumulated
in his office, he noticed a paper that he had passed over in his rapid skim-
ming of available research. The study that caught his eye bore Goodenough and Mizushima's bylines on the title page. "As soon as I started
reading that paper, I realized that [Goodenough's] material was the perfect material for me," Yoshino told me. He had understood that lithium
cobalt oxide (LCO) could make for batteries that were much safer and
less liable to catch fire.

In his lab at Asahi Kasei, Yoshino mixed lithium carbonate, cobalt
oxide, and stannic oxide (a form of oxidized tin), then roasted them at 650
degrees centigrade (just over 1,200 degrees Fahrenheit) for five hours.
After another twelve hours of firing the mixture at over 1,500 degrees
Fahrenheit, he pulverized it in a ball mill. Then he mixed it with a de-
composed form of carbon and an acrylic resin solution in a liquid solvent.
When he finally applied the mixture to one surface of a very thin sheet
of aluminum, the positive cathode was ready to be made by nipping the
aluminum foil between stainless steel mesh. "I just followed the way he
described it in the paper, step by step, and fortunately, I was able to make
it without too much trouble."

By the end of December 1983, Yoshino had applied for a patent for his
new polyacetylene-LCO battery, but polyacetylene turned out to be less
conductive than he had first hoped. He didn't have to look far for an alternative: Scientists at another wing of Asahi Kasei had created a form of
carbon that showed promise. The material was known as vapor-grown
carbon fiber, or VGCF, "grown" in a laboratory by heating a hydrogen gas

in a furnace to 1,200 degrees centigrade. He filed for another patent in 1985, but he thought he could still build a safer and more powerful battery, so he started to experiment with other forms of carbon, including petroleum coke, a byproduct of oil refining, for his anode. Later developments saw him settle on graphite as the best material for battery anodes—this decision would come to be questioned in the 2020s, when China controlled the production and processing of graphite. In late 2024, China placed export restrictions on graphite and other critical minerals, and scientists looked to build anodes out of materials like silicon.

One summer morning in 1986, Yoshino traveled to Nobeoka, a town on the southern Japanese island of Kyushu, where Asahi Kasei had a testing facility. He had brought with him two batteries: one with a cathode containing lithium metal, the other containing lithium cobalt oxide. To test them, a heavy-metal slug was dropped onto the two batteries.

When the lithium-containing battery was struck, flames leapt from each side, eliciting *oohs* from the scientists who were watching. An alarm began to sound as the cell burned for almost half a minute. Yoshino thought the sparks looked like fireworks. When the slug came down on the battery containing lithium cobalt oxide, however, nothing happened. There were no flames.

Cobalt oxide had made the battery safe enough not to explode when whacked and crushed (or at least if they had been made properly: Faulty batteries and ill-conceived batteries that cut corners for price, including the kind of cheap Chinese e-bike batteries that would start fires in cities around the world, would still be incredibly dangerous). Lithium-ion was ready for its world debut.

CHAPTER 12

THE MILKING COW FALLS ILL

As a boy, Augustin Katumba Mwanke once won a scholarship for being the "nicest student" at his school. In the mid-1970s, he moved from his native Pweto, a Congolese town on the shores of Lake Mweru, to become a boarding student at an elite high school in Lubumbashi, the country's copper capital, following the well-worn tracks trod by successful men like Kufi Kilanga. "He was very quiet, actually," Mwamba Wanzala, a classmate of his, remembered. "Not a very talkative chap, very quiet, but very friendly. We played soccer together."

Katumba had initially wanted to become a priest, but in Lubumbashi he had become fascinated by Gécamines, Mobutu Sese Seko's giant copper-and-cobalt mining company. "Life at that time was supremely dominated by Gécamines," he would later write. "We breathed Gécamines. We lived Gécamines. We dreamed of Gécamines." When Wanzala and Katumba chatted with their classmates at school, they all shared their life plans. "He wanted to study mechanical engineering, just like I did," Wanzala told me. "He didn't plan on becoming a politician."

For boys in Katumba and Wanzala's time, studying engineering was a surefire way to end up working for Gécamines. The company seemed to be a haven in a country where the number of going concerns was dwindling

under Mobutu's corrupt rule; it supported vast swaths of the South financially and socially. In the Gécamines neighborhood of Lubumbashi, there
were smart canteens, football fields for workers' children, schools, hospi-
tals, even Gécamines orchestras. *"Gécamines djo baba, djo mama,"* the saying
went. "Gécamines, my father, Gécamines, my mother."

Katumba would later remember that he wanted to emulate Gabriel
Umba Kyamitala, the director general of Gécamines, and rise to the
top of the mining agency. "The idol, the model, the sphinx of my fantasies was him. He was God, in his time, of the biggest mining giant
in decolonized Africa." Over the next few decades in Congo, Katumba's childhood fantasies would eventually coalesce, but the ensemble
they formed would more closely resemble the charred wreckage of a
nightmare.

DURING THE 1980S, IN KINSHASA, KUFI LIVED THE LIFE OF A BIG VEGETABLE.
He had cars, villas, staff, multiple wives, and girlfriends. "He was a
great polygamist," Gaylord Kilanga, Kufi's son, said of his father. "He
left behind twenty-something children." But he was also, by all accounts,
devoted to studying solutions for the betterment of the Congolese people. He served briefly as minister for primary, secondary, and professional education, and he traveled overseas to represent his country with
UNESCO. "He was a Mobutist in his deepest essence," Kilanga said. "A
true believer."

Kilanga can claim that his father was never involved in mining, but in
effect, he and the rest of the Big Vegetables depended on it just as much
as Katumba and his cohort in the South. Other companies, like the tin
mining company Géomines in the town of Manono, had been stripped
bare by Mobutu's vultures by 1982. But Gécamines was still healthy;
some 70 to 80 percent of Zaire's foreign earnings came from the export
of copper and cobalt from Katanga, and the firm was, de facto, the dictator's personal piggy bank, what the Congolese call a *vache laitière*, or
milking cow. From the late '70s onward, the money that the company
made from exporting copper and cobalt went into an account he controlled. During the 1980s and early '90s, according to an analysis of World

Bank documents, the dictator skimmed somewhere between $150 million and $400 million a year from the company's ledgers. This practice, known in Congo as *bouffer l'argent*, or eating money, is one that the chairmen of Gécamines, as well as local and national politicians, continued to emulate into the 2020s.

As Gécamines crumbled in the late 1980s, the company, once strictly a state concern, turned to private groups like the Entreprise Générale Malta Forrest, the firm headed by George Forrest, to try and save itself. In late 1986, Gécamines decided to hire Forrest's firm to help it remove arable topsoil so that it could reach ore deposits buried underneath. "Layer by layer, the engines descended toward the ore and fashioned the landscape so particular to open-pit mines, in steps," Forrest later wrote. Private companies like Forrest's were also burrowing into the state, first finding contracts to prospect for minerals on Gécamines's behalf, then operating some of the mines through "public-private partnerships," the first of what would come to be the dominant model for extraction in the Democratic Republic of the Congo, once Mobutu left. "Gécamines was facing real financial difficulties, all of its contracts were suspended, it was on the brink of asphyxiation, it was only just staying afloat," Forrest remembered. "Little deposits, little gain—just surviving." Soon, Forrest was in charge of several mines, one of which produced five thousand tons of cobalt a year, just over 10 percent of the world's total production in the late 1980s.

Other illegal networks of Gécamines workers, organized by Greek and Lebanese traders, would steal cobalt from the Shituru and Luilu factories. Using teams of fifty to one hundred workers, and heavily bribing officers within the Zairean hierarchy, they were able to steal ten to seventy tons of refined cobalt a week. These traders, known as "cobaltists," financed their operations by preselling cobalt on international markets or through grants from within their communities. They would export the material out of Zaire using double-bottomed trucks and then sell it to Chinese businessmen in Johannesburg. The "cobaltists" could not have realized it at the time, but they had begun to create a parallel supply chain for Zaire's critical metals that would become of vital strategic importance well into the next century.

THE COLD WAR WAS ENDING, AND THE U.S. AND EUROPE BEGAN TO CUT their aid to Mobutu, whose profligacy and ruthlessness were regularly making headlines. In 1987, Mobutu signed an agreement with the Chinese to build a giant stadium in Kinshasa. He decided to name it Kamanyola, after a decisive victory that he had won against Simba rebels in 1964. After the Berlin Wall fell in 1989, the desire in Washington to support Mobutu waned, especially as his human-rights abuses became more and more widely known, and China sensed opportunities to build bridges with Zaire.

Opposition leaders like Étienne Tshisekedi wa Mulumba were making their voices heard. Tshisekedi, a Luba-Kasai, was a former Kasaian separatist who had worked with Mobutu and then turned against him. His stubbornness was the stuff of legend. He was a recognizable figure at rallies, making fiery speeches from beneath the brim of his signature flatcap.

Starting in the early 1990s, partly to punish Tshisekedi, Mobutu and his officials began to tacitly support a Katangese movement against the Luba-Kasai, stoking interethnic hatreds that had been encouraged by Moïse Kapend Tshombe's ministers during the Katangese secession. For many years, Mobutu had used the Luba-Kasai as the Belgians had—to run the mines. Now he began to orchestrate a campaign of ethnic cleansing against them, purging the Luba-Kasai from Gécamines. "The Kasaians are foreigners. Katanga is certainly a hospitable land, but the foreigners must not forget their status," Antoine Gabriel Kyungu wa Kumwanza, Mobutu's governor of Katanga, said in 1991. Some one hundred thousand Luba-Kasai were also forced to flee their homes in the South. Many were pushed into squalid displacement camps near the railway stations in Likasi and Kolwezi.

For Gécamines, the purges were an unmitigated disaster. Everyone who was in Lubumbashi and in Kolwezi at the time remembers the two *pillages* of the early 1990s—mass nocturnal ransackings in which the businesses, and especially the state-owned mining businesses, were stripped bare. For Forrest, they brought back eerie memories of the 1978 Kolwezi attack. "These events were disastrous for the country because a part of the economy was destroyed by these rebellions that pushed thousands of people to flee, leaving everything behind them," he later wrote. He instructed his foreign staff to evacuate Luba-Kasai families. At least 661 Luba-Kasai people died of exposure or hunger, and scores more were killed in massacres. Some reports put the death toll between 50,000 and 100,000.

CHAPTER 13

A BATTERY AND A BUBBLE

In the late 1980s, as Mobutu Sese Seko was laying waste to his country and Akira Yoshino was completing his experiments at Nobeoka, Sony, the Japanese tech colossus, was also on its own path to creating a lithium battery. Sony had broken ground with handheld video recording during the decade. But they had a problem: The available batteries were either too big and heavy or didn't have enough power for the company's devices. According to a corporate history, "Sony's yearning for a battery that could be reused again and again was growing stronger every day."

Spearheading this initiative was a charismatic man named Keizaburo Tozawa, the head of Sony's battery division. Tozawa worked on the problem with a singular focus. He employed a management style that he based on a naval target-practice technique. "By shooting three guns at once, all aiming at the same target, the chance that one of the guns will actually reach the target is greater. With this approach, the target is reached faster and more accurately," the corporate history contends. "In the same manner, Tozawa decided to start several research efforts simultaneously using slightly different approaches to reach the target of developing a lithium rechargeable battery."

A lingering question about who first discovered the dream battery hangs over the history of lithium-ion. Sony had almost certainly seen

Yoshino's 1985 patent and had, in fact, been in contact with Asahi Kasei executives, who had demonstrated their battery to the corporation on January 21, 1987. Sony was far bigger and better resourced, but Asahi Kasei was worried that it could be gobbled up by the device behemoth, which had become, after the release of the Walkman, one of Japan's most visible companies. The two firms forged a "joint work team," which meant that their scientists could work together, and then Sony announced that it had come up with its own battery, which looked suspiciously identical to Asahi Kasei's. "The question of how the two companies would end up with identical chemistries would always linger," Charles J. Murray would later write in *Long Hard Road*, a history of the lithium-ion battery.

When Yoshino won the Nobel Prize, Asahi Kasei's supporters took it as a sort of nod from the science gods that his had been the first workable battery. Sony, meanwhile, engineered its battery far better, and it would end up reaping more of the rewards. "The debate over who was first," Murray writes, "would eventually be lost to history."

Either way, Sony and Asahi Kasei both rushed to get their batteries into production. The peculiar mechanics of capitalist innovation meant that the race was now on to secure patents—John B. Goodenough's for the LCO cathode, for example—and gear up manufacturing capability. The companies needed to dominate the field and produce as many batteries as possible. Sony, which under Akio Morita especially drew more on American corporate culture in the move-fast-and-break-things mold, rapidly began to engineer its battery. Asahi Kasei dithered, much to the frustration of the teams that had been working on lithium-ion and the new anodes.

On February 4, 1991, Sony announced that its new technology would be commercially introduced. The battery would be called a lithium-ion battery rather than a lithium battery—a necessary rebranding, as the latter was known to catch fire and all sorts of prohibitions on carrying them on commercial aircraft existed, set forth by the U.S. Department of Transportation. It would be a long time before Samsung Galaxy Note 7 smartphones began catching fire and the current prohibitions on stowing lithium-ion batteries in checked luggage would come into effect.

The battery's popularity blossomed, and Japan became the epicenter of the new lithium-ion rush. The battery found its first use in Sony's

CCD-TR1 camcorder. Technologists quickly saw the far-reaching pos-sibilities of the small, lightweight, and powerful cells: Laptop makers re-alized that lithium-ion batteries remained charged for longer, and the nascent cell-phone industry began testing out lithium-ion in its phones. By 1994, models powered by the new batteries, such as the Motorola MicroTAC Elite, began to pop up.

But no one seemed particularly concerned about where the metals to make these batteries were coming from. They seemed abundant, never-ending, even.

That's not to say there were not warning signs: The price of cobalt shot up in the early 1990s, primarily because of instability in Zaire (other in-flationary factors included an economic crisis over the border in Zambia, and the collapse of the U.S.S.R.). The pillage of the national mining firm following the purge of Kasaians from the country's mining provinces made international traders uneasy, and they noted that the mines in southern Congo had suffered from chronic underinvestment. People began to stock-pile the mineral, but then, when the U.S. decided to sell its strategic reserve of twelve thousand tons of cobalt starting in 1993, prices fell. With cobalt selling for around $8.50 a pound in 1999, the device-makers were not yet concerned about the source of metals that went into their products.

The same was true of lithium, another one of the ingredients in the new batteries. Everywhere from South America's Lithium Triangle, which stretches across Argentina, Bolivia, and Chile, to the United States, the metal that these days is known as "white gold" was dirt cheap. When the Soviet Union collapsed, the U.S. government began negotiating to sell a thirty-six-million-kilogram stockpile of lithium that it had squirreled away in order to build nuclear weapons. Bombs aside, the material was mainly used in the aluminum, ceramics, and glass industries, and few people foresaw that it would become vitally important—the U.S. Geological Survey wrote that in 1994, lithium was thought to be so abundant that supplies couldn't possibly be exhausted by demand. Until the 2010s, in fact, ac-cording to the authors of one paper on the rise and fall of this metal in the U.S., "lithium prices were generally not sufficiently high to justify ex-tensive exploration, nor encourage production from more than a limited number of high-concentration / low-cost operations."

THE 1980S HAD BEEN A TIME OF IMMENSE FINANCIAL SPECULATION IN JAPAN. Anything seemed possible. But that bubble had burst in 1989; the 1990s was, for most Japanese, the hangover after the party. Morale plummeted. "The idea of Japanese economic supremacy would be forever discarded," the journalist David Pilling writes in a history of Japan's stagnation. There was one exception, however: For the first years of the 1990s, Japan was the center of all things to do with rechargeable batteries. Sony sat at the white-hot core of the boom.

By 1995, the conglomerate had released its third-generation battery, which was twice as powerful as its original model, and later that year announced that it would begin developing a car; by the spring of 1996, Sony had agreed to develop a battery pack for Nissan's FEV-II, the first-ever lithium-ion-powered concept vehicle. Even though the FEV-II would never get beyond the conceptual stage, it was portentous, heralding a new era of battery-powered locomotion:

Akira Yoshino, John B. Goodenough, and M. Stanley Whittingham
after winning the 2019 Nobel Prize in Chemistry

CHAPTER 14

GETTING RICH IS NO SIN

B y the mid-1980s, Robert Aronson, the early electric-car evangelist, was becoming frustrated. He had wanted to open an EV business in a tax-free zone in the Bahamas, but government bureaucracy
had stymied him. So he set his sights on other markets. Aronson began
closely following developments in the East. "That was just about the time
that China was opening up," remembered Barry Iseard, a British environmental scientist who serves as vice president and director of an Aronson-
founded company, Apollo Energy Systems. Iseard met Aronson when the
inventor showed up in a class he was teaching in the Bahamas in 1980:
They had worked together ever since.

Deng Xiaoping, China's paramount leader, had assumed power in
1978. A disciple of Zhou Enlai, the leader who had defined China's aid
framework during his 1964 Africa tour, Deng took to heart Zhou's ambition to transform the country into a leading industrial power by the beginning of the twenty-first century. To do this, he began to move China
toward a market economy. By 1984, China had started pursuing a contract system whereby the managers of state-owned enterprises would sign
contracts with the state, and the employees of those enterprises would
sign contracts with the firms. As Deng told the CBS correspondent Mike

Wallace in 1986 during his first major interview with a Western broadcaster: "To get rich is no sin."

The contract system was hardly unfettered capitalism, but U.S. businessmen sensed an opportunity—if the Chinese premier was saying it was no sin to get rich, why not give it a go? Aronson had friends who were keeping him abreast of opportunities in China, not least among them executives at the aviation and defense company McDonnell Douglas. Encouraged by their reports, he took a meeting on June 5, 1985, with some representatives of the state-owned Shanghai Aircraft Industrial Corporation at McDonnell Douglas's headquarters in Long Beach, California. SAIC had been involved in lengthy negotiations with the U.S. firm and, in 1985, entered a coproduction agreement with McDonnell Douglas to assemble MD-80 aircraft. "It's very slow but very positive," a McDonnell Douglas executive told a reporter from *The Washington Post* in late 1984. "The Chinese invented the word negotiation. I may be retired and my successor may be retired, but I think we'll make it with China."

Aronson began to negotiate opening a factory in China with SAIC. After his June 5 meeting, he told his associates that they could make a huge profit by moving his operation to China. "In those days, labor was really cheap in China," Iseard told me. "You could pay a Chinese person a month's wages, which would be not much more than the hourly-rate wage for an American worker doing the same thing."

In 1988, after three years of negotiations, Aronson signed a contract with a subsidiary of SAIC called the Shanghai Far East Aero-Technology Import & Export Corporation, or SFAIC, to begin producing batteries at a fixed price. His firm would provide the raw materials, the equipment, and the expertise, while SFAIC would provide the factories and the manpower. In September of the next year, the plant was up and running, but three months later, the state-owned enterprise said it could no longer guarantee the price of the batteries.

Aronson, like Elon Musk and many other battery entrepreneurs who would come after him, discovered that doing business in China was far from straightforward. Investors may have been lured in by Deng's assurance, but they had ignored the second part of Deng's comment to Wallace:

"What we mean by getting rich is different from what you mean. Wealth in a socialist society belongs to the people."

They would soon find that the business climate in China was confusing and opaque, especially as politicians in Beijing laid down policies of retrenchment in the early 1990s in reaction to massive pro-reform protests. Dissidents were silenced. Guardrails were placed around foreign investment, even as more and more overseas companies poured money into the country. The transformation of state-owned enterprises into capitalist models became layered with complexity as the government moved to protect strategic industries and firms at which layoffs threatened labor unrest.

SAIC, which had close ties to the Chinese military-industrial complex, refused to negotiate with Aronson. Something else was also afoot. "The Chinese government decided to commandeer his factories to make submarine batteries, and they wanted to take over his factory," Iseard told me. "And Bob didn't want to allow that. He wanted to continue to manufacture for electric cars and other vehicles and so on." The battery factory was seized by SFAIC, and Aronson took the firm to arbitration court in Stockholm, where he was awarded $4.5 million. SFAIC declared bankruptcy and refused to pay. "We began learning that China has a policy of non-enforcement of arbitral awards in favor of foreigners," Aronson said when he testified to Congress about the case in 1997.

Even more worrying, the Chinese government was using SFAIC as a tool for spying on the U.S., and as a way of diverting technology to Beijing's military-industrial complex. "They had a liaison office at McDonnell Douglas in connection with their agreement with us, so we were really a cover for an espionage operation which went bad," Aronson testified. Five Chinese spies were eventually arrested at McDonnell Douglas's headquarters in California. The U.S. firm found itself caught up in a web of accusations that it had allowed vital machine tools—tools that could be used to build military aircraft—to fall into the hands of China, and it was fined $2.1 million by the Department of Commerce in 2001. *The Wall Street Journal* called it "one of the largest civil penalties ever in an export-control case." The company was acquired by Boeing, which paid the fine. SAIC's parent company, the state-owned Aviation Industry

Corporation of China, is still around. In 2023, it was ranked 150th on *Fortune*'s Global 500 list of corporations.

As for Aronson, he never received the arbitration money, Iseard told me. During his congressional testimony in 1997, at a hearing on whether China should be allowed to join the World Trade Organization, Aronson didn't mince words. "China's disgraceful behavior . . . is not the behavior of a country mature enough to join the World Community of Nations," he said. "Therefore, at this point in time, I could not recommend that China be allowed to become part of the World Trade Organization."

Despite the admonition of Aronson and others, the corridors of power in Washington and Europe were awash with enthusiasm for globalization, and China joined the WTO in 2001. Beijing would quickly build up the capacity to become a world leader in battery production and the processing of critical minerals. Aronson and his dream to build smokeless cars would be forgotten. And lithium-ion, a technology that the pioneering American had overlooked—and that China understood it had to control—steamed to the vanguard of power storage.

CHAPTER 15

TWILIGHT OF THE
BIG VEGETABLES

As the West moved out of Zaire, Beijing rushed in to capitalize on the relationship it had built with Mobutu Sese Seko. The Chinese-built Kamanyola Stadium, which was finished in 1993, was the fourth biggest in Africa, and around the same time, China agreed to mine at the marquee Katangese copper mines of Étoile and Ruashi. For a brief while, it appeared that Mobutu might be saved by his alliances in the East. But then, even that money dried up: The Chinese had seen the writing on the wall for the dictator. They had better things to do than stand alone with an isolated kleptocrat.

The crumbling of Congo in the 1990s and early 2000s and the way it was divvied up are key to understanding today's complex lithium-ion battery supply chain. "The country's formal economy," a U.S. State Department assessment from 1996 read, "has virtually collapsed. . . . Zaire's public sector is insolvent and unable to provide even the most basic public services at satisfactory levels. . . . The vast majority of Zaïrians live in poverty."

Congo's mining industry had also collapsed. Katanga's huge Kamoto mine, which followed a huge deposit of copper and cobalt deep through underground tunnels, had literally collapsed six years earlier, burying the jig concentrator and fifteen million tons of ore. At the nearby Kamoto

Oliveira Virgule, one of the country's largest mines, the pit had been flooded. Copper production was just under a fourteenth of what it had been in 1988. Cobalt production had almost halved. When I asked Bruce Jewels, a banker who had dealt with Gécamines before Mobutu's fall, what the company was like in those years, he laughed: "Corrupt." Top executives stole wantonly. "These guys were being kept in positions of power and houses bought for them by Belgian friends in South Africa so that the status quo on the delivery of cobalt and copper could continue," he went on. "It was a complete mafia."

In desperation, Mobutu turned to his own countrymen. He invoked what is sometimes called "Article 15" of the Congolese Constitution: *Débrouillez-vous*—fend for yourself. He gave a speech to soldiers who had not been paid. "You have guns," he told them. "You don't need a salary."

As the State Department report surmised, "The informal economy, while dynamic and increasingly active in providing essential public services, consists essentially of subsistence activities, and cannot arrest the country's overall economic and social degradation."

Mobutu encouraged the people of the South to *fend for themselves* as well. In 1982, he had legalized small-scale mining, mainly of gold and gemstones, but as Gécamines collapsed in the early 1990s, so-called artisanal miners became a common sight on the ailing giant's plots. During and after the Belgian era, the practice had existed on the periphery in Katanga, often for traditional copper smelting by the descendants of precolonial smiths. The difference was that now, for the first time, the small-scale mining of copper and cobalt ore—often carried out by unpaid Gécamines agents and achieved using tools no more sophisticated than sharpened lengths of metal pipe—had become a business. Copper and cobalt mined in this manner began to find its way onto international markets; ore was sold to traders at small makeshift trading houses known as *maisons* and then to bigger fish who would ship it out of the country.

ODILON KAJUMBA KILANGA'S FATHER, JOSEPH KUFI KILANGA, WAS A COUSIN of Kufi Kilanga, the minister, but he had enjoyed none of his relative's success. In the 1980s, he had moved from northern Katanga to the outskirts of Lubumbashi and settled near a copper-cobalt mine called Kakama. Joseph had come to find a better life for the family he was starting with his wife, Beatrice, and he began building a small business selling tires. He talked about his illustrious cousin every now and then, but he was starting from scratch.

As Gécamines folded, people started to enter the concession near Joseph's shop to mine for copper and cobalt. Joseph remained wary: He wanted no part in the business. Besides, nobody was making that much money. The pits people were digging were getting deeper and deeper and more and more dangerous. Kajumba remembered that during these early days, he and his brothers would help their father take off tires, fitting and refitting them onto the wheels of passing cars and trucks. Their father didn't want them working too much, though: School was the most important thing, he used to say. Joseph worried for his young sons, as the country seemed to be sailing further and further into dire straits. "He was a great example," Kajumba would later remember. "He was the best dad in the world."

When other kids in the neighborhood started to go into the mine site to earn money scrounging for ore, Kajumba's father forbade him from joining them. He was the kind of father whose instinct to protect overrode his children's wishes. "He didn't want me to go to the mines, or even to venture too far from home," Kajumba said. "He wanted us to study."

———

THE COUNTRY SEEMED TO HAVE SUNK AS LOW AS IT COULD. BUT WORSE WAS to come. In 1995, Mobutu's defeat was being actively plotted by rebels to the east, in Rwanda and Uganda. After the 1994 Rwandan genocide, the marshal had made enemies among Rwanda's new Tutsi-led government. Mobutu had allowed ethnic Hutus to flee into the country from Rwanda after Hutu supremacists slaughtered up to eight hundred thousand Tutsis. Many of the people responsible for the killings were among the refugees and now threatened Rwanda from Congolese refugee camps. Rwanda's Tutsis mustered a force to unseat Mobutu and install someone who would be loyal to them.

In the past, Mobutu might have counted on Western support to quell an all-out assault, as he had during the Shaba I and Shaba II attacks on Kolwezi. But this was a new world. Besides, the dictator's China policy, which had seemed so wise when the money was flowing in, hadn't exactly endeared him to his erstwhile allies in Washington, Paris, and Brussels. They didn't discourage or encourage Kigali's plans. Mobutu asked Hassan II of Morocco for assistance, just as he had during the Shaba rebellion, but this time none came. A mine entrepreneur of the era told me that the CIA was content to sit on the sidelines and watch a victor emerge.

IN THE 1990S, LAURENT-DÉSIRÉ KABILA, A FORMER LEADER OF THE ANTI-Mobutu Simba rebels, was selected by the Rwandan government to become one of the leaders of the insurgency. After one of his fellow rebel leaders died, he quickly maneuvered himself into becoming the only leader of the campaign to oust Mobutu. Kabila, who was in his late fifties, was in many ways an odd choice to lead Congo. He was known as the M'zee, after a Kiswahili honorific for an older gentleman. He was a Marxist doctrinaire and a wily operator. A French report of the era described him as "like an ebony Buddha with a polished skull . . . a plump ghost."

Kabila was born to a Luba father and a Lunda mother in northern Katanga, on the shores of Lake Tanganyika, not far from where Kufi Kilanga had spent his early school years at the mission. Kabila's first taste of politics had come in the early 1960s, during Katanga's fractious secession period. A Balubakat militia he had joined at independence killed his father, an educated man who had been elevated to the status of a local functionary by the Belgians. Kabila never publicly expressed any remorse for his old man's death.

Soon after, Kabila joined the Simbas in rebellion in eastern Congo. There he liaised with Che Guevara, who had come to spread revolution in Africa. He constantly frustrated Che, standing him up at remote guerrilla camps for weeks. Che wasn't entirely critical—he said Kabila was a fine leader—but he noted that "it is essential to have revolutionary seriousness, an ideology that can be a guide to action, a spirit of sacrifice that accompanies one's actions. To date, Kabila has shown that he possesses none of these qualities. . . . I have very great doubts about his ability to overcome his defects in the environment in which he operates." After a few months, Che packed up and left.

In the years after Che's disappointment, Kabila had continued to underwhelm. When he'd tried to run a Marxist rebel republic in eastern Congo, it had fallen to pieces. In later years, he had enjoyed better luck with a brothel and a series of fisheries in Tanzania. Kabila certainly seemed an odd choice for leader of the new rebel movement, but only if one thought that the Rwandans wanted him to actually *run* Congo. He was less of an odd choice, however, if one considered that the Rwandans saw him as a puppet leader. They really wanted to control Congo through him and control the giant country's resources themselves. And as Sony's production of lithium-ion batteries went global, those resources would become ever more valuable.

CHAPTER 16

BUILDING DREAMS

Wang Chuanfu, the founder of BYD

In February 1995, the same year that Sony introduced its powerful third-generation battery, a company called BYD began doing business in the backstreets of Shenzhen, one of China's manufacturing boom-towns. A picture from the time shows the entrance to the firm, housed in a grimy concrete building, behind a red sign that reads, in Mandarin, "Metallurgical Compound." There was scant indication that, before long, the firm it advertised would become one of the world's largest battery makers and, soon after, one of the world's largest automobile companies.

In the spring of 2024, I watched as a black BYD Seal, the company's midsize sedan, silently rounded a bend in Paris's Second Arrondissement. Later I logged on to the company's website. "Dynamic and intelligent," the site gushed. "Ultra-rapid charging is no problem." Through a chat window, I asked an associate (or a chatbot) named Charles if he knew where the company sourced the primary materials it used in its lithium-ion batteries. "It's a BYD Blade battery," Charles said, a novel technology that orders cells in a unique way, making for a more powerful, longer-lasting battery.

But where did the material in the battery come from? BYD's salesman didn't want to get into the specifics of the supply chain. "I'm very sorry," Charles replied. "I can't dispose of that kind of information directly."

Back on the street, I stopped to snap a photograph. It was the first time I had ever seen a BYD. Around me, nobody seemed to notice the pebble-shaped vehicle as it slipped into the afternoon traffic. It was fitting: The company had built its reputation on a certain lack of flash, eschewing grand public proclamations like those of Elon Musk and Steve Jobs for a slow and steady grind. In early 2025, BYD would surpass Musk's Tesla to become the world's largest producer of electric vehicles.

———

THE BYD SEAL WAS EVEN MORE EXTRAORDINARY IN LIGHT OF WHERE ITS founder had come from. Wang Chuanfu was from a rice-farming town called Wuwei, in China's Anhui Province. Translated, *Wuwei* means something like "Do nothing" or "Let it be" in Mandarin. Wang's father, a carpenter, died when he was thirteen, and his mother passed two years later, leaving his elder brother and sister to raise him. The tune of his upbringing seemed to consist of a single, somber note, one that tolled crushing poverty for hundreds of millions of rural Chinese in the 1970s and '80s. His family sometimes had to beg to pay for basic necessities, and a storm once blew down their small home. Wang's older brother dropped out of school to work, and his family gave one of his five sisters up for adoption because they could not support her.

Wang was bright and worked incredibly hard, wasting little time on leisure. As a young man—thickly bespectacled, on the smallish side, with a round, jovial face—he began his studies at the Central South Industrial University (now the Central South University). There he joined the youth league of the Chinese Communist Party and became known for his moves at cafeteria dances. On weekends, he helped his older sister and brother run their shop, declining to take time away from work to join his friends on hiking excursions.

In 1987, Wang went to study for a postgraduate degree at Beijing's General Research Institute for Nonferrous Metals. Within three years, he had acquired his degree, and he became an associate professor. Two years after that, he was promoted to deputy director of the entire institute. He married and prepared for a stable career in academia. Such jobs, known as "iron rice bowls," had assured a stable salary for Chinese citizens as the country settled into its Communist era, and they were part of the state's implicit promise to furnish work to those who applied themselves. A year later, Wang was promoted again, this time to run Big Power Nickel-Hydrogen Battery Co. Ltd., a joint-venture battery company that the institute had established with Baosteel, a state-owned metals concern.

Taking on the new role meant that Wang had to move with his young family to the southern city of Shenzhen. As China had opened up for business, Shenzhen had led the way. The city was originally a fishing village next to Hong Kong, an Asian finance hub, then still under British control. In 1980, Deng, the Chinese premier, designated Shenzhen a special economic zone (SEZ), relaxing controls on private industry. Businesses could be set up in a heartbeat, and fortunes were there to be seized.

When Wang arrived in 1992 to run Big Power, he was transfixed by the currents of business and success running through the city. A picture of him from the era shows a man who had come a long way from Wuwei: Wang sports a low-cut, double-breasted gray suit and seems to be in a hurry, his gaze fixed on some point out of the frame.

Wang had noticed something else too. It was the early '90s, and more and more of the new business elites he saw marching through the streets were spouting off into large mobile phones with long antennae.

Wang understood that there was a gap in the market: The batteries were incredibly pricey, and this contributed to making the phones scarcely affordable for most people. Wang thought he could produce them in China at a far lower cost. "Importing batteries from Japan was very expensive," he later told an interviewer. "There were import duties, and delivery times were long." One day in 1994, Wang read an article suggesting that Japan was moving away from nickel-cadmium technology in a bid to minimize pollution; he knew, though, that the technology was still needed to power many devices, especially the cell phones that he was beginning to see everywhere.

At first, Wang tried to make the idea stick at Big Power, but infighting between the joint-venture partners made the work environment chaotic. He decided to strike out on his own. In November 1994, he rented a small workshop on one of the top floors of the metallurgical academy in Shenzhen's Buji neighborhood. As a book in praise of Wang's business acumen later put it: "He renounced a comfortable life to become an ascetic." Wang, who had spent his life studying chemistry, began to deconstruct Japanese batteries to see how they could be made quickly and cheaply. He took aim at the large battery firms across the Sea of Japan. Two months later, just shy of his thirtieth birthday, Wang incorporated BYD and started searching for funding. The initials didn't mean anything at first— Wang reportedly chose the character that corresponded to the Roman *B* so his company would be high up on lists of companies at trade fairs.

It was difficult for Wang to secure funding for the new business. When he went to banks and asked for a loan, he found that they required real estate as collateral for their cash. Wang didn't even understand the meaning of the term *real estate.* But by June 1995, he managed to secure a $300,000 loan from Lu Xiangyang, a cousin of his who had made a small fortune working in banking in the SEZ, as well as some more seed funding from an associate who had worked in the insurance sector. Their stakes would later propel Wang and the other investors into the ranks of China's billionaires. The batteries were reverse engineered—copied— from Japanese and Korean models. In the absence of mechanical equipment, Wang hired a thousand low-salaried workers to build electronics on the cheap and on the quick.

Though he didn't have advanced machinery to put together batteries, Wang figured that China's cheap labor was his advantage, so he divided production into bite-size stages, splitting his employees into teams of two people who formed a rapid production line. Conditions at the Buji workshop were grueling: A dozen people lived and labored there around the clock. Wang lived with them, working, it seemed, all the time; when his daughter was born, he stayed at the office instead of accompanying his wife to the hospital, and he only visited his newborn a few days later.

Wang was wary of spending too much money on machinery for BYD, even as his company grew out of the space in Buji: Every week, more migrant labor would sweep into the country's coastal cities from the countryside, so he had a steady supply of people willing to work for BYD. According to *The Creative Wisdom of Wang Chuanfu*, a business-strategy book by Li Daqian, Wang only slowly integrated machines into his company. He "invented a production pattern of 'equipment + workers = mechanical hands' and carried out production with a 'human sea' strategy. He separated a production line into several processes. The core process was controlled by automation, while other links were finished by laborers. In this way, Wang Chuanfu only spent millions achieving the effects on what others had spent tens of millions."

The hard work paid off. In the subsequent decades, China and BYD grew like lightning. The country was doubling the size of its economy every eight years. In short order, the firm was able to produce four thousand batteries a day, and Wang's emphasis on cost cutting and repetitious work made his batteries 40 percent cheaper than Japanese ones. From 1999 to 2001, BYD grew sales by three times to around 1.3 billion renminbi ($157 million in U.S. dollars). Meanwhile, some two hundred battery start-ups had taken fast root in China. "Our competitors were all local," Wang later told the authors of a Harvard Business School case study. "Their quality was inconsistent. Our aim was to improve quality while keeping the price low so we could compete in the high end of the market. So we started to invest in process improvement."

AT THE SAME TIME, JAPAN HAD SLOWED TO A CRAWL. AT THE END OF THE 1980s, the Bank of Japan had raised interest rates in a bid to slow inflation; the market, opiated by cheap government money, was at first slow to react. In the 1990s, the withdrawal symptoms truly kicked in, and Japan started to lag. The 1990s would be known as Japan's "Lost Decade." Growth slowed to about a quarter of its rate in the previous decade, and Japan went through three recessions. Even by the early 2020s, the largest Japanese companies had barely grown since 1989.

Sony was initially able to ride out some of the aftershocks of the rate rises. Batteries certainly helped offset the stagnation affecting other economies. Thanks to innovations like the lithium-ion battery, Sony recorded $63 billion in sales in 2000. But the company was not immune to the slump that had stopped the clocks in Japan, and it was suffering from its own lack of impetus.

The mass adoption of the mobile phone intensified the race for lithium-ion, but Japanese companies seemed unable to hang on to their early-mover advantage. Executives from Japanese firms and from around the world scoured the earth for a place where they could build batteries as cheaply as possible at scale. Sony turned to China: In 2000, the company issued a press release trumpeting the creation of a new factory that created high-tech lithium-polymer batteries in Wuxi, a city close to Shanghai. "Sony believes in the great potential of the Chinese market," said Hiroshi Shoda, president of the firm's China subsidiary. "Its high-tech and new technology fields, represented by the communications industry, are growing rapidly." Soon, Chinese companies were making their own lithium-polymer batteries—and all at lower prices than the Japanese versions.

Japan was reduced to making parts of devices, like batteries—and even that market would soon evaporate. Across the Sea of Japan, the rest of Asia was looking at Tokyo's demise as an opportunity, one that would come to be of paramount importance as the world began to decarbonize and turn to electric cars.

Korean firms like Samsung, and later LG, began to develop lithium batteries. At the beginning of the 2010s, Samsung was producing 110 million battery cells a quarter and held more than one-fourth of the market share of a global lithium-ion battery business that Reuters valued at $18 billion. Sony's share at the time was 7 percent. At the turn of the millennium, Japan had controlled 90 percent of the lithium-ion battery industry; by 2012, Sony's value had dropped to one-ninth of Samsung's.

The economics in Korea worked better than in Japan, but according to the *Financial Times*, Samsung and LG enjoyed only a 10 percent operating margin on their battery production at the best of times. The margins in China, as Wang's work at BYD had proved, could be much higher. In a globalized world that promised frictionless trade, more and more firms began looking at China, a country that had followed the Japanese and Korean models of development but had an almost endless supply of cheap manual labor. In 2022, an executive at a Japanese battery company explained to me how Chinese companies like BYD had leveraged low wages and looser safety standards. "In China, the mentality is different," he said. "The mentality is: 'We go all in; we can fail; we can even fail the safety tests—the government will support us, and even if the batteries blow up from time to time, they will look the other way.' Now they have surpassed us."

IN THE ADVANCE RANKS OF CHINA'S BATTERY-BUILDING MACHINE STOOD BYD. Wang was, by all accounts, an eccentric character, a workaholic who lived and breathed his company. He rarely appeared in anything but a business suit, though he sometimes sported a silky tie that shimmered in the light, its fabric patterned with his company's monogram, stitched in miniature.

Wang never really seemed to rest. Unlike his Japanese and U.S. counterparts, he continued to be frugal, even as money flowed in. He insisted that his executives fly economy, and at the North American International Auto Show, in Detroit, the firm rented a cheap suburban house for its presenters rather than fork out more for a hotel.

Helpful, too, was that at the outset, the firm didn't have to invest in research and development. "China and Korea are good at copying Japan," the Japanese battery executive told me. "Sony built their factories to build batteries in China in the 1990s. We gave them the keys to copy us." BYD reverse engineered Sony's battery designs, leading the Japanese firm to file a patent-infringement lawsuit in Tokyo against Wang's company in 2003. Sony withdrew it when BYD invalidated other patents held by Panasonic: It was a win for BYD, which showed it could defend itself against larger legacy firms, although the company did set up an intellectual property department of its own.

BYD had begun creating lithium-ion batteries by the turn of the new millennium, and China was on the way to becoming one of the world's largest producers of the technology. Rival firms, like Robin Zeng's Amperex Technology Ltd., or ATL, also began mass-producing batteries. Zeng bought a U.S. patent to build a lithium-polymer battery that was safer and more flexible than other lithium-ion technologies; at first, the firm's technicians couldn't seem to make the technology work, but then they employed a technique similar to Wang's, working long nights until they had created a usable battery. Zeng's factory was located in Dongguan, which soon became one of the cheapest places in the world to produce lithium-ion batteries.

By the early 2000s, BYD was making cheap cells for Motorola and Nokia, which profited from the massive savings Wang's supply chain entailed. Mobile phones provided an important growth segment for Chinese battery makers. In 2002, an estimated 95 percent of mobile phones were powered by lithium-ion batteries, and over the next decade, most mobile phone batteries came to be made in China. "It's a volume game," Jeffrey Char, a Tokyo-based tech and energy investor, told me in 2023. "Economies of scale and large volume basically provide you with the opportunity to lower your cost and therefore be more competitive."

WITHIN A DECADE OF ITS FOUNDING, BYD HAD MUSCLED ITSELF INTO BEcoming the world's second-largest battery maker. Competition was steep, but Wang had mastered a work cycle that brooked little rest. He was soon mooting the building of "a green-energy car," a plug-in vehicle. In Wang's conception of BYD, "price would be the greatest driving force of sales"— relentless cost cutting would create cheaper products for consumers, especially Chinese consumers.

The company was listed on Hong Kong's stock exchange in 2002 and started to produce cars a year later. "We successfully sustained remarkable growth, as we have done in the past eight years," Wang wrote in his first annual report. The company had grown yearly revenue by 76 percent.

BYD, the report boasted, had enjoyed "enormous success." But BYD had also gotten a helping hand from the Chinese state: At first, funds were available through the government's 863 Program, which ran from 1986 until 2016, spurring domestic high-tech development in China and birthing supercomputers and spacecraft. Then, at the turn of the millennium, the Chinese state prioritized the commercialization of battery systems during the period of the Tenth Five-Year Plan, a national development blueprint that set economic policy, and BYD profited from government financing and tax breaks. More and more, Wang was becoming part of the fabric of the Chinese state. In 2000, he was elected deputy of the Shenzhen Municipal People's Congress, and he "enjoyed special allowances from the State Council," according to one of the company's yearly statements.

Soon, Wang was joking that his company's initials stood for "brings you dollars."

Zeng's ATL had also enjoyed wild success, thanks to the same concoction of state capital and hard work. Its early batteries, created after a patent Zeng had licensed from Bell Labs, would swell and occasionally explode if charged repeatedly. ATL's engineers changed the original chemical formula to stabilize the battery and filed for a new patent, which, in 2003, caught the eye of Apple as it was launching its iPod media player. Zeng's new batteries were now powering the Walkmen of their era. In 2005, the company was bought by TDK, the Japanese electronics manufacturer, but a spin-off, Contemporary Amperex Technology Limited, or CATL (pronounced as "cattle" by some), would grow to become the world's largest electric-vehicle battery maker. Wang and Zeng had the production capacity, the know-how, and the sheer engineering muscle. Elon Musk may have been thought of as the battery king of the West, but Zeng and Wang were power-storage emperors. In 2024, the dichotomy was underscored by Zeng. "He doesn't know how to make a battery," the CATL head said of Musk. "It's about electrochemistry."

BYD AND WANG WERE SHOUTING THEIR AMBITIONS TO HIGH HEAVEN, BUT at first few people in the West took any notice. In January 2003, BYD bought 77 percent of Xi'an Qinchuan Automobile, a struggling state-run automaker in the interior of China. (The next year, the company upped its stake to 92 percent, bringing BYD's investment to $38.9 million.) "Due to the limited oil resources and increasing environmental protection consciences, the growth potential for electric vehicle and automobile is enormous," Wang wrote six months later. By 2004, it was marketing the Flyer, Qinchuan's boxy $4,000 petrol-powered car, at the Auto China show, in Beijing, at a booth that featured two models clad in gold bikinis and mermaid tails.

Most Western auto executives scoffed: The Flyer had a tiny engine and doors that would only fully open some of the time. But in China, the cheap vehicle was attractive to a new generation of workers who, on the back of the country's industrial success, were able to afford cars for the first time. Wang had achieved further savings by bringing his supply chain in-house as he created "a whole-industry chain of low-cost components," Li explains in *The Creative Wisdom of Wang Chuanfu*. "At least 70% of BYD's auto parts were produced by internal business units."

From around 2005 on, BYD made everything but tires, windshields, and some "universal auto parts." It even made things that were usually outsourced, like CD players and wires. This allowed the company to maximize its cost competitiveness. It was perhaps inevitable that the "resource advantage" talked up by companies like BYD would lead Chinese businesses—and the Chinese government—to try to bring the raw materials that went into producing lithium-ion batteries under their aegis.

It wasn't exactly difficult to figure out what would come after the Flyer: a BYD car powered by batteries. Part of Wang's strategy for his firm was neatly summed up in one of the company's mantras: "When others have something, I have a better one; when others don't have something, I have it."

BYD was also hard at work replicating its original model: what the company called "innovating through imitation"—and what foreign firms were quickly realizing was simply copying. "After a year and a half in business, we've realized there's a lot of technology that's not patented in China," Wang told *Forbes* in 2004. "We can learn a lot from that."

BYD WAS NOT THE ONLY FIRM THAT WAS GROWING IN CHINA. THE COUNTRY was "no longer content to be the home of low-skilled, low-cost, low-margin manufacturing," a writer for *The New York Times* opined in 2008. "Chinese companies are trying to move up the value chain, hoping eventually to challenge the world's biggest corporations for business, customers, power and recognition." Communist Party planners were also beginning to realize that electric vehicles represented a China-shaped space in the global market. That year, at the Beijing Olympics, Chinese companies showcased some five hundred new energy vehicles, including ones powered by batteries.

By 2008, the writing was on the wall, and a few wealthy Westerners had begun to read it: That year, Warren Buffett bought 10 percent of BYD's stock, and the company unveiled the world's first plug-in hybrid: a lithium-ion-powered car with a petrol-range extender. In Shenzhen, Wang presented the car from beneath a banner that read TECHNOLOGY FOR A GREEN TOMORROW. The car would combat pollution, which had become an acute health crisis in the nation's cities during China's push to become a manufacturing powerhouse. That year, it was also widely reported that the upstart company's name had taken on a new significance: Its initials, BYD, stood for "build your dreams."

CHAPTER 17

FIRE SALE AT THE KARAVIA

Augustin Katumba Mwanke, the young Katanga native who had grown up idolizing Gécamines, headed to South Africa after he finished his studies. It was the early 1990s, and he took a job with HSBC Equator, a subsidiary of the British multinational bank that did its business in sub-Saharan Africa. "He was a kind, thoughtful, intelligent, quasi-religious person," Bruce Jewels, the banker, told me. Jewels was Katumba's former boss at HSBC Equator. "A churchgoing man and churchgoing family, you know? And that resonated with me because I'm also a churchgoer." At the time, Katumba, a slight man, wore a thin mustache and double-breasted gray suits with wide lapels, the style that Wang Chuanfu was also sporting over in Shenzhen. He showed up one day at Jewels's office looking for a job. Katumba was a young accountant with no experience in banking, but Jewels took him on as an assistant.

By the mid-1990s, Mwamba Wanzala, Katumba's old classmate, had moved to the U.S. to study engineering and was starting out in a logistics role for the U.S. Army. Wanzala, who knew Katumba from the time he was fifteen or sixteen through their university studies together in Kinshasa, traces a change in Katumba's attitude to when he became involved in finance. He told me that Katumba used to call him when he came to the States on a training course with HSBC. During their long conversations, they shared their experiences in the U.S., and Katumba told Wanzala of his excitement over his new role and the financial possibilities it offered. "He was more interested in what he was doing. He was doing some training to become a banker," Wanzala said. "He liked it. He thought that was one way he was going to make money."

WHILE KATUMBA WAS LEARNING THE ROPES IN SOUTH AFRICA, MOBUTU SESE Seko's Zaire finally collapsed. The rebel insurgency led by Laurent-Désiré Kabila and sponsored by Rwanda and Uganda invaded the country from the east, easily routing Mobutu's troops and quickly capturing land. The Big Vegetables, people like Kufi Kilanga, faded into the woodwork or fled.

Kabila, badly in need of investment, decided to promise rights to Congo's mineral assets in return for cash to fund his insurgency. At first, the type of state-backed investment that China espoused did not look likely to win out in Congo. Congo's economy was in tatters after decades of Mobutist profligacy, and the country had embraced a kind of turbo-war-capitalism that was surely distasteful to Beijing's planning-obsessed economists. Inflation was hovering somewhere between 500 and 600 percent, and Gécamines—the mining giant that Katumba had so admired in his youth, and that had supported Mobutu's empire—was a laughingstock.

Kabila's decision to go to the market and sell off mining assets would shape Congo for years to come, and it would create a marketplace where only those with the fattest pocketbooks would survive. In retrospect, looking back on a world that was globalizing and consuming more and more products derived from mining, a world in which Beijing was flexing more muscle and the West was becoming more averse to risks, the path to China's dominance in energy storage takes on a certain sense of inevitability. But in the 1990s, China was still a haven for things being churned out on the cheap.

And where Congo was going was anyone's guess. Even as their war against Mobutu continued to rage, Kabila's rebels had begun to run fire sales of Congo's most prized copper-and-cobalt assets. Kabila, formerly a Maoist, was prepared to go to the market to fuel his political ambitions, and there were those on the market who were all too willing to comply. On April 9, 1997, just over a month before Mobutu fled, Kabila's forces seized Lubumbashi and made it the de facto rebel capital.

The next day, Bill Turner, CEO of Anvil Mining, flew in. Anvil was an Australian mining firm that had just submitted the year before an application to work in Katanga under the Mobutu government. "We want to know how the new authorities intend to honor the old contracts signed with the Mobutu regime," he told the French newspaper *Libération*. "We will not renegotiate our contracts—it took us years to snatch them from Mobutu." The vultures, he said, were circling, looking for deals with the rebels. He seemed to be counting himself as a vulture too. The visit paid off: Anvil was allowed to explore for minerals in Katanga. On February 27, 1998, Kabila personally ratified the Dikulushi Mining Convention with the firm. The convention gave Anvil a 90 percent stake in the Dikulushi copper-silver mine on the shores of Lake Mweru, which the firm said would produce 20,000 tons of copper and 1.8 million ounces of silver a year.

THE SALES OF MINES BY KABILA'S REBELS BEFORE THE FALL OF MOBUTU'S government formed the germ of the mining regime that now provides the world's battery and electric-vehicle industries with so much of their copper and most of their cobalt. The day after Turner arrived, more international businessmen flew into the capital of Congo's copper-rich South. When their planes landed, company officials were driven through Lubumbashi's wide streets to the twin concrete slabs of the Karavia Hotel, nestled next to a small lake and Lubumbashi's golf course. In faded rooms that had originally been designed by Gécamines for visiting dignitaries, they went over their pitches with Kabila's men. Kabila himself arrived on the fourteenth, by way of a corporate jet owned by the mining entrepreneurs Max and Jean-Raymond Boulle. The rebel leader set up in one of Mobutu's palaces.

But Mobutu was still at the head of government in Kinshasa, and the deals that were being struck at the Karavia were not valid under the Zairean

Constitution, which declared that the subsoil belonged to the state. In the blanched luxury of the hotel's halls, nobody seemed to care. The old order was dissolving, and a new one was being constituted before their eyes. "This is the first time that a developing country recovering from economic devastation and civil war has gone straight to the private sector for help," *Africa Confidential* noted. Bechtel, the U.S. engineering and construction behemoth, even began to provide the rebels with satellite imagery, and according to *The Wall Street Journal*, Robert Stewart, an executive at Bechtel, began to travel around Congo with Kabila, instructing him on how to put down "ethnic uprisings." The investors at the Karavia were told by a rebel who was advising Kabila on financial matters that they had better commit capital or risk losing out. "I think those who trust us today will have a jump-start on you," Mawampanga Mwana Nanga, the finance adviser, told the investors. The foreign businessmen understood that the money would go directly to Kabila's rebel army.

The air crackled with tension between sparring suits at the Karavia. The atmosphere was so thick that even the breakfast buffet felt like a battlefield. Boldface names, such as South Africa's Anglo American Corporation, were present, as well as smaller mining companies and "junior miners," firms that prospected for mines in the toughest-to-access parts of the world. Among the latter were representatives of Anvil Mining; Consolidated Eurocan Ventures Ltd., a firm founded by Adolf H. Lundin, a Swede (family motto "No guts, no glory"); and the Boulle brothers' firm, American Mineral Fields, Inc (AMF). Junior miners are often hard-bitten, larger-than-life characters who have grown accustomed to living under pressure. As Max Boulle told *The Philadelphia Inquirer*'s Andrew Maykuth, "You need strong nerves to work here."

Jewels, the HSBC Equator banker, also traveled to Lubumbashi that May. He invited Katumba, his mild-mannered apprentice, to join him; he thought the young man might be able to learn something. But Katumba was also a local, and local knowledge can sometimes be the difference between doing a deal and going home without a contract. Besides, Katumba was eager to learn. When they arrived at the airport, they noticed a group of men speaking among themselves. "One of them recognized Katumba and called out to him," Jewels recalled. He asked Katumba who the people were. "He said they were his old school friends, but they turned out to be Kabila's people."

Jewels had come to Lubumbashi to talk to Gécamines about a $30 million loan for a concession in Kolwezi called Tilwezembe. HSBC Equator was involved in arranging the funding for the project. Soon after arriving, Katumba advised Jewels that one of his friends wanted to meet them later that evening at the Karavia for a drink. Mwana, the rebel finance adviser, arrived in his T-shirt and running shorts. "He had been a waiter or something in the States," Jewels remembered. He made few promises. "It wasn't for me to tell whether they were going to win the coup or not. You know, I recollect being quite cautious in what I was saying because I didn't know them."

The HSBC bankers met with the rebels several times at the Karavia. At one point, according to Katumba's memoir, Mwana turned to Katumba and asked, "Why are you speaking to me about other people before your country, our country?"

Katumba replied that he was employed at the bank, and that his country had never offered him anything. "You are here," Katumba said. "Mobutu is over there."

Mwana retorted that there would be a new government in Kinshasa in seven days. "Think of your country. You ought to put your talents at the service of your country," the finance adviser said, snorting. "That's what I propose to you." With that, Mwana abruptly ended the meeting.

Mwana's words, however imperious, spoke to a secret longing that Katumba had long nurtured. The banker did feel called to public service, but he demurred that day in Lubumbashi. "That which is most important to me," he later wrote, "that which I truly like, that which lives in my most profound depths, I rarely exteriorize, and then only by barely noticeable signs."

At the Karavia, someone pointed out a young man to Katumba. To the banker's reckoning, the man, who was in his twenties, looked "calm and serene." It was Joseph Kabila Kabange, Laurent-Désiré's son, a man Katumba would soon come to know as "the Boss."

ONE NIGHT AT THE KARAVIA'S BAR, TWO OF THE ATTENDEES ALMOST GOT into a fight. On one side was an official from Anglo American; on the other, someone who worked for the Boulle brothers' AMF. In the year before, AMF had signed a deal with Mobutu's government for what the Belgian colonialists had christened the Prince Léopold Mine, a huge underground zinc-and-copper complex in Kipushi, on the Zambian border. But when Kabila's rebels took the river port of Kisangani in March 1997, AMF decided to go all in and back the rebels, letting Kabila use the company jet. "Do you wait until everybody gets here and be last or do you get in early?" Max Boulle said at the time. "We've made a conscious decision to get in early."

With Kabila's men looking set to overthrow Mobutu at any opportunity, it was looking like Anglo American, a behemoth that controlled more than half of private business in South Africa, might lose the rights to several diamond mines to the upstart Boulles. The company was left looking a little stupid. Tired and overwound at the Karavia's bar, Anglo's officials finally hit a boiling point when it was revealed that the Boulle brothers had scored a contract with the rebels to reprocess the tailings at Gécamines's sites in Kolwezi. (Tailings are rock scrap left over from already worked minerals.) In earlier generations, this material had been cast aside, but now it could be processed again with modern techniques to squeeze even more metals from the scrap. Some of the tailings in Kolwezi dated all the way back to the colonial era. And that was what so irked the South Africans at Anglo: Just like the beginning of Congolese independence marked the downfall of many a Belgian firm that had dominated in Congo, Kabila's arrival seemed to portend the end of the firms that had established themselves under Mobutu. They believed that at the Karavia, under their noses, the Boulles had just signed a billion-dollar deal that gave them access to more than a million tons of copper and 275,000 tons of cobalt.

But all was never as it seemed in Congo. The Belgian social scientist Erik Kennes has written that although dark money flowed freely at the negotiations, not a single contract was truly signed, sealed, and delivered. That December, around the time Kabila went to China, he reneged on the deal he signed with the Boulles, claiming it had been falsified. In Texas, the Boulles brought a $3 billion suit against Anglo American for wrecking their deal, but Anglo denied the charges. The case ended up being dismissed.

And so it would go for the next decade and a half. Caveat emptor. The history of Congo's mining after Kabilist deregulation is littered with such stories: Welcomed with open arms when the money was flowing, small and large companies were cast aside when a better offer was on the table. Hard-nosed businessmen like the Boulles and the Zimbabwean entrepreneurs Billy Rautenbach and John Bredenkamp—people who have all, at one time or another, been criticized by the international press for their lucrative exploits in some of Africa's poorest nations—got a once-over from Congolese leadership.

Even George Arthur Forrest, the Katanga-born businessman who was used to double-dealing under Mobutu's regime, got duped. Due to his deals with Gécamines, Forrest was in some ways known as the progenitor of Congo's mining privatization. "For international companies wishing to invest in the mining sector in Katanga, George Forrest has become a must-have operator," one European academic wrote in 2004.

When I visited Lubumbashi, from 2019 onward, one of the town's central squares was called Square George Forrest, and he had donated much of the town's art museum. Nevertheless, in a rare interview with a Belgian newspaper in 2023, Forrest would grumble that "Kabila took over our mines." Forrest's family had been in Congo since 1921. "He didn't compensate us, and the Belgian government did nothing to help us. We lost these mines and a lot of money. This is an area to avoid."

But few people seemed to mind that Kabila was, for the moment, reneging on his promises. It just wasn't a priority to many international observers. The war in the East was developing into a veritable bloodbath, and Kabila was becoming ever more erratic and authoritarian, locking up dissidents and silencing freedom of expression. The huge profits that could be imagined in Congo, however, meant there were always people who would jump at the chance to fill the investment gap if they thought the price was right.

Stewart, the Bechtel official who had traveled with Kabila, reportedly supplying him with satellite data and helping him put down uprisings, was furious when he learned that the mine concessions promised to him had been signed over to a Rautenbach company, Wheels of Africa. The journalist Howard W. French later wrote that while working in Congo during the period, he saw "an impressive, bound briefing book" that purportedly outlined Bechtel's plan. "Bechtel would shepherd in the construction of roads, dams, airports, and other projects on a massive scale, all of which would be collateralized or paid for by Congo's mineral wealth." The plan sounded a lot like a resources-for-infrastructure deal that China would negotiate with Congo—and that U.S. commentators would criticize—only a few years later. "For reasons that have to do with the erratic policies of the Kabila government but also with the United States' chronic inability to invest much focus on Africa, they all came to naught," French concluded, "and China, with a greater appetite for risk and a longer-term vision about the continent, ran with the ball."

CHAPTER 18

PLANS ON THE BACK OF A COMET

Dan Gertler claims to have fallen in love with Congo while smoke was still rising over Kinshasa. It was May 1997, and Mobutu Sese Seko had just fallen. Gertler was twenty-four. He had just finished two years of military service in Israel, and he was out to prove himself in Africa.

At the beginning of the twenty-first century, no international businessman in Congo's copper-and-cobalt mining industry would receive more scrutiny than Gertler. At first, however, the young man had no interest in critical metals. He was focused on diamonds. The precious stones were part of his inheritance, after all: His grandfather Moshe Schnitzer was known as "Mr. Diamond," and the square in front of the Israel Diamond Exchange was called Moshe Schnitzer Square. Moshe had created the exchange, and Shmuel Schnitzer, Moshe's son (and Gertler's uncle), was one of its fastest-rising stars.

The name Schnitzer came from the Yiddish word for "to cut"—and cutting diamonds is how Moshe rose to prominence after he arrived in Mandatory Palestine from Romania in 1934. During the 1940s, working as a diamond cutter, he joined the Irgun, a Zionist terrorist group agitating for the creation of an Israeli state. In 1967, Moshe became the president of the Israel Diamond Exchange. He held the post until 1993. Under his tenure, exports of polished diamonds from the country increased from $200 million to $3.4 billion.

Diamonds—and Israel—were everything to Moshe. "There was no real separation between diamonds and family life," Shmuel told a journalist in 2002. Gertler remembered getting up at five every morning as a child to learn how to polish stones. Moshe's Irgun comrades had made their way to the top of Israel's defense and political apparatus—powerful people like the right-wing prime ministers Menachem Begin and Yitzhak Shamir. These deep connections between Moshe and the Israeli military and state apparatuses would occasionally shine through as Gertler built his network in Congo.

Gertler was the son of Moshe's daughter Hanna and Asher Gertler, a onetime professional soccer goalie turned diamond trader. At first, the son wanted to follow his father onto the playing field, but he was asthmatic, so he focused on his other talents, excelling in mathematics and science. He was also personable, which helped—boisterous, endlessly talkative, and, to some people, very likable, with a broad smile that flashed a set of gapped teeth. (The charm wasn't universal, however: As one mining executive who later worked with him told me, "Dan was annoying as hell.")

Under the charming facade, Gertler quickly grew to nurture an ambition passed down to him by his grandfather: He wanted to control the supply chain for diamonds, thereby cutting out middlemen and miners—people like the Boulles and companies like De Beers and Anglo American—and keeping more of the profits for himself. While speaking to two journalists from the magazine *Jeune Afrique*, a childhood friend of Gertler's recalled that the young man had a "particular" understanding of the world: "When he looks at a wooden chair, he ponders not only how it was made, but from which forest it came, and what infrastructure made it possible to exploit the forest."

Gertler decided to strike out on his own to try and see whether he could make some diamond deals away from the family business. He soon began to focus on Africa, the continent where the overwhelming majority of the world's diamonds came from. Vast sums could be made with the right connections and know-how.

By the time Gertler reached Congo, he had already negotiated deals in Angola and Liberia, but he hadn't yet made the big score that would cement his reputation.

Watching the news coming out of Kinshasa as Laurent-Désiré Kabila's troops took the city, Gertler saw an opportunity, one that he, like the miners at the Karavia, understood would accrue to those who got in early. He took a flight from Tel Aviv to Germany and then to South Africa: After twenty hours, his plane finally touched down in Kinshasa.

KABILA'S REBEL ARMY HAD SEIZED THE CAPITAL FROM MOBUTU ON MAY 17, 1997, only a few days before Gertler's arrival. The streets were patrolled by child soldiers. In the aftermath of its capture, the city was anything but safe. The Big Vegetables had vamoosed with whatever they could take—to Belgium, to the countryside, or to nearby African countries—and the capital felt ransacked and raw. Kabila, the new president, spent his first nights in the city on the sofas of the VIP lounge at the Chinese-built Kamanyola Stadium, accompanied by the Rwandan commanders who had managed his rebellion for him.

Outside, the city, a symbol of everything Kabila had ever fought against, was a skeleton of its former self. The remnants of Mobutu's troops and loyalists were rounded up and incarcerated. The nights spluttered with the noise of bullets. Kabila realized that Congo faced the same problem it had encountered after decolonization—a lack of skills—so he began to make job offers to the former dictator's bureaucrats. Kufi Kilanga, who was thought of as a professional, was offered amnesty and a job with the new administration, but as his son told me, he just couldn't bring himself to work with the people who had unseated Mobutu.

Kabila, in an echo of the dictator he had just unseated, began to rename things. The country became the Democratic Republic of the Congo. Kamanyola Stadium became the Stade des Martyrs, rebaptized in memory of a group of politicians killed by Mobutu. And the sprawling mining province in the South became Katanga once again.

Very soon, Augustin Katumba Mwanke, the HSBC banker who had attended the Karavia fire sales, would find himself a beneficiary of Kabila's "erratic" decision-making. As Kabila seized Kinshasa, Katumba received a call from one of his bosses at HSBC. The boss's tone was "excited and gluttonous," Katumba recalled. "Words came out of his mouth, telegraphically, as if he was firing joyous shots or hammering out truths," Katumba would later write, "not only to convince me, but also to exult, to enjoy and to project himself into the vast future that he was already dreaming of: he was dreaming up plans on the back of a comet." The vast future that his boss was dreaming of was one in which well-connected insiders could make huge profits from Congo's gigantic stockpile of natural resources.

Before long, Katumba was seconded to the new Congolese government by HSBC. He wrote that he wanted to "keep his bank salary and dedicate himself to his country," so he hashed out a deal with the new Finance Ministry in which he worked for both the bank and the government. "Magic, no? It's what you could call a win-win deal," Katumba later wrote.

He quickly became the face of Congolese finance to the outside world. "Despite the fact that Augustin hadn't had much banking experience, he was very bright and in fact much brighter than the guy who was the minister of finance," Jewels recalled. "Within three or four months, he was effectively running the ministry."

Katumba started work in July 1997. In September, four months after Kabila seized power, Mobutu died of prostate cancer in a military hospital in Morocco. The marshal's era was definitively over, and Kabila was building a new Congo, one in which there were vast opportunities for international finance to prosper from the country's stores of mineral wealth—or so it seemed.

———

GERTLER WAS ALSO ENJOYING THE FRUITS OF HIS EARLY-IN GAMBLE IN Kinshasa. Upon arriving, he enlisted the help of Shlomo Bentolila, the chief rabbi of Kinshasa's Chabad-Lubavitch center, to make introductions. In short order, the rabbi arranged a meeting between Gertler and the president's twenty-five-year-old son, Joseph, at the InterContinental.

The two hit it off almost immediately. Gertler and Joseph were separated by only two years, and both men carried the burden of their family legacies. Joseph, taciturn, with an angular face and the shadow of a mustache, was a commander in Kabila's military. He had presided over the fall of the jungle city of Kisangani just two months earlier. "We have a reason to fight," Kabila had told a *New York Times* journalist as the city fell. "That's what kept us going all this time." His father would soon send him to the People's Republic of China for military training.

In the meantime, Katumba had so impressed the president that Kabila wanted to bring him into government definitively. On a Saturday morning in mid-March 1998, just under a year after the Karavia meetings, the president called Katumba out of the blue. At first, Katumba thought it was a friend of his on the line, and the president had to make himself clear, summoning Katumba to the presidential palace. "You ought to work for your country," Kabila told him, echoing Mawampanga Mwana Nanga's order the year before. "I want to name you governor of Katanga."

Katumba was thirty-three and had never imagined himself as the governor of the province. He froze for a few seconds, then mumbled an excuse about his lack of experience.

But Kabila had taken his decision. The president would soon call Jewels to tell him that Katumba would be working for him as governor. "If that's what you want, Mr. President," Katumba said, "I shall do my best."

CHAPTER 19

THE NIGHTMARE

A bust of Joseph Kabila Kabange in Malemba-Nkulu, Congo, 2022

Gertler and Kabila Jr.'s friendship developed quickly. They would chat until the early hours at military bases. They became, in the businessman's later telling, "like brothers: We talked about life, religion, the development of the country." They did not know it at the time, but the conversations contained within them the framework for a critical metals dilemma that, twenty years later, would preoccupy a world that was increasingly reliant on Congo's resources, a world that was increasingly rent between the twin poles of Washington and Beijing.

The Israeli entrepreneur was also flying around Congo to diamond mines, which were owned by a plethora of small and large traders whose numbers had only multiplied in the post-Mobutu chaos. By 1998, he was meeting with ministers and showing them letters indicating that his firm had "the full backing of M. Schnitzer & Co"; that he was doing business in Russia, Guinea, and Sierra Leone; and that he wanted to do $2 billion in diamond deals over the next two years. There is no hint that he had any interest, at this point, in the country's copper-and-cobalt mines.

In any case, those mines had been promised elsewhere. In mid-1998, the Rwandans and Ugandans, frustrated that the older Kabila was not going to rule as their puppet, began a second war, this time to replace the president whom they had put on the throne.

The ensuing conflict would kill more people than any other since World War II. The new president turned to Robert Mugabe's Zimbabwe for help; as collateral, he offered Mugabe and his associates the mines that he had so quickly signed away earlier in the war. "It was a way to thank them for their logistical support during his taking of power," George Arthur Forrest, the Katanga-born Belgian businessman who had contracted for Mobutu, would later remember. In 2002, when they were no longer useful to the regime, the Zimbabweans, too, would be kicked off the mines.

Wars are expensive, and Kabila Sr. needed money up front. He began to tap miners using novel taxes and renegotiations of contracts. To lure investors in, he dangled deals and monopolies that would be swiftly overturned. The Congolese diamond industry was a shambles. De Beers had left Congo after the government ended its contract to buy Congolese diamonds, and the country's diamonds were being smuggled out, then sold on international markets to the highest bidders. The diamond trade was pushed onto the black market.

Kabila Sr. decided to enact even more radical plans. In a move that might have seemed, at first blush, rather like giving a lion the key to the zoo, Forrest was made head of Gécamines in 1999. The previous head of Gécamines had been Billy Rautenbach, the Zimbabwean businessman who had close ties to his country's dictator, Mugabe. Kabila Sr. now wanted the Zimbabweans out of the Katangese mines, and he charged Forrest with making Gécamines profitable again. "He didn't have much knowledge about administrative matters, let alone experience in public governance," Forrest wrote years later. "When he arrived in power, he still thought Gécamines was the big Gécamines that he had known back in the 1960s. . . . He had a vision far removed from reality."

Forrest insisted that although he did not want the job, Kabila Sr. had given him no choice. To avoid conflicts of interest, he claimed, he left the room whenever his companies were being discussed. Nevertheless, Forrest was successful. He spoke Kiswahili and was able to converse directly with the workers when they went on strike, promising to pay them regularly and restart industrial production. By the end of Forrest's term as head of Gécamines, the firm was making a modest profit for the first time in years.

KABILA JR. WAS EAGER TO INTRODUCE HIS FATHER TO HIS NEW FRIEND, AND not just because the two had a good relationship: Gertler could provide much needed liquidity, and Kabila Sr. had talked about how he wanted to establish a diamond monopoly in order to ensure that tax money was paid directly to the government, and that they wouldn't have to deal with traders and miners who were ripping the country off. Gertler was now spending 90 percent of his time in Congo, and he might be just the man to manage such an enterprise.

By the turn of the millennium, the Israeli had managed to make some deals in the country, but he was still looking for the centerpiece, the venture that would justify his big African gamble. Initially, Kabila Sr. snubbed Gertler. At their first rendezvous, meant to take place inside the presidential palace, the guards turned him away.

Kabila Jr., however, was keen to strike a deal with Gertler. By the year 2000, his father had promoted him to deputy chief of staff of Congo's armed forces, and he understood the country's need for cash better than almost anyone in the country. He arranged another meeting, at which his father asked for $20 million up front in exchange for a diamond monopoly. A UN report would later peg the deal as part of "a pattern of miscalculated decisions taken by the cash-strapped Laurent-Désiré Kabila, whose main interest was the immediate cash flow."

Gertler returned to Israel to celebrate. Twenty million dollars was a minute sum compared with the profit he could expect to make from the diamond monopoly. Two days after his meeting with the president, however, he received a call from one of Kabila Sr.'s associates: Perhaps he hadn't understood. The president needed the money immediately. The young Gertler liquidated stock and turned to banks and his family for the cash. His grandfather personally guaranteed a loan from the Bank of Israel. The money went to the Swiss account of Congo's central bank and, presumably, toward fighting the war.

At the end of July 2000, Gertler's company, International Diamond Industries, signed a deal worth $700 million that gave it control over the entirety of Congo's diamond output for eighteen months. It promised to provide certificates that showed where the diamonds had been mined. Gertler's gamble had paid off in ways that most people could only dream of. He was twenty-six years old. "This is the optimum way for the Congo diamond production to be marketed in a transparent manner that will inspire trust and confidence in the country's certificate of origin, which will accompany each and every parcel to be exported by IDI," Anatole Bishikwabo Chubaka, Congo's acting mining minister, said when the deal was signed.

But despite the minister's assurances, much in the deal was not transparent. The certificates he described, for example, were never issued. Later, Kabila Jr. would also claim that only $3 million was paid from the $20 million that was demanded.

Gertler maintained that he paid the full price, and Congolese government reports do indicate that the country gained $6.6 million in revenue from the monopoly, more than under the previous system of independent buyers. An April 2001 UN report, however, said that the "deal turned out to be a nightmare for the Government of the Democratic Republic of the Congo and a disaster for the local diamond trade." In the end, the contract would be canceled after just eight months.

Kabila Sr. wasn't just looking for money; he was also casting about for military support wherever he could find it. In 1999, it was reported that North Korean military trainers had surfaced in Congo, close to the Shinkolobwe uranium mine in Katanga, the same mine that was the source for the fissile material used in the atomic bombs dropped on Hiroshima and Nagasaki. (It was never clear if the North Koreans received uranium for their troubles, but two years later, on the day he was assassinated, Kabila Sr. had also arranged to meet with Iranian officials to negotiate a uranium deal.) Gertler had also apparently promised Israeli military training and support to Kabila Sr., leaning on the close contacts between his family and his nation's defense-and-intelligence apparatus.

According to the UN report, the deal contained "unpublished clauses, in which IDI agreed to arrange, through its connections with high-ranking Israeli military officers, the delivery of undisclosed quantities of arms as well as training for the Congolese armed forces." The report alleged that diamonds were then flown to Tel Aviv by "former Israeli Air Force pilots." In Israel, according to the report, stones would be cut and sold at the Ramat Gan Diamond Center. Gertler's network was apparently far-reaching: A group of retired Russian generals called "the Russian Military Brotherhood," people he referred to in a 2002 interview as "good friends," reportedly wanted to do a deal in Congo with him.

It is unclear whether the trainers ever came to Congo—Gertler's company has always denied any military dealings, and he would not speak to me for this book. Melissa Sanderson, a chargé d'affaires at the U.S. embassy in Kinshasa during the 2000s, told me that Gertler worked with Israeli forces, including after the war in Congo.

"We knew for a fact that there were times that the Israeli military was trolling around down in Mbuji-Mayi [Congo's diamond capital], and we knew he got them there, and why they were there," Sanderson said. "But, of course, it was the U.S.-Israel relationship—we're not going to get upset about that."

Sanderson came to know the country and its players in granular detail. She made a point of traveling the length and breadth of the country during her time at the State Department. "I was dragged kicking and screaming to the Congo for the first twenty minutes," Sanderson told me. "I hated it. Then I fell in love with it." She came to know the country inside out, traveling as much as she could, investigating the stories she was told, and, crucially, meeting all the major players.

In a 2002 interview with *The Christian Science Monitor*'s Nicole Gaouette, Gertler made light of the allegations that he helped Kabila out militarily, sarcastically retorting that "he put together a consortium involving the FBI, the CIA, the Russian secret service, and the Israeli police, army, and foreign ministry, adding that he controls the last three." When Gaouette pressed him on the rumors, he exploded, asserting that his "philosophy" was never to hide anything. Sanderson told me she knew for sure that such training had been provided. "What was discouraging," she said, "was that they didn't make any headway in trying to professionalize the Congolese military."

And Gertler would certainly continue to have close contacts with the Mossad, Israel's secret service, for some time to come. In 2019, it was reported by *Haaretz* and *Bloomberg* that Yossi Cohen, then head of the Mossad, had traveled to Congo with Gertler at least three times. In Congo, Cohen and Gertler had met with the president, to whom Cohen, in an echo of the late 1990s, offered defense technology.

ON JANUARY 16, 2001, LAURENT-DÉSIRÉ KABILA WAS KILLED BY ONE OF THE child soldiers who had been assigned to guard him. His son was in Lubumbashi with Augustin Katumba Mwanke as news about the assassination began to filter out of Kinshasa the next day. Rumors abounded: People were evacuating Kinshasa; the president's body was being transferred to Zimbabwe; the borders were sealed. Before Kabila Jr. left on the presidential jet, he chose his words carefully, telling Katumba, "Hold on, Augustin, watch your province, stay ready for everything. I will keep you informed."

By January 26, Kabila Jr. had ascended to become the fourth president of the Democratic Republic of the Congo. He was just twenty-nine. As he took power, the country descended further into war and ruin. In the East, the country was checkerboarded with warlords and roadblocks: Kabila Jr.'s father had used armed militias known as Mai-Mai to fight Rwandan and Ugandan invaders. They were now fighting among themselves.

Kabila Jr. concerned himself with how to bring some kind of stability back to the fractured country. Katumba stepped into the breach. Sanderson would meet him when she was working at the U.S. embassy. He was, she recalled, "very bright, very funny, always tired, very frustrated." The first time they met, Katumba told her, "You know, I hate Americans." But she managed to win him over, and they became close.

By that point, Katumba was being called the King of the Congo on the Kinshasa circuit. People said he was the brains behind Kabila Jr.'s administration. "He would say, 'No, I'm the cockroach of the Congo. All I do is sit outside the president's office and wait for him to yell,'" Sanderson told me. Katumba was "a pseudo-father" to the young president, she explained. "M'zee was never father to Joseph in any positive way." It wasn't that Katumba was so much brighter than Kabila Jr. ("Joe's nobody's idiot"); it was just that he knew how to explain things to the president.

Once again, the government began reviewing the contracts it had executed under the older Kabila's rule. Katumba was made a minister of state, and it fell to him to examine Gertler's monopoly. The Belgian government had been pressuring the Congolese government to cancel the deal, and Katumba was also of the opinion that it should be nixed. In April, Kabila Jr. withdrew from the contract his father had agreed to.

At first, Gertler was livid: How could his friend have double-crossed him? He called Kabila Jr. over and over but got no response. Then he began to visit Katumba's office, demanding answers. He threatened to leave Congo.

But the new leadership "saw Gertler as an important asset to use," Sanderson told me. Kabila Jr. had a problem with Lebanese diamond smugglers who were making a killing in the province of Kasai: They would pay artisanal miners with used blue jeans in return for the valuable stones. Even worse, the diamond smugglers had started to throw their weight around in Kinshasa, and Kabila Jr. felt threatened. "The best possible asset for countering a Lebanese problem is an Israeli," Sanderson said, "particularly one with connections to the Israeli government and, most especially, to the Israeli military." That message, Sanderson noted, was clear: "Remember that you are guests in this country: This is not your country." Kabila Jr. understood that he needed Gertler.

In the end, Kabila Jr. agreed to a compensation deal. "Our termination of his contract was, in effect, abusive: we did not respect the procedures in this area, in particular regulatory deadlines," Katumba would later write. "We rightly decided to compensate him."

Compensation or not, Gertler was still left hanging. He lobbied the government. "Dan would come and he would leave without getting anything," Katumba wrote. "He was neither reimbursed for his expenses, nor did he obtain new contracts. During all this period of bitterness, he hated me, blaming me for all the blockages he suffered."

After a couple of months, Kabila Jr. found another use for Gertler: He asked the businessman to connect him with powerful people. Gertler, by all accounts, threw himself into the task. He set up meetings with the U.S. government, and he also arranged for Katumba to meet officials from the China Development Bank, a state-owned bank that had begun to give overseas loans to developing nations. It wasn't always easy for Katumba to manage his president. When the chairman of the bank visited Kinshasa a few years later, Kabila Jr. was too busy to meet him. "The Chinese official had a bitter taste in his mouth after his trip," Katumba remembered. "I had to apply myself to dissipate any ill will."

Gertler, who had become more outwardly religious, also pushed Kabila Jr. to negotiate the end of the war. He became a representative of the Congolese president, traveling abroad to carry out diplomacy. Foreign officials would get a shock when meeting with Congolese delegations. They would expect a group of African diplomats and instead Gertler would appear, in the traditional garb of a Hasidic Jew, sometimes with a rabbi in tow.

In April 2002, Gertler arrived in Washington, D.C., bearing a letter signed by the Congolese president and addressed to President George W. Bush. It described Gertler as "a respected and well known international businessman" and an "old and trusted friend." Kabila Jr. explained that although "the Democratic Republic of Congo is rich in many natural ressources [sic]," it now faced "occupying nations that are exploiting these ressources." He was pleading with the U.S. administration to rid Congo of its occupiers—Rwanda and Uganda. "Mr. Gertler has conviced [sic] me to put trust in you and join the forces of good, rather than succumb to help offered by other nations, whom we doubt are on the good side," he wrote.

The letter was the prelude to a series of meetings where Gertler told U.S. officials that "President Kabila believes that the rebel movements and foreign armies are using the cloak of war to disguise what has become a blatant exercise in self-enrichment through the illegal plunder of scarce resources." He warned that Kabila Jr. could "go to war" with Rwanda, which he said was backing the rebel groups and which he called "a disruptive force in the Democratic Republic of Congo." This scenario—and the threats that accompanied it—was almost exactly the same as the one that threatened the battery supply chain in the 2020s, when Rwanda supported a rebel group that took swaths of eastern Congo.

One trip led to another, and Katumba was sent to Washington with Gertler. The young minister met with Condoleezza Rice and explained to her the situation in Congo.

At one point, Gertler turned to Katumba and said, "Why are you blocking me?"

Katumba stayed silent. He understood that Gertler wanted him to promise something. But it was not in his gift to make any assurances; Kabila Jr. was biding his time before promising Gertler any more of Congo's rich resources.

On July 21, 2002, Katumba wrote Gertler a note: "I got to know you better, thank you for what you did behind the scenes." From that point onward, Katumba would later write, Gertler and the minister forged closer and closer relations. "This friendship developed, strong and beautiful, over the months and years," Katumba remarked, "like a meeting between two men with profoundly similar spiritual convergences, despite their dissimilarities."

IN WASHINGTON, PRESIDENT BUSH WAS KEEN TO SCORE A FOREIGN-POLICY "win" as the first year of the war on terror drew to a close. He brought together Kabila Jr. and Rwanda's Paul Kagame at New York's Waldorf-Astoria, on the sidelines of the UN General Assembly, in September 2002.

"Paul, when are you leaving Congo?" Bush asked Kagame.

The Rwandan president assured him that he would be gone in three weeks.

Almost seven months later, in April 2003, the "final act" of the peace accord would be signed at Sun City, a South African resort and casino. But Gertler once again horned in on Congolese diplomacy: According to him, he put together a team that came up with some of the provisions, including a stipulation that Kabila Jr. would rule with four vice presidents from the different rebel factions. "Dan Gertler, through his actions, saved the lives of Congolese," the Congolese politician Barnabé Kikaya Bin Karubi professed in a slick advertorial on Gertler's website. "He helped stop a war in which DRC lost 6 million people. Through his actions, the war ended." Gertler was clearly of the opinion that he had been in large part responsible for the peace agreement.

In reality, although the Sun City peace pact might have slowed the worst atrocities of the Second Congo War, it did not end the violence in the country. Rwanda, in particular, would continue to meddle in its neighbor's affairs, sponsoring militias and exporting purloined minerals for decades.

Kabila Jr. had blocked Gertler from the precious-stones market, so the businessman started to look to the South, where fortunes were being made mining copper and a metal that Congo had in overwhelming abundance, a metal that more and more companies demanded for their lithium-ion batteries: cobalt.

———

IT WAS ONE OF THOSE COINCIDENCES OF HISTORY: JUST AS MORE AND MORE lithium-ion batteries were being produced in Japan, the mines had come roaring back to life. Unlike in the 1970s, when the two Shaba wars spiked the price of cobalt so high that it made commercial sense to ship it out via airplane, Congo's wars didn't seem to affect the southern mines. In 1999, even as the older Kabila began to arm the Mai-Mai and war raged in Congo's East, the copper-and-cobalt hubs of Lubumbashi and Kolwezi, as well as the road that connected them, seemed to be immune from the violence. Cobalt production, the U.S. Geological Survey's Andrew L. Gulley found, "increased seven-fold from 1996 to 2003 despite two African wars over the DRC and its resources." Thanks to this development, and the sale of the U.S. stockpiles, the global price of cobalt began falling at the end of the Mobutu regime, and prices crashed.

In 2002, with the help of Katumba, Congo instituted its first mining code since Mobutu's fall. This was also the year that Gertler was shuttling back and forth between Congo and Washington on Kabila Jr.'s behalf. Gécamines, the mining company, had been destroyed in Mobutu's downfall—the government was looking to round up international financing, and, of course, ensuring that money continued to flow to Kinshasa. Copper was booming on the back of a massive amount of demand from

China, and cobalt prices had also begun to creep up. "Dan then moved away from diamonds and began to be interested in copper and cobalt in Katanga," Katumba later remembered.

The jewel in the Katangese copper-and-cobalt crown was the Kamoto mine in Kolwezi, but it had already been acquired by Forrest, who stepped down as head of Gécamines after the assassination of Kabila Sr. Forrest had spent the past half decade rehabilitating the mine with money borrowed from banks. He thought it could be a landmark project in a country torn apart by war.

Gertler set his sights on Forrest's mine, but he would have to cool his heels for the time being. In the meantime, he took a large stake in a neighboring project in Kolwezi, a giant agglomeration of some of the country's highest-grade copper deposits, known as KOV (the initials standing for the ore bodies Kamoto, Oliveira, and Virgule), as well as the copper-cobalt mine of Kansuki, near the entrance to the city of Kolwezi. By the end of 2012, Gertler would be invested in companies that controlled 9.6 percent of global cobalt production. "This was approved by the transition" government after the peace accord had been signed, Katumba wrote. "No one found fault with it."

As he developed his mines in Congo, Gertler discovered a talent for luring in investors, even as he and his companies took on loan after loan, becoming, in one businessman's words, a "debt monster." To convince people to place money with him, he deployed a network that included Beny Steinmetz, a businessman who was later convicted in Swiss court of bribing Guinea's former first lady in order to acquire mining rights; Yaakov Weinroth, a high-powered lawyer who had represented Israel's right-wing prime minister, Benjamin Netanyahu; and a series of rabbis who explained to potential investors that Gertler had an almost mystical power to yield profits "beyond the natural order."

But the reason for Gertler's success was hardly supernatural: In fact, it was mainly thanks to Katumba, whom he paid handsomely for his services. For Gertler, Katumba was more than just a fixer; he was an essential part of the Israeli businessman's operation. "If there is no Katumba, they also take away the [mining] license," Gertler would later say. The Congolese politician came to Gertler's house and behaved like he was an owner of the mines, and Gertler explained during the arbitration proceedings how he split his companies with him. "Katumba was the one who dealt with government officials, company CEOs, all the tax authorities who come and give tax bills all the time to the local community," Gertler said. The Congolese politician held incredible power over the president and his ministers. As Gertler put it, "I think everyone who was in Congo understood that Katumba's word was law."

Forrest, who would be maneuvered out of his ownership of the Kamoto mine by Gertler, saw things differently from Katumba. "Despite the massive sacrifices that everyone made, and which we made to relaunch this flooded and therefore mothballed mine, they took the Kamoto mine away from us to give to Dan Gertler," Forrest would later fume. He mused that Gertler's strings were being pulled by powerful foreign multinationals. "It goes without saying that we lost lots of money, but that didn't stop us from continuing, from believing in our country, and from pursuing our work there."

Part 3

BATTERY BOOM

"Globalization takes place only in capital and data.
Everything else is damage control."

—GAYATRI CHAKRAVORTY SPIVAK,
AN AESTHETIC EDUCATION IN THE ERA OF GLOBALIZATION

CHAPTER 20

ACCELERATING
THE TRANSITION

The Tesla Gigafactory Berlin-Brandenburg, May 2024

The group of investors had formed fast friendships over the past couple of days, gathering into knots. A New York bank had arranged for them to visit automakers in Sweden and in Germany and France. It was late March 2024, their final day in Germany before heading to London. Europe was trying to show its best face: Among the sparring Chinese and U.S. elephants, European leaders were trying to position their countries to best take advantage of the green transition. A light drizzle washed over the fields and parking lots outside the Berlin airport hotel.

Two of the investors were sitting down for breakfast, having bowls of oatmeal before their final site visit. They sprinkled sugar into their coffees and jargon into their sentences—*BEV* for "battery electric vehicle," *ICE* for "internal combustion engine." They were people who wanted to make it very clear that they *knew* what they were talking about. "Two takeaways: We know electricity is the way. We know electricity is the best type of power, but the best power infrastructure is not there," a Belgian fund manager said. "Second takeaway: The Chinese are also coming after the market."

"They aren't so vertically integrated," his colleague, a New York–based investor, replied.

"One question: Where is the Tesla factory?" the Belgian wondered. "Can I get easily from downtown Berlin to there?"

"It's rural enough that some people in the countryside attacked it," the American said. "What would you call it? Environmental protests? Environmental terrorism?"

It turned out that the Tesla factory—the Gigafactory, to use the moniker that the company's CEO, Elon Musk, had given all his plants after he built the first one in Nevada in 2013—was not far. It was just under half an hour away.

Part of Tesla's raison d'être, Musk would tell his growing legion of online fans through the 2000s and 2010s, was altruistic. He wanted the large auto companies to adopt lithium-ion and thus reduce carbon emissions. Plus, Musk found it cool.

At a meeting for Tesla employees in early 2009, Musk read from a slide: "What's cooler than making a hot product, a sexy product, that also saves the world!"

Now Tesla was being critiqued for its own deleterious impact on the environment. Since February, a group of protesters had been camping out near the factory in the damp forest around the community of Grünheide. They were protesting the more than three hundred thousand cars a year that Tesla pumped onto the streets of Germany and Europe. "An electric car creates a huge ecological footprint through the consumption of resources and thus drives the global climate catastrophe even further," one of the group's manifestos read. The "neocolonial supply chains" that fed the Tesla machine were, they argued, contributing to polluting the world more than they were saving it from global warming. "Whole swathes of land and ecosystems are devastated as water is either used for mining or returned to contaminated groundwater. And the people who live in these areas, often in the global South, are being robbed of their livelihoods by mining," they went on. "Instead of perpetuating car capitalism in a green guise, we are fighting to end it!"

Just over a month after the bankers visited, eight hundred protesters would attack the Gigafactory, clashing with police who prevented them from crossing the boundary into Musk territory. "Why do the police let the left-wing protestors off so easily?" Musk griped on the social-media app Twitter (which he had renamed X) after the release of video footage showing masked disruptors, clad in black, charging toward the factory.

AT THE CHECKPOINT OUTSIDE THE FACTORY, THE BUS CARRYING THE IN-vestors was turned around. The driver, whose phone ringtone was set to the 1990 hit "Wind of Change," a song by the German rock band Scorpions that presaged the fall of the Soviet Union, grumbled a bit and then pulled into a parking lot a little way off from the main build-ing. Outside, some of the investors dashed through the fizzling rain; oth-ers turned their collars up and flipped the cameras on their phones to snap selfies. The Gigafactory would probably be the most iconic stop on their trip.

Tesla's facility had been built at an astonishing speed in a country with a reputation for red tape. After Musk's November 2019 announcement of the factory's construction, Tesla had started to clear a swath of pine plan-tation so that it could begin building. The company had constructed the factory using temporary permits up until March 2022, when the first full permit was issued. Musk personally delivered the first thirty cars to their new owners later that month.

These were facts that a rep sporting a leather bomber jacket à la Musk did not gloss over as he welcomed the investors into Tesla's steel-and-glass building, then led them up to a conference room with Tesla car doors hung on the walls like abstract sculptures. On a wall at the back of an office space was a sign whose last line had been partially obscured by a television monitor. It read oddly, like a clipped haiku:

Our mission:

To Accelerate the world's transition

[To Sustainable Energy]

The razing of the pines had angered local environmentalists even before Disrupt had begun its campout. A proposed battery factory next door, which Tesla promised would create ten thousand new jobs, stoked their rage even further. Environmentalists filed a lawsuit against Tesla in early 2020 to try and halt construction. Musk's lawyers pointed out that the felled pines had been planted there to make cardboard, and a German court allowed Tesla to proceed. Ramona Pop, a senior member of the Green Party, publicly backed the project. "We need to keep some perspective," Pop said. "Tesla's future investment should be allowed quickly for clean mobility and climate protection."

The investors donned hard hats and protective goggles ("Sometimes sparks fly up") and moved through a netherworld of loading bays. Men with tattooed arms took long hits of nicotine-laced vapor and stared blankly into space. They were, in Teslaspeak, "production associates." Twelve thousand people worked at the factory. Some were Ukrainian refugees. Others chatted away in Turkish.

Inside was a nightclub for roughshod demons. Techno beats competed with the clanging of machinery. Gargantuan robots named King Kong and Godzilla lifted vehicles onto conveyors. The factory is meant to produce the body of a Tesla Model Y, the company's compact crossover SUV, every forty-five seconds. The production line moved like a river along the factory floor, with workers rushing up to cars and going to town on them, installing lights, computers, upholstery. The pace was frenetic. Electric trucks whirred about and missed whacking into the investors by mere inches.

BE NEAT, read signs on the wall. BE ON TIME. And also: LEAVE BEHIND YOUR COMFORT ZONE.

The velocity of Tesla's production comes at a cost. In 2023, reporters from the German magazine *Stern* investigated a trove of safety documents from the Gigafactory in Brandenburg and found that "serious accidents happen almost daily." At the factory, Tesla had also installed an illegal diesel station, hidden under a party tent, and leaked feces into the water supply. The magazine accused the local government of selling out to Tesla. In response, local politicians and Tesla denied the claims and said there had not been a significant increase in workplace accidents. But Tesla has also been accused of fostering a "toxic" culture in its U.S. factories; in *The Nation*, Bryce Covert reported that at one Gigafactory, workers said they were "putting ourselves at even more danger" to meet quotas.

AS THEY AMBLED ACROSS THE FACTORY FLOOR, THE INVESTORS PEPPERED the rep with questions: *How many, how much?* The rep smiled and fended off the queries—especially the ones about Tesla's lithium-ion batteries. Articles proliferating online claimed that the cars made at the factory used batteries from BYD, the Chinese battery company, and Musk had publicly mused about producing a compact car there that included batteries from CATL.

The group passed into the casting section, where molten aluminum was pushed into molds by machines capable of injecting the metal at a pressure of six thousand pounds per square inch. Using the machines, Tesla can trim seventy production steps down to just three.

"We haven't done a tour like this in a long time—we need you," a Tesla rep told one of the investors. The investors nodded. In fact, they needed one another. The investors had cash to spend, and the company had positioned itself, and the lithium-ion-powered cars it had pioneered, as the solution to clean transport. Everyone—from the Green politician Ramona Pop, to the European Parliament, to the U.S. Congress, to the titans of finance, to the legions of fans who bought Model Ys and rhapsodized about Musk online—had accepted that lithium-ion would be *the* solution for mobility, for buses, for the vaguely annoying scooters that were zipping noiselessly down the streets of every major city by the early 2020s. Such people had not factored in the pollution of production. In fact, when all emissions were taken together, Tesla cars were more polluting than hybrid vehicles, but everyone seemed to have bought into the company's promise of greener transport.

Still, Tesla wasn't moving as fast as its competitors in China. There, companies like BYD had been steaming ahead with government support long before Tesla even built its first Gigafactory. Beijing was cheering on its domestic companies, but in the U.S. and Europe, businesses such as Musk's were still regarded with suspicion. In Brandenburg, thanks to environmental regulations, recent plans for an expansion of the Gigafactory and of the battery-production facility had been stymied.

On the investor tour, as molds were pressed and machines clanked, a tattooed Tesla worker connected his phone to the sound system. The music began with a howl. The rep's explanation of the die-casting process was drowned out by the sound of Fat Joe rapping:

> *I did it all, I put the pieces to the puzzle*
> *Just as long, I knew me and my peoples was gon' bubble*

One of the investors—a slim British-public-school type with crew-cut red hair burst into a smile. He was sold. "I mean, Fat Joe and Terror Squad—*legends*," he said, laughing. "I mean, you just have to admire the *hustle*."

CHAPTER 21

BREAKING THE ICE

Electric vehicles experienced a turbulent liftoff during the first two decades of the twenty-first century. Even in the mid-2000s, with consciousness around emissions surging, battery-driven cars were still something of a cautionary tale, and almost all the major auto manufacturers were betting on internal-combustion-engine, or ICE, vehicles.

General Motors, the U.S. auto giant, was still smarting after the negative fallout from the EV1, an electric car it had debuted in the decade before. EV1 had been launched by GM in 1996 to much acclaim, but due to lack of demand, it was recalled in 2003, when the company crushed the vehicles for scrap. Environmental activists blamed oil companies, but GM said the cars had lost it large amounts of money.

As late as the second half of the 2000s, hybrid vehicles, especially the flagship Prius, made Toyota into the world's biggest automaker, a position GM had occupied for seventy-seven years. "Toyota was the darling of the media, the American public," Bob Lutz, a legendary GM executive who assumed responsibility for global product development in 2005, would later say. One of the driving forces behind the company's renewed interest in electric vehicles, he remembered the sentiment at the time: "Toyota is light-years ahead. General Motors people are all brain-dead; all they keep doing is internal combustion engines. If only we had the smart people in Detroit [the headquarters of the U.S. auto industry] that Toyota has, the noble Japanese who are not motivated by profit. They're motivated only for the good of society," he said. "It was enough to make you sick."

———

IN THE EARLY AUGHTS, A LOS ANGELES ENGINEER NAMED ALAN COCCONI was telling anyone who would listen that he could make a car that would travel three hundred miles on lithium-ion batteries. "This could be really a very viable solution and could meet most people's needs for transportation," Cocconi said. In 2001, Martin Eberhard, a newly minted e-reader millionaire out of Silicon Valley, invested some money in Cocconi's concept. The tZero, an electric vehicle priced at $200,000, was born.

To build the tZero battery, Cocconi used one hundred blocks of sixty-eight cells, each as big as a standard car battery. The cells were called 18650s because they were eighteen millimeters wide and sixty-five millimeters tall. They contained something that was essentially the original LCO formulation that John B. Goodenough had discovered, as did most of the EVs that followed in those early years.

After the money ran out, Cocconi's funder, Eberhard, teamed up with a friend named Marc Tarpenning, and they founded Tesla in 2003. Musk invested $6.35 million a year later and quickly took control, ousting the original founders. It would turn out to be unquestionably one of the best investments ever made. By 2014, it was estimated that Musk had made around $2.79 billion from Tesla. In August 2024, the company's shareholders approved a pay package of $46 billion for him.

In the early days, Musk made sure the Roadster, the car that had grabbed Lutz's attention, was sleek and sporty and *sexy*. Moreover, he made sure that it was driven by famous Hollywood actors. By 2009, celebrities like George Clooney, Matt Damon, and Leonardo DiCaprio would all own one. These cars could go from zero to sixty in less than four seconds flat, and the battery was good for 244 miles before it needed recharging.

Innovation tends to happen in two stages: the scientific eureka moment followed by the scaling of that triumphal discovery industrially. Tesla, and Musk, had changed the narrative around electric cars. Now they needed to master the production of those cars. Musk began to plan the Gigafactories that would produce Teslas quickly and cheaply, in a style he would dub "Ultra Hardcore" in a 2012 email to staff. "People put their heads down, worked through the problems, and ramped up that factory," said Ryan Melsert, a founding member of Tesla's Gigafactory design team who has gone on to build a battery-recycling business. "I think it's hard to picture that time, because it wasn't financially successful; now it is one of the biggest corporations in the world."

The Gigafactory in Brandenburg was the end result of all that grafting. "Industrial scale is an innovation ladder all unto itself," Mujeeb Ijaz, founder of Our Next Energy, a billion-dollar battery company in Michigan, told me. "Way beyond technology, industrial scale creates the motivation to do things differently, to do it better, do it faster, to make it cheaper, to localize the supply chain and create the opportunity to drive cost down."

By the late 2000s, Detroit was starting to wake up. Not only was Tesla innovative, but it was also mastering the art of car production. "I mean, this is outrageous, here is a small start-up car company on the West Coast, obviously, very confident about lithium-ion batteries, is going to go into production with this car," Lutz said on *Charlie Rose* in 2011. "Whether Tesla is ever hugely successful or not, I will always owe them a debt of gratitude for having, kind of, broken the ice."

———

LUTZ, A NATTY DRESSER FAMOUS FOR HIS CRIMPED-COLLAR SHIRTS AND EL-egant ties, had noticed Musk's moves at Tesla early on. He understood the need for the U.S. auto industry to innovate. He was no environmental bleeding heart; when he had headed Chrysler in the late 1980s, he had pioneered high-octane vehicles like the Dodge Viper. He had also served as CEO of Exide Technologies, the battery company, so he knew quite a bit about energy storage. He realized that if GM was going to win against Japanese automakers, as he put it, the "only way to stop them was to leapfrog them and do something beyond what Toyota had done." Lutz was constantly being told that General Motors needed a game changer "to give people a new perception of the company." The firm turned to its batteries.

Instead of coming up with a hybrid or purely electric car, GM designed something between the two: the Chevy Volt, a "plug-in hybrid" that utilized an "E-Flex" system. Cost was not a strong consideration, Lutz told me. "I thought, *Well, there's no point of just doing another hybrid*," he said when we spoke in 2022. "If we are going to do a green car, I thought we should be doing something that is manifestly and demonstrably better than what Toyota is doing with the Prius."

The car had a forty-mile range, but to cure the "range anxiety" that had supposedly scuppered the EV1, it had a 1.4-liter engine that would kick in when the car's battery had drained. When the Volt debuted at the 2007 North American International Auto Show on a stage whose backdrop fizzed and crackled with blue lightning bolts, Lutz stepped out of the vehicle beaming. "The GM electric vehicle is an inconvenient truth," Lutz quipped, referring to former Vice President Al Gore's film warning of the dangers of global warming. (Lutz himself was a climate-change skeptic, calling global warming a "total crock of shit" in 2008.) "This is not a PR exercise or a pure show car," he continued. "This is a real program with real money behind it that is heading for production."

CHAPTER 22

LE PETIT

By the mid-2000s, people may have been turning to lithium-ion-based batteries to power their vehicles, but the major demand for lithium and cobalt came from handheld devices. Mobile phones had become ubiquitous; by 2002, most of these devices used lithium-ion batteries. "Lithium-ion batteries are the foot-soldiers of the digital revolution," *The Economist* contended.

In Congo, more and more people were heading to the southern mines to find the riches by which these devices were powered, especially copper and cobalt. They were banding into loose groups, or cooperatives, in order to sell ore to traders. Copper was used in all kinds of electrical devices, and cobalt was increasingly being used in batteries. On the outskirts of the Kakama mine, business had picked up at Joseph Kufi Kilanga's tire shop. (As noted, he was Odilon Kajumba Kilanga's father and the minister Kufi Kilanga's cousin.) "My father wasn't rich, but he still had means," Kajumba told me. "When we wanted something, my father always did his best to get it for us. He was obliged to always satisfy us. Even if he didn't have money, he would try to satisfy you. In short, he spoiled us."

Even though production levels for copper and cobalt remained fairly high, the South's once proud mining industry had become a shell of its former self. Murray Hitzman, the geologist who worked in Congo in those days, saw how Gécamines had been devastated firsthand. "There were some of the best geologists I've ever met in my life still working for Gécamines, and they hadn't been paid for three years," he said. "Really good people, really good workers, too, at the mines. I couldn't believe that they just—they were working for nothing. Literally for nothing." Hitzman continued, "It was just sad, to be honest. It was just sad as hell."

The scramble for wealth was leading people into ever deeper pits, which became unstable during the rainy season and were wont to collapse, burying miners alive. Artisanal *creuseurs* even went into the uranium mine at Shinkolobwe, which had provided the material for the first atomic bombs, and carried out radioactive material. "They're digging as fast as they can dig, and everyone is buying it," John Skinner, a mining engineer, told the Associated Press at the time. "The problem is that nobody knows where it's all going. There is no control." Moïse Katumbi Chapwe, Katanga's governor after 2007, began efforts to prohibit anyone from entering the Shinkolobwe mine (it had officially been closed since 1960). "I spoke to a lot of foreign ambassadors. I said, 'This mine remaining like this is a danger for all the world,'" Katumbi told me when we spoke in 2019. "A lot of people are going there illegally because it has a high cobalt content and a high uranium content."

IN AN ATTEMPT TO HARNESS MORE PROFIT FROM THE MINES, KABILA JR., along with his main adviser, Katumba, and economists from the World Bank, crafted a mining code that was enacted into law in 2002. The new code liberalized the mining sector, provided for Gécamines to sell off its assets, and enshrined artisanal mining into law.

Artisanal miners had to belong to cooperatives and seek permits from a body that was charged with educating them in the ways of mining and organizing them into effective teams. It also established "artisanal exploitation zones" where permitted artisanal miners could work. In theory, there was a slate of rules that provided for best practices, including around safety and a minimum age for miners.

In reality, though, there was little about the artisanal mining sector that suggested the rules were being followed. Children worked in mines, and pregnant women washed minerals. After years of colonialism and dictatorship, there was a desperation to the way that people worked on sites like those around Katanga. The artisanal miners, the *creuseurs*, were mining to make ends meet. They were often from the very poorest stratum of society in a country that consistently ranked as one of the poorest in the world.

Amid the poverty, a few people smelled opportunity. Atop the heap, reaping gains from mining, were politicians in Kinshasa, most prominently Katumba and Kabila, who were busy creating a system by which the revenues from mining went into the coffers of the elite. Then there were the enablers, the "fixers." They came from every corner of the globe. One of them was Dan Gertler, who was beginning to buy more mining concessions in Katanga through a company called Fleurette at bargain prices. There was Shiraz Virji, an entrepreneur of Indian extraction who, having started life as a wood and spice trader on the Kenyan coast, founded a business in 2001 called Chemaf, or Chemicals of Africa, which not only bought mines but also built a cobalt-processing plant. There was Billy Rautenbach, the Zimbabwean businessman who had served as head of Gécamines and now ran a company called Boss Mining. And, as always, there was George Arthur Forrest.

Congo's fixers understood they needed fresh money for their projects, so they looked abroad. Moises and Mendi Gertner, London-based investors, were two people who decided to allocate money to the country's burgeoning mining scene. They had been told at a wedding sometime in early 2006 by Yaakov Weinroth, the attorney, that Gertler was a "charismatic child prodigy with extensive and unusual connections, who is capable of making a fortune," so they decided to invest large sums of money with the young Israeli even after he did not repay a loan they issued to him. The connections Weinroth was talking about included Kabila's sister, Jaynet, and Katumba. "Dan and Katumba have a tremendous relationship," Mendi Gertner later said during an arbitration resulting from a later disagreement over money. (The 1,200 pages of arbitration documents were provided to me by the Platform to Protect Whistleblowers in Africa.) In early 2006, the Gertners met Katumba in Congo and when he traveled to Israel. "We knew from the outset that Mr. Katumba was the one helping to make things happen."

Few people understood the power of the artisanal mine, however, like Alex Hayssam Hamze. Often described as a "logistics genius" by the people who knew him, Hamze was the son of a martial arts instructor from the South of Lebanon. With his homeland embroiled in civil war, Hamze's father went to teach martial arts in Congo. "They left Lebanon when he was two," an associate of Hamze's told me. His father soon lost his job and set himself up as a merchant in a small town near the outskirts of Kolwezi.

By the mid-1990s, Mobutu Sese Seko's wane had ushered in the era of artisanal mining: The Lebanese community of traders in Kolwezi, who had already gained some experience as "cobaltists" during the 1980s, began to buy hand-mined cobalt from Gécamines sites. "It was the Lebanese community who started the market, who started buying and selling, and in the mines that were abandoned by Gécamines," Norbert Nawiji, a longtime copper and cobalt trader in Kolwezi, told me. "People were pouring in and doing *handpicking*—collecting of stones from the surface—and they sold to the Lebanese."

HAMZE AND HIS BROTHER BEGAN BUYING AND SELLING MINERALS MINED by artisanal *creuseurs* from a mine called Mutanda ya Mukonkota, near the eastern entrance to Kolwezi. He was still in his teens, but he was determined to make money trading like his father.

The Mutanda mine lay on a nature reserve called Basse Kando, established by the Belgians in 1957. During Mobutu's rule, black antelope and elephant lived in its forestland, and hippopotami bathed in the Kando River. But then gold was discovered at a site called Kawama, or Shabara, and artisanal miners rushed to begin digging there: Rich seams of copper and cobalt were unearthed, the forests were cut down and made into charcoal, and the earth was ripped open. By 2024, according to Jonas Kiriko, a Congolese investigative reporter, around 77 percent of the Basse Kando Reserve would be allocated as mining concessions.

When Hamze began working at Mutanda, it was becoming a desolate strip of wasteland. The gold miners had flushed mercury into the rivers, killing off wildlife. People were scraping together a few dollars a day from the ocher-tinted soil. But amid the devastation, Hamze saw opportunity. "He's *very* focused. He understands logistics and numbers innately," his associate told me. "The guy didn't even graduate from the Lycée, but he's a fucking *genius*, a computer genius, a logistics genius." In 2002, Hamze set up a trading company called Bazano. He was small in stature, and the Congolese miners began using a sobriquet for him that stuck—Le Petit.

At one point, the miners at Mutanda revolted—against the mining conditions and their earnings, as traders were paying them a pittance for their minerals. Le Petit stepped into the breach, hashing out terms with the miners in their native Kiswahili.

In May 2001, Gécamines finalized an accord with a firm called Southern African Metal Refiners Congo, or Samref, to exploit Mutanda's minerals. Gécamines technically possessed the mineral rights to Mutanda but was not mining there. At the beginning, the state company retained 40 percent of the new company's shares, and Samref owned the rest. The new venture was christened Mutanda Mining, or MUMI.

Little was public about Samref and the investors behind it. One of them was Charles Brown, an American adventurer who had seen an opportunity to profit in Congo as the Kabila regime took the reins from Mobutu. Samref was a junior mining firm. Success for such junior companies is based on not only the richness of the reserves they have developed but also how much money they can eventually convince large operators to pay them.

After a few years, Brown was worried that his bet on MUMI had not paid off. When Hamze offered to buy some of Brown's shares in the company for $300,000 in 2006, Brown agreed to sell.

As the mine became profitable, Brown began legal action, alleging that Hamze and his future partner, Glencore, a commodities-trading firm based out of Switzerland, used coercion to wrest away his rights to the mine. (Hamze has denied any coercion; Glencore maintains that it never met with Brown, who died shortly after filing suit.)

By 2005, Hamze was operating MUMI using a hybrid between artisanal and industrial methods. Bazano was able to exploit the rich seams of cobalt and copper at Mutanda. At the time, it was difficult to refine the rock into sheets of copper cathode and cobalt hydroxide, a blue powder that is further refined into battery precursor material. Since the Benguela Railway through Angola was closed, Bazano began using trucks to export large quantities of copper and cobalt across the Zambian border. Much of the material from Mutanda began arriving for smelter refinement at a mine in Zambia called Mopani.

The rock beneath the russet hills of northern Zambia contains ores that are less concentrated than those on the Congolese side of the border. The Mopani mine, whose giant Nkana concentrator had been operational since the 1930s, during the days of British colonialism, had been acquired in 1998 by a consortium majority-owned by Glencore. By 2003, people at the company had noted "the potential sources of 'imported' concentrates and copper bearing material produced from new mining developments in northern Zambia and the Democratic Republic of Congo," as one report put it. Glencore committed itself to building a new smelter and sulfuric acid plant (acid is used to leach cobalt out of the ore), which went operational in 2006, but it still wanted more.

Glencore began to expand into Katanga, and as its officials crossed the Congolese frontier, they heard tell of a young Lebanese trader whose business was booming. "Alex is a very smart guy," a former Glencore cobalt trader told me. "He probably knew the market for cobalt and copper in the DRC better than anyone at the time."

Exporting most unprocessed cobalt and copper ore from Congo had technically been illegal since 2003. Like the cobaltists, the cobalt thieves during Mobutu's era, traders like Hamze were now known for heterogenite, the name of the copper- and cobalt-rich ore they smuggled. They were dubbed "heterogenists."

For the Congolese state, this posed a problem: Millions of dollars a month in revenue were being lost as minerals were smuggled across the Zambia-Congo border at Kasumbalesa. "A comparison of official Congolese export and Chinese import statistics for cobalt and copper trade in April and May 2004, for example, showed a greater than ten-fold difference," a 2006 report from the International Crisis Group explained.

As the 2000s wore on, some politicians, such as Katumbi, vowed to stop the minerals flowing over the border. But the 2002 Mining Code was increasingly showing its deficiencies. It gave leeway to traders. It allowed for minerals to be exported with special permits from the minister of mines (and, according to some interpretations, by companies that could prove they were building processing plants in Congo). The discretion of the Ministry of Mining was an easy target for under-the-table dealings, but Hamze was also constructing refineries in Likasi and Kolwezi, so he could legitimately say that he was building processing plants.

The Katangese mineral-smuggling situation may have been problematic for Congo, but for Congolese politicians, who ensured loyalty through payments to their supporters, it was a gift. Like Mobutu's "big vegetables," many of them had developed a taste for luxury goods and expensive homes abroad. Accusations were rife that the powerful, right up to the president's office, were profiting from the trafficking of minerals in Katanga.

BY 2007, BAZANO BEGAN TO SIGNAL THAT IT DID NOT WANT TO BUY FROM artisanal miners. It was clear that the artisanal trade was for small-timers, and Hamze was moving up the ladder. Hamze invested in Mutanda, carried out a feasibility study, and reassessed the size of the mine's deposits. When the contract had been signed between Gécamines and Samref in 2001, there was no solid estimation of the size of the deposits. After Hamze's companies finally did a size assessment, they found that the mine held around 958,000 tons of copper and 85,375 tons of cobalt, numbers that continued to grow with time. He had hit the mother lode. Soon, Hamze brought his operational genius to bear on the site: Mutanda basically became an industrial mine. "It was run like a Swiss operation," one of Hamze's associates recalled.

But Bazano also continued to buy artisanally mined cobalt from other mines, despite its protestations to the contrary. In 2012, researchers found that Bazano was buying hand-mined cobalt from Tilwezembe, a site that was technically owned by Glencore but had been abandoned to a massive artisanal operation. Tilwezembe was known for its rampant use of child labor. It was one of the sites that had been bought by Gertler and flipped to Glencore, but the company had decided it wanted nothing to do with the site after its employees were threatened by people with artisanal mining interests there. Artisanal mining at Tilwezembe, it argued, was "to our disbenefit. However of even more concern is that their operations may well be unsafe and creating environmental problems."

Approximately sixteen hundred artisanal miners, including Kajumba, were working on the site by 2012. They were surveilled by at least fifty-five armed guards. A Lebanese man named Ismaël, a former associate of Hamze's, bought the site's cobalt. Ismaël was described by one NGO as "an extremely tough man in business, who has no respect for the miners." Ismaël's firm processed the ore at his smelter and at the Mopani concentrator in Zambia. Bazano insisted that it wasn't involved in mining at Tilwezembe, but the firm had signed a contract with a firm named Misa Mining to purchase ore from the site.

On a visit to Kolwezi ten years later, in 2022, I visited makeshift shafts dug into the red earth of people's back gardens outside Tilwezembe. Bazano was long gone, but the *creuseurs* and child laborers were still there.

CHAPTER 23

INTO THE PITS

S ometime in the early 2000s, Odilon Kajumba Kilanga's father, Joseph, the tire salesman, was bitten by a malaria-carrying mosquito. The parasite began to wear him down. "He was suffering," Kajumba said. "His health began to get worse." Malaria, it can be said, is the most lethal side effect of the country's crippling poverty. In Congo, the disease is the principal cause of early death; every time I visited the country, people whom I'd come to know would tell me how they or someone they loved had just recovered from a nasty bout of the disease.

Joseph couldn't work as hard as he used to, and he didn't have money to send his children to school. Education was supposed to be free, but in a country where the state had all but crumbled, it was up to parents to pay teachers' salaries. Kajumba had to drop out. His father "couldn't support his family any longer," Kajumba told me.

Joseph became weaker still when he contracted typhoid fever. He tried to continue working to support his family. "When he started to get treatment, it was too late," Kajumba said. "Typhoid had already started to eat away his insides. He didn't realize how far it had gotten." As Kajumba remembered his father's last days, he shook his head. "He was a good guy," he said. "The best father in the world."

Kajumba was now alone with his mother and four siblings, unable to pay for school and watching everything his father had built fall apart. His dream was to open a little restaurant, a bar or a nightclub maybe, but he didn't know where to start, and he didn't have the funds. His oldest brother went to Kinshasa to see if he could find a job. As his father had become sicker and sicker, Kajumba had begun to venture to the mine site to moonlight as a *creuseur.* "Even if it was against my father's will, I said to myself, 'You're a man—you have to work,'" he told me.

Kajumba soon began to get sucked in by the work, lugging rocks across the burned and pitted craterscape of the Kakama mine. They were sold to intermediaries. He didn't think about where they went. He learned how to lower himself into narrow, deep, hot holes in the earth; how to chip away at the rock face with the crudest of tools, usually an iron bar whose ends had been ground to points; and how to bring scraps of precious ore back to the surface before he fainted from lack of oxygen. "I had to work out how to take care of myself and my family," he said. He was sixteen, maybe seventeen.

One day in 2006, when Kajumba was eighteen, he got a call from a friend who had moved to Kolwezi. This friend urged him to join a *creuseur* cooperative that was roaming from mine to mine, sharing profits. He told Kajumba that Kolwezi's mineral seams were rich in not just copper but cobalt too, and that Kajumba would be able to turn his family's fortunes around. "There were good sites that you could just turn up to and work," Kajumba remembered.

In those days, it took eight hours to get from Lubumbashi to Kolwezi by bus. The country was still settling down after years of civil war, and the thickets on either side of the rutted two-lane road crawled with outlaws who would occasionally hijack vehicles using weapons they'd leased from impoverished soldiers. Once, bandits stopped a bus and ordered its passengers to strip; the hijackers took everything, even people's underwear. Kajumba knew that the journey to Kolwezi had its risks, but of the *creuseurs*, he said: "If they tell you to come, you come."

———

GROWING UP IN KATANGA, KAJUMBA HAD ENJOYED WATCHING SOCCER WITH his father. He grew to love Joseph's favorite team, Tout Puissant Mazembe (All-Powerful Mazembe), Congo's most successful club. TP Mazembe, which was founded by Benedictine monks during the colonial era, had as its logo a crocodile clenching a ball in its long jaws. The team's owner, Moïse Katumbi Chapwe, the suave Congolese businessman who would soon become governor of Katanga, was a protégé of Augustin Katumba Mwanke. "This man was a brother to me," Katumbi said of Katumba when we first met in 2019, at a lodge in Zambia. "I can't deny it."

Katumbi's father was a prominent Greek trader who had fled anti-Jewish purges during the Axis occupation of the island of Rhodes during World War II, and his mother was a princess who traced her heritage to the eastern Lunda kingdom of Kazembe. Katumbi's father had made a fortune from a set of family businesses, including the importation of fish from Lake Mweru. His brother Raphaël Katebe Katoto had political ambitions: He reportedly supplied UNITA rebels in Angola, planned a coup against Kabila, and even backed anti-Kabila rebels in the East of Congo.

Katumbi, by contrast, jumped on the Kabila bandwagon and supported the M'zee. He also profited from the mining industry; a business run by his family was the monopolistic food supplier to Gécamines workers, and he owned mining and trucking companies. One of them, Mining Company Katanga, or MCK (the initialism recalled Katumbi's own initials), would reportedly make him tens of millions of dollars, much of which he spent on his political career. In 2006, Katumbi became a parliamentary deputy.

In 2007, soon after Katumbi's foray into politics, he was elected governor of Katanga. He ran the province like a business and became popular because he handed out large sums of cash in poor areas, built roads, and outfitted schools. Katumbi supported education for girls and tried to stop children from working in dangerous mines. The province was soon faring well. "Fighting corruption took us from third place in contributions to the national budget to number one after six months," Katumbi told me proudly. When he became governor, 3 percent of Katangese people had access to running water. By 2013, that number had climbed to 67 percent.

Katumbi was concerned about the effect of rapid industrialization on his province, even as he profited from it (MCK was signed over to his wife). He saw how the number of artisanal miners had swelled to hundreds of thousands; how, every day, tons of ore was crossing the border into Zambia; and how the people of Congo were receiving little, if any, benefit.

Shortly after taking office, Katumbi, who said that mining companies were "keeping the money and sending it overseas," banned the export of raw minerals across the frontier to Zambia. "Show me even one toilet that has been built with that money," he thundered, insisting that an export ban would begin to bring benefits back to Congo. All of a sudden, the traders who had profited from the minerals couldn't sell their wares.

KATUMBI WANTED BIG WESTERN COMPANIES TO INVEST IN KATANGA. GLENcore, the Swiss trading firm, had been on an asset-buying spree since around 2003. It was headed by Ivan Glasenberg, a former champion racewalker for South Africa and Israel. Glasenberg's intensity and intellect made him unusually suited to the world of commodities trading, which required a combination of hardheadedness and an ability to sink oneself into the minutiae of deals that spanned continents and were often renegotiated at the last moment. Glencore employees of the Glasenberg vintage often tried to replicate IG, as he was known among his workforce, getting up before sunrise to indulge in physical activity, especially cycling through the rolling hills around the company's headquarters in the Swiss town of Baar, one of their boss's favorite pastimes.

But Glencore was not all fresh air and freewheeling. It had been born out of the ashes of Marc Rich and Co., a commodities firm whose eponymous founder had been served "the biggest tax-fraud indictment in history" in 1983. (Rich was later controversially pardoned by President Bill Clinton.) In late 1994, Rich's traders booted him out and formed two rival firms: Glencore (for "global energy commodities and resources") and Trafigura, which was based in Geneva. Both firms would soon find themselves on the front lines of the global rush for battery minerals, in Congo and across the world.

Glasenberg had struck it large by betting on coal and other commodities, and the firm was looking for investment prospects. Copper, which was booming in China, seemed like a good bet in a rapidly electrifying world. Cobalt was an afterthought, a last wring of the profit rag; cobalt is a *byproduct*, several current and former Glencore traders were at pains to explain to me while I researched this book. They meant that the market for cobalt was a volatile one, dissimilar from the big-time futures markets for metals like copper or even nickel, and that cobalt was stockpiled, especially because trading volumes for the metal were comparatively low.

In the mid-2000s, Glasenberg was looking for a source of copper, and he was willing to take on risk. More risk, in fact, than large established firms like BHP Billiton, which had examined several sites in Katanga, including those owned by Dan Gertler, but then withdrawn because it feared running afoul of financial authorities. Glencore officials began scouring the world for deals. In Congo, Glasenberg did not want to purchase shares from Gécamines because Glencore wasn't seeking a relationship with the Congolese government. In July 2006, Nikanor, a mining company founded by Gertler, was floated on the London Stock Exchange, and the firm soon came to Glasenberg's attention.

There was substantial back-and-forth as Glencore tried to acquire Nikanor. An early offer of some $1.6 billion was not accepted in May 2007. It was at this point that Gertler and other investors in Nikanor appeared and began meeting with Glencore officials: Samie Monderer, then a young Glencore employee, helped introduce the company to Mendi and Moises Gertner, who by that time held about 22 percent of Nikanor. When one of the businessmen negotiating the sale came to the office, he "would come, spend half the day there—it was important," a Glencore employee recalled.

In June 2007, Glencore obtained its first shares through a share placing, whereby Nikanor offered new shares directly to the company, and the company ended up taking a 12.5 percent stake in Nikanor. According to several Glencore employees, Gertler wasn't a major figure in the early stages, but he began requesting meetings more and more persistently. In the next half decade, the Gertners' relationship with Gertler would go south and their arguments developed into the arbitration proceedings. In documents submitted at the arbitration, Gertler revealed that he had paid some $360 million in bribes to Katumba "to ensure 'goodwill and influence' on his part concerning continued activity in Congo." The Gertners' lawyers would later insist to *Bloomberg* that they had no role in running their mine investments in Congo and that they were "run exclusively by Gertler, causing the Gertner brothers significant financial damage."

In his quest for copper, Glasenberg wanted a stake in Katanga's other big mine, the giant Kamoto Copper Company (KCC) mine outside Kolwezi, which was owned by Katanga Mining Ltd., a Toronto-listed company in which George Forrest held a stake. The world was in financial crisis, and everyone was looking for financing, which also meant it was a good time to make strategic acquisitions. In 2008, Glencore obtained its first shares of Katanga through a merger with Nikanor; then, in June 2009, it obtained more shares in the mine when it executed on a convertible loan it had given to Katanga; and the next month it bought even more shares in Katanga. Through his involvement in Nikanor, Gertler had also become involved in the mine. In a separate transaction, in 2013, he arranged to purchase the royalties from the mine's production from Gécamines through a firm named African Horizons, meaning he would have a steady stream of income from the mine's production of copper and cobalt. Glencore would buy out more than $37 million of a Gertler company's shares in Katanga in 2017, but he continued to receive royalties from the mine well into the next decade.

The culture at Glencore was intense, and the company focused single-mindedly on making money—sometimes, it seemed, at any cost. In 2023, a New York federal judge would order the firm to pay $700 million after it pleaded guilty, in 2022, to a long-running bribery scheme involving foreign officials in eight countries, including Congo. "In the DRC," the judge said, "Glencore admitted that it conspired to and did corruptly offer and pay approximately $27.5 million to third parties, while intending for a portion of the payments to be used as bribes to DRC officials, in order to secure improper business advantages." The "third parties" were known as agents. "Almost everybody used agents," a former Glencore trader told me.

BY THE MID-2000S, GERTLER HAD, FOR A FOREIGNER, AMASSED ENORMOUS power and assets in Katanga. George Arthur Forrest still controlled what Katumba once described as the "flower" of Katangese mining—the Kamoto deposit—but Gertler was quietly buying up more and more shares of Kamoto, and he owned two giant concessions: KOV and Kansuki, which abutted Mutanda. Money was essential to keeping the deals flowing. The record of the arbitration proceedings between Dan Gertler and the Gertner brothers does not show explicit bribery, but it does illustrate how Gertler cycled huge amounts of cash and diamonds to Katumba and others at the apex of Congolese politics in an extensive program of penumbral transactions. "What does he [Katumba] get in return for all this?" a lawyer asked Mendi Gertner during the proceedings. "Mr. Gertler's friendship?" Gertner said that he and his business partners had agreed for 15 percent of the shares of the business to go to "local people." Shell companies and code words were used, and payments were obscured. During the arbitration, Gertler was not forthcoming on this matter, but his answer to one of the questions hints at the web of transnational companies being used to extract wealth from Congo and aggregate it into a few pockets. "It's none of your business whether it was in my account in Panama, an account in Cyprus, or an account in Gibraltar," the Israeli businessman said when questioned about payments to Katumba.

Glencore officials saw working with Gertler as a matter of expediency, and they wondered if the Congolese had been a little taken in by the foreigners operating in the country, people like Forrest and Gertler whom they considered big talkers with little to back themselves up. Frankly, they seemed a little amateur to people with knowledge of international finance. "Suddenly Kabila has these mines—what are you going to do with them?" a former Glencore official later said. "And you have these two monkeys [Gertler and Forrest] who say they can find investors who want to put money into them." Gertler kept asking for money and was extremely litigious. Forrest was better liked by Glencore officials, but he was also looked at with some skepticism: He would make a habit of insisting that investors from far-flung places were on the cusp of putting money into his projects, and then the deals never materialized.

There was no denying, however, that Gertler had something unique: both the president's ear and the ear of Kabila's éminence grise, Katumba.

In his autobiography, Katumba reveals how Gertler once invited him and his wife, Zozo, on a yachting holiday around the Israeli port city of Eilat: "Me, the little Congolese from Pweto, I looked at this world with wide, marvelling eyes." Early on a Friday evening during the trip, the celebrity magician Uri Geller, of spoon-bending fame, visited the boat. Katumba was impressed, but Zozo got the heebie-jeebies. Gertler advised Katumba to use his time in Israel to get medical treatment for the aftereffects of an old car accident. As he recovered from an operation at a medical center in Tel Aviv, Katumba fell into a coma, then awoke to see his friend Gertler along with a rabbi, whom the businessman had brought in to pray for him.

Gertler, who'd also gotten doctors to come in from London, had "saved my life," Katumba writes. When he woke, he was paralyzed and could not talk or eat. The rabbi spoke to Katumba for a long time about good and evil, urging him to give freely. As he recovered, he writes, the message resonated. "Thanks to Dan and his Rabbi, I found a new religion, a new family," Katumba effuses. "Dan, my friend, my 'twin brother,' despite everything that appears to make us different, I am proud to be the brother that you didn't have. Let us be at one with one another, for ever."

When Katumba returned from Israel, people in Kinshasa noticed that something had changed. "After that, Dan became Augustin's number-one guy," Melissa Sanderson, the former U.S. chargé d'affaires, told me. "Because at that point, it wasn't just about money anymore."

Alongside Gertler and figures like him, Katumba and a coterie of politicians in Kabila's government helped construct a sort of parallel payment system that aggregated money not into the state's coffers but into private pockets. As the Carter Center, a human-rights and transparency NGO, later documented in a 2017 investigative report, Gécamines was again being used as a *vache laitière*, echoing the use to which Mobutu Sese Seko and his elites had put the parastatal. The authors of the Carter Center report argued that corruption through Gécamines, which usually retained 20 or so percent of mining projects, was hard-baked into the Kabila system of government, in which backdoor deals proliferated. "Wolves don't eat among themselves," a popular Congolese saying went, meaning that the corrupt had to cooperate in order to steal. It was, the Carter Center report opined, "a far cry from the competitive, well-organized licensing system that the World Bank had envisaged at the adoption of the 2002 Mining Code."

Katumba, in particular, seemed to be surrounded by scandal, but nothing stuck. "He became the shadow of the chief," a Congolese journalist once told me. When a UN report accused Katumba of corruption and suggested that he be banned from travel because of his involvement in corrupt deals, Kabila relieved him of his position. But Katumba wouldn't be out in the cold for long.

CHAPTER 24

OOZING EVIL

By the 2010s, Katanga had developed a fully-fledged parallel economy that revolved around minerals, deals, and occasionally violence. Powerful figures profited from instability. State offices were split along ethnic fault lines that recalled those that divided the Congolese people during the days of the Katangese secession. The Lunda formed the backbone of the provincial mining office in Kolwezi, the Luba in Lubumbashi; people from the Chokwe ethnic group ran the mine regulator; and the Hemba—Odilon Kajumba Kilanga and Kufi Kilanga's people—predominated at the atomic-energy regulator, which had to sign off on potentially radioactive ores.

In Congo, as the online newspaper *Africa Intelligence* pointed out at the time, "ethnic allegiances take precedence over everything, especially in natural resources management." The country's mines would provide "the manna that will finance the country's presidential and provincial elections and military operations to repel foreign forces." And the political and economic elites who controlled them would control Congo.

Augustin Katumba Mwanke sat at the apex of the shadow hierarchy in Katanga. Katumba also had a particular interest in Bill Turner's Anvil Mining, which operated the Dikulushi copper-and-silver mine. The Australian company was one of the few groups that had managed to operate relatively unscathed since the days of the Karavia fire sale. From 2001 to 2004, Katumba even sat on Anvil's board; the company paid him for attending meetings and leased its office in Lubumbashi from him. The group was well connected to other powerful figures too—Raphaël Katebe Katoto, Moïse Katumbi's brother, leased machinery to the company.

But then, in October 2004, a young fisherman named Alain Kazadi Makalayi gathered a handful of men and a smattering of weapons, denounced Katumba and Kabila for "pocketing money from the mines," and seized the nearby town of Kilwa, declaring a new separatist state of Katanga.

Anvil provided trucks and aircraft for the army's counterattack: According to a UN investigation, once Kazadi and his forces were neutralized, soldiers from the Congolese army's 62nd Infantry Brigade went house to house, rounding up civilians for supposedly abetting the uprising. At least seventy-three people were killed, and the UN documented some twenty-eight summary executions. The soldiers also ransacked the town, raping and torturing scores of civilians.

In their report, UN investigators hinted that the uprising and subsequent killings were somehow orchestrated by figures in the Congolese establishment, perhaps as part of a higher-level struggle to control resources. Anvil denied wrongdoing and said that the army had coerced it into supplying matériel, but the company was sued in Canada for abetting war crimes; the case was eventually dismissed for lack of jurisdiction. After the incident, Katumba himself visited the area to convince traumatized civilians that it was safe to return home.

———

THE USE OF VIOLENT GROUPS TO SEIZE POWERFUL RESOURCES WAS A FREquently employed tactic of powerful figures in Congo. In the 2000s and 2010s, for example, Nkambo Gédéon Kyungu Mutanga Wa Bafunkwa Kanonga, a Catholic altar boy turned schoolmaster turned warlord, ran a brutal insurgency in northern Katanga, not far from Dikulushi.

Gédéon had originally been part of the Mai-Mai—fighters mobilized by Laurent-Désiré Kabila who sprinkled themselves with blessed water in the belief that it would ward off bullets—and he had risen to power through brutality and by claiming that he was in possession of powerful "gris-gris," or magical charms, that he had collected near the town of Malemba-Nkulu. Mines for valuable minerals like tin and coltan, an ore of tantalum, lie in and around this area. Tantalum is in demand for use in capacitors, which store electricity in modern devices; John B. Goodenough had looked at using it for a lithium-ion battery cathode, but concluded it was too heavy.

Gédéon's former spokesman, Thierry Mukelekele, told me that Congolese army officers and politicians also profited illegally from the mines.

John Numbi, a general who, under Kabila, was, at least publicly, one of Gédéon's chief antagonists, often got blamed. "I can tell you that Gédéon is not in any mine," Numbi told me from exile somewhere in southern Africa in 2023 (he had fallen out with Félix Tshisekedi, Kabila's successor). "I know Gédéon well." He said that he himself had not profited from the illegal export of minerals, but he pointed a finger at other figures in the Congolese government for taking advantage of the situation.

Between 2002 and 2006, Gédéon's men progressively became more and more violent, battling rival militias and carving out of northern Katanga an area that was known as the "Triangle of Death." The group was estimated to be around two thousand fighters strong at its height. UN peacekeepers were deployed; Numbi and the Congolese army ran a brutal counterinsurgency campaign; and Gédéon's militia killed and tortured people. Gédéon and his men were accused of cannibalism and mystical practices. Eva Gilliam—an information assistant with the UN who at the time witnessed the disarmament of Mai-Mai fighters in the area where Gédéon operated—remembered the significance of the gris-gris for the fighters. "We would arrive for disarmament ceremonies by helicopter, in the bush, maybe in a mud hut, and they would be there," she told me when we spoke in 2022. "They sometimes had human fingers and ears hanging from their necklaces." The UN report she authored at the time reports that she was told fighters would "nibble pieces of flesh" during combat, although she noted that the specter of cannibalism was also a perfect boogeyman to scare people away from an area rich in minerals. An estimated 165,000 residents of the Triangle of Death fled their homes.

DESPITE THE VIOLENCE—OR PERHAPS BECAUSE IT DIVERTED ATTENTION away from more mundane matters, like business—people in the highest echelons were profiting. So too were businessmen like Dan Gertler. In public, the Israeli businessman hailed Kabila's efforts to run the country. "He's the most promising new president in the world—a new Mandela," he told *Newsweek* in 2003. But Gertler also began to use the same kind of language to describe himself: "I should get a Nobel Prize," Gertler would tell journalists from *Bloomberg* in 2012. "They need people like us, who come and put billions in the ground. Without this, the resources are worth nothing."

By 2006, with Congo lurching toward presidential elections, a pattern emerged: Key assets, including mines, were being sold at bargain prices to fund political activity by the ruling party. Congo planned to hold its first democratic multiparty presidential vote in four decades (Mobutu Sese Seko used to run himself as the only candidate). The elections were blisteringly expensive for everyone involved: The bill for the vote came to around $500 million, almost half of which was funded by the European Union and its member states, making Congo's elections the costliest in African history. They were expensive for the candidates too: Kabila needed money, and Gertler was an instrument for procuring cash. Shortly before the elections, an American diplomatic officer cabled Washington from the embassy in Kinshasa: "Key political and economic actors are currently dismantling Gecamines [*sic*] to assert control over the mining sector and reduce Gecamines' political influence in Katanga and in the DRC generally. Proposed joint-ventures—with Dan Gertler International, Group Forrest and Phelps Dodge—would take the most valuable of Gecamines assets and leave the company with few options for further development." At the center of this web of transactions was Katumba. Michael J. Kavanagh, a journalist for *Bloomberg* who has assiduously followed the twists and turns of the Gertler story, has written that the Israeli businessman used his connections with Katumba to horn in on the best deals.

Then, just before the 2011 elections, *Bloomberg* reported, Gertler bought five mining ventures at below-market prices from state enterprises and sold them, at least in three instances, for a hefty profit. The implication was that Gertler was involved in "grabbing and flipping" mines, as the *Mail & Guardian*, a South African paper, put it. He would then funnel the money back into Kabila's purse, and the president would spend it on his costly reelection campaigns. (Gertler vigorously denied these stories at the time they were published.)

In private, however, Gertler would represent himself as uniquely powerful, the only person who could get things done in a chaotic country. The Israeli businessman began to live, as he described it, "the Congolese way." During the arbitration with the Gertner brothers, Gertler at one point turned on a lawyer who was questioning him. "My power in Congo is that my word is my bond," Gertler said. "You don't know what trouble is. . . . Do you know what it's like to deal with a chief who has to support ten thousand families, each with five children, people who are hungry for *fufu*, the local food there? You'll never understand that." Gertler boasted to the lawyer, "I am a king in Congo to this day."

Investors keen on gaining exposure to the world copper market were given road shows: Meetings with Gertler and George Forrest in Europe were followed by junkets to Congo, where the investors would fly into the country for the day, then tour a mine where they would be sold on "very good mineralization" and how they could "create value for the shareholders"—and profit for themselves. One such junket would be shown in the 2009 film *Katanga Business*, by the Belgian documentarian Thierry Michel. Such visits to "assets" were considered normal occurrences in the world of globalized finance (not dissimilar to the visit to Tesla that the New York bank had arranged). And, because finance had linked asset management firms with pension funds, the financiers might not just be investing the money of the wealthy but also their retirement savings.

A London financier who went on some of the junkets and began investing in Congolese mining, including in a Gertler-linked project near Kolwezi during the mid-2000s, told me that he had been looking for a risky investment that might make him and his firm big money. From the early stages, the financier sensed that things in Congo were not as stable as his interlocutors had promised him. "I think we always managed to go in and out on a day trip and so, you know, deliberately avoid spending the night in Congo," he told me in 2024.

The financier invested in a mine named Mukondo, expecting a healthy profit, but things soon started to go wrong. The name Dan Gertler often came up; it was he, the financier was told, who was using his political connections to create problems at the mine. Problems that only he could fix—for a price. "We all knew he was a scumbag," the financier said. "He was someone who was malevolent."

At one point, as his investment crumbled in value, the financier went to visit Gertler in his suite at the Ritz in London. A fund manager accompanied the financier to the meeting and remembered large sums of money being discussed. The meeting with Gertler was contentious, the financier recalled. "He oozed evil."

AMONG THE DEPLOYERS OF FOREIGN CAPITAL EAGER TO INVEST IN THE boom years before the 2008 financial crisis was the giant New York hedge fund Och-Ziff. Before a U.S. court, it would later plead guilty to paying bribes in several African countries, including Congo. In the documents that emerged after Och-Ziff pleaded guilty was the allegation that the fund had used a foreign "agent" to funnel at least $21.5 million to an unnamed Congolese official whose detailed résumé matched Katumba's.

The "agent," on the other hand, was a fixer whom the U.S. Justice Department described only as an "Israeli businessman" with significant interests in Congo's copper and diamond mining sectors. Subsequently, media outlets and NGOs identified the "agent" as Dan Gertler, and a subsequent civil lawsuit in the Southern District of New York named him explicitly. (Gertler has denied paying bribes in Congo.)

The road to prosecution was paved with public and private deals. In 2008, Och-Ziff bought $150 million worth of shares in a Congolese mining company owned by the agent and then, over the next few years, gave him $254 million in loans that were funded out of the hedge fund's investment capital. Prosecutors from the Eastern District of New York would later contend in a lawsuit against two Och-Ziff employees that they violated the Foreign Corrupt Practices Act (FCPA), because they were certain the money "would be used to bribe high-ranking government officials in connection with the acquisition of assets on behalf of Och-Ziff."

"The DRC landscape is in the making and I am shaping it—like no one else," the agent wrote in an email recovered by federal prosecutors. "You see there is a bigger picture in all of this." When I spoke to a former Och-Ziff manager in 2021, he told me that the fund had changed. "I wasn't on that desk," he insisted. "But it was pretty clearly an open-and-shut case of bribery."

IN 2007, GLENCORE BOUGHT ITS FIRST SHARES IN MUTANDA, THE MINE that Alex Hayssam Hamze had managed to take control of. Hamze was a plainspoken man, and said that Glencore was wasting money and could act more efficiently in Congo; Glencore's engineers disagreed and frequently squabbled with Hamze's team.

An associate of Hamze's told me that Ivan Glasenberg "randomly ran into" Hamze while prospecting another copper-cobalt mine during a visit to Kolwezi. Glasenberg, who liked scrappy rags-to-riches stories, was impressed. "Glasenberg recognized that when he came to Kolwezi. He said, 'You guys aren't running the show—it's this kid here,'" the Hamze associate said. Still, an engineer who used to work for Bazano told me that it was mostly an act. "Alex bought a lot of brand-new machines," the engineer said. The equipment painted a picture, and Glencore understood that MUMI could be a world-class mine if managed correctly.

A Gertler company soon acquired a strategic parcel of land next door at a knockdown price from Gécamines. By 2011, Glencore realized that MUMI was running out of space to develop and needed the land next door to develop the mine, so the Swiss resources company decided to enter into yet another joint venture with a Gertler company. "Gertler had the other mine with them," Hamze's associate said. Hamze "didn't get on with Gertler. He's a powerful local guy. Dan didn't like that." Finally, Le Petit called Gertler out on the price he was offering: "Hamze said it was too low," the associate remembered, referring to the price offered by

Gertler for MUMI shares. "Why should the guy get in at a fraction of the price?"

HAMZE FELT THAT GLENCORE HAD BEEN STRONG-ARMED INTO THE MERGER with Gertler's company: He understood that the mine next door was worthless. He wrote Glencore a short note protesting the sale, and though it resulted in his asset price rising, he remained incredulous. "How can you force a merger," Hamze's associate remembered him asking, "with a company that is literally worth zero for thirty percent of the world's best cobalt deposits?"

In 2012, Glencore signed a deal that would see it pay $480 million to take control of Mutanda by buying shares owned by Bazano and another Hamze company, High Grade Minerals. The next year, it bought the rest of Hamze's shares for almost as much. Le Petit had been paid out almost $1 billion for an asset that he had acquired for $300,000 only seven years earlier.

According to Hamze's associate, Glencore insisted that Gertler be allowed to enter the Mutanda joint venture. "They did it to please the president and to please Dan Gertler." (Glencore said it was not a matter of pleasing Gertler, but rather that it needed the next-door space; the merger was completed in July 2013.) "We are delighted this investment is now starting to bear fruit," Gertler said publicly.

Glencore officials from the time remember Gertler being less cheery in private. The Israeli businessman's shares had been diluted during the Katanga Mining deal, and he was upset at having lost family money on the deal. He was still very wealthy by most standards, but he was suspicious he would lose money again on the MUMI deal; in a separate transaction later on, he would successfully buy the royalty stream for the mine from Gécamines, ensuring him a steady income from the sale of MUMI's minerals.

By the time of the merger, Hamze, on the other hand, had become seriously wealthy. "He was making millions," Hamze's engineer told me. "There weren't enough trucks to get the minerals out." He built a hospital and a park, refurbished a golf course, and even created the town of

Likasi's central marketplace. Bazano, his company, had a farm with cattle and vegetables. Hamze remained devoted to his mother and would always make sure to pick her up from the airport when she returned to Congo after traveling to Lebanon, her suitcase stuffed with Levantine flatbread. For fun, Hamze's family Jet-Skied on Lake Nzilo. One day, a Bazano employee was hurt in a crash. One of Hamze's family members pulled $300 from his pocket, handed it to the injured worker, and told him to go on leave until he felt better. "They were very generous people," the engineer remembered.

But through the stormy negotiations, Hamze had also made a powerful enemy: Gertler. Soon, the government was creating problems for Hamze everywhere he turned. He left the country, traveling to South Africa, Malta, and Dubai. "I said, 'Just look at how many zeros you have on your account,'" the associate told me.

Hamze hadn't fully retreated from Congo, though: Bazano and affiliated companies still do business there, and I was put in touch with a cousin of his who was running an element of logistics for him. Le Petit also owned permits to mine at the Shabara mine, which abuts Mutanda. It was crawling with artisanal miners who had formed a collective called COMAKAT. Hamze was buying copper and cobalt off the artisanals using another company. In 2015, Glencore bought that mine too, although by the end of 2024, it had not managed to expel COMAKAT.

Hamze had been popular in Congo. Jeef Kazadi, for instance, remembered him calling in journalists and offering them money to write stories about his operations. Hamze's associate remembered visiting Kolwezi and Likasi with him. "It's like a riot when he's on the street. It's like—I don't know—being with Eisenhower or something," he told me when we spoke. "He is now sitting bored in Dubai."

One of the interesting things about the Hamze story is that it shows just how unconcerned the United States was with critical metals during the 2000s and the first part of the 2010s. The presence of Hamze—who had family connections to a part of Lebanon where Iran held sway, in close vicinity to a metal that was used in power storage for almost all cell phones and computers—didn't seem to raise an eyebrow. (Several well-connected Lebanese sources told me that Hamze had nothing to do with politics in his home country.) U.S. officials at the embassy in Kinshasa didn't seem to care who he was.

After Hamze left Congo, however, the U.S. government did appear to take an interest in him. He constructed a lavish home near the town of Naxxar on the island of Malta; in 2022, a summary of Maltese planning applications showed he was having an issue with his boundary wall. "I saw him out one night with these two security guards," a Malta-based U.S. official told me in 2024. "One of them was very pushy."

During the early part of the Cold War, U.S. spies had fought hard to keep Congo's uranium out of Soviet hands; during the Mobutu era, the CIA had backed a coup and provided logistics support to keep the dictator in power; now, under Kabila, no one seemed to believe that cobalt was important. The world's largest economy was still run on hydrocarbons—it was even fighting a losing war for them in Iraq. Perhaps that was why the U.S. government barely seemed to notice when China began to move into the southern Democratic Republic of the Congo.

CHAPTER 25

A NEW CATHODE

Denise Gray was a company woman: She had grown up in Detroit and risen in the ranks at General Motors since her college days, in the 1980s. As an engineer, she had worked on the electromechanical systems and transmissions of cars. She was even married to a GM executive. After Bob Lutz directed his staff to create an electric vehicle, they had to figure out what type of battery to use. The task fell to Gray, who, in 2006, was promoted to lead GM's energy-storage solutions. "Lithium-ion is the reason we can do a Volt," Gray would say. Journalists typically didn't ask for more specificity on things like cathode composition.

There were, by the mid-2000s, several choices, in fact, when it came to battery chemistry. John B. Goodenough and other scientists had spent the last two decades figuring out how to make more powerful and more cost-effective batteries. What this meant was that Gray's team wasn't bound to the LCO cells that powered cars like Teslas and portable devices like laptops. And executives at GM were not advertising the makeup of their batteries—they were playing their cards close to their chests. Gray had quickly alighted on a battery chemistry called lithium nickel manganese cobalt, or NMC.

As with the Japanese lithium-ion batteries, the origins of NMC are contested between a few different laboratories, including Pacific Lithium in New Zealand, Dalhousie University in Canada, and Osaka City University in Japan. The first patents for the battery, however, were filed in the U.S. by researchers at the Argonne National Laboratory in Lemont, Illinois.

The scientists at Argonne were led by Michael M. Thackeray, a sandy-haired South African marathoner who had worked at John B. Goodenough's Oxford laboratory in the 1980s. There he had discovered that metal spinels—a type of crystalline structure composed of tetrahedron-shaped ion arrangements linked at their corners—were good hosts for lithium ions. (When Thackeray first suggested this to Goodenough, the older professor queried whether there would be enough space for the lithium ions to be intercalated during discharge: "By all means try, but I suggest that you look around the laboratory to see what other projects are going on.") After a few false starts, Thackeray found that a compound of manganese, a brittle element used in alloys, might provide a suitable receptacle for the lithium-ions in the cathode. "This discovery," Goodenough remarked to Thackeray in his typically understated manner, "may have commercial significance."

After a stint in South Africa, Thackeray took a job back at Argonne, arriving during "a blizzard to pale the skin" in February 1994. He would spend the next few years quietly working with a tightly knit team of twelve battery scientists to fashion new cathode materials. One such project involved tests on a material that layered lithiated nickel manganese cobalt oxide with spinels of lithium manganese oxide, a structure that would support the cathode as the battery was charged and discharged.

In 2000, when Thackeray attended a conference in Como, Italy, the birthplace of Alessandro Volta, he got wind of a project by a lab in New Zealand that used lithium-manganese-chromium oxide. Anxious not to be outmaneuvered by the Kiwi team, Thackeray quickly phoned Argonne. His team threw together a patent application, which was provisionally filed in a matter of weeks. The patent gave the team the exclusive right to a cathode of nickel, manganese, and a third metal—cobalt. NMC was combined with the spinel technology to get around the latter's lower energy density. This new material allowed 60 or 70 percent of the lithium ions to migrate out of the cathode, 10 or 20 percent more than from lithium cobalt oxide, resulting in a more powerful battery. The battery also had a longer cycle life, meaning it would last longer.

NMC batteries allowed electric vehicles to go farther and faster than Goodenough's original batteries. These days, they are especially popular in the United States and in Europe. As Steve LeVine writes in *The Powerhouse*, a book about the lab at Argonne that produced the new material, NMC was a "superior" cathode material, and perhaps even a supreme one when it came to electric-vehicle design. Nickel and manganese "really made the price come down for cathode materials," Shirley Meng, a materials scientist, told me. When we spoke, Meng was working at the University of California, San Diego. She has since moved to the University of Chicago and taken the position of chief scientist at the Argonne Collaborative Center for Energy Storage Science. In the years since the Volt's release, NMC batteries have been adopted by companies like Audi, Ford, and Tesla as they compete to produce better electric vehicles.

LIKE TESLA, GENERAL MOTORS NEEDED TO MASTER THE ART OF INDUS-
trial scale: It needed a place to make the batteries. Gray, the engineer, had winnowed down a list of battery-cell producers. High on her list was A123, a U.S. start-up. The company, where the battery innovator Mujeeb Ijaz worked, was in dire financial straits. It had invested too much in production capacity for orders that never materialized. In many ways, the company's bosses had repeated the mistakes of Robert R. Aronson, the '70s electric-car innovator: They had scaled up production too early and too quickly. Only, this time, they had used federal loans to do so. (The Republican presidential candidate Mitt Romney's campaign would later use the company as a prime example of how the Obama administration was "gambling away" taxpayer money on pie-in-the-sky schemes.) Beijing, on the other hand, was making the gambles it needed to create more powerful batteries and cheaper electric cars.

As A123 crashed, Gray took a job with LG Chem, part of LG, the Korean chemicals behemoth. The subsidiary for which she worked—Compact Power Inc., or CPI—began producing the batteries. It had conveniently placed its offices in Troy, Michigan, about half an hour from GM's headquarters on the Detroit River. CPI was positioning itself to the U.S. automotive industry as a "one-stop shop" for battery solutions. Compact Power had also received federal funding to build a battery-production facility in Holland, Michigan. That money, like the A123 loans, had come out of President Barack Obama's American Recovery and Reinvestment Act, a stimulus package created in the wake of the 2008 financial crisis— in many ways, a ringer for President Joe Biden's 2022 Inflation Reduction Act, which also had as its goal the spurring of battery development.

GM was careful with what it released publicly. The veil of secrecy extended up and down the supply chain. It was broadly known that the Volt used a laminated flat-pack battery, but information about the chemistry used in the cathode was patchy. A website called Green Car Congress reported that "CPI is using manganese-spinel ($LiMnO_2$), which features high stability and resistance to thermal runaway." The author of the article was apparently unaware of the battery's NMC component.

Finally, on January 6, 2011, GM held a conference call with Argonne and Mohamed Alamgir, the research director of Compact Power. GM had licensed a new battery technology from Argonne called NMC, Alamgir announced, but he also let slip that the license covered technologies already being used in the Volt. On the call, GM Ventures' Jon Lauckner noted that NMC was "probably the most capable cathode material we have seen out there."

The technology took off. By 2019, NMC battery output made up 69 percent of total lithium-ion battery production. "Due to its quality uniformity and high energy density, NCM batteries have become the most widely used component in the battery industry in total," an LG Energy Solution report informed its readers. (The nickel-rich cathode material is known as both NMC and NCM, depending on whether the cathode contains more manganese or more cobalt.)

Using the stable and thermally resistant spinel technology, LG was able to reduce the amount of cobalt in the cathode. Cobalt formed only 20 percent—and later 10 percent—of an NMC cathode, far less than Goodenough's original LCO cathode, which contained about 60 percent cobalt. Battery makers in the 2000s had already become concerned about cobalt, not so much for its association with child miners as for its high price. (According to the U.S. Department of Energy, in 2023, cobalt, followed by nickel, was the costliest material used in cathodes.) "There's a huge drive to cut the cost of lithium-ion batteries. By going from one-third cobalt to, let's say, fifteen percent cobalt, it cuts the price of that particular material by about forty percent," said M. Stanley Whittingham, the creator of the first lithium-ion battery. "It's a huge incentive to get the cobalt out."

Whittingham, speaking in 2020, also pointed out cobalt's association with human-rights abuses in Congo. During the 2010s, those concerns would begin to surface not just in Africa but in the U.S. Some people took to calling NMC cathodes "low-cobalt cathodes," and they became the standard for high-performance electric vehicles. "Cobalt—I mean, that problem—we *solved* that problem," a senior scientist with the Argonne National Laboratory confidently told me at the Battery Show in Novi, Michigan, just outside Detroit, in 2022. "And it was scientists at Argonne who did it."

But cobalt was still needed in NMC, just in lower quantities. And lithium cobalt oxide wasn't going away anytime soon. It still had many advantages: Batteries made with it, for example, could maintain a high voltage even when low on charge. Well into the 2020s, tech companies that made smaller devices like laptops and cell phones—Apple, for instance—would still use LCO batteries. But there was another problem with NMC: To produce it at scale, battery makers were going to need a whole heap of nickel.

CHAPTER 26

NICKEL FROM THE FOREST

Morning in Bahodopi, on the Indonesian island of Sulawesi, begins with *fajr*, the dawn prayer. It crackles out of clapped-out speakers over a town of corrugated roofs, muddy streets, and tens of thousands of men pulling on blue-and-green smocks and fumbling for hard hats in the morning's half-light. When the prayer is over, new sounds commence, as if summoned by the incantations to Allah: the crash of hammers and the whir of drills. The town, which sits in a jumble of humid forest, is still being fashioned.

Sadip, a farmer in his sixties who arrived in Sulawesi from East Java in 1993, told me when we met in 2022 that it hadn't always been like this. (Like many Indonesians, especially those from Java, Sadip uses only one name.) For Sadip's first twenty years, life in the town was calm, unhurried. The farmers spent their time corralling wild pigs and digging irrigation tunnels between their fields. They dammed a local river. According to an article in the Chinese magazine *Caijing*, until 2013 "the only sign of modernity in Bahodopi was a couple of dated motorcycles."

Since the time of Thomas Edison, who proposed nickel-iron batteries as an alternative to lead-acid ones, the battery world has been fascinated with nickel. In 1901, Edison traveled to Sudbury, a Canadian town abutting some of that country's largest nickel deposits, to look for the metal.

He had little success: In the town of Falconbridge, he dug into quicksand and missed a mother lode of the metal by a few feet. Now NMC batteries were spurring a new rush for nickel, buried deep in pits outside Bahodopi.

Much of the nickel mined in Indonesia is used in the Chinese steel industry because the ore is less pure than that from other sources, like in Russia, but NMC cathodes such as the one used in the Chevy Volt have also become a big driver of growth in the country in recent years. As a report by the Center for Strategic and International Studies, a Washington think tank, noted, "Indonesia's nickel strategy has become part of the country's goal to create an integrated EV supply chain." One of the epicenters of this rush was located at the giant Indonesia Morowali Industrial Park in Bahodopi, known by locals as IMIP.

In 2019, a $4 billion battery project that included Chinese investors began construction in Morowali. Green Eco-Manufacture, or GEM, a Chinese colossus in the recycling of metal and electronic waste, said that it was planning to make battery-grade nickel chemicals there. "Indonesia will become the main player in lithium batteries," said Luhut Binsar Pandjaitan, an Indonesian minister, as the project was unveiled. "We will control the world market." In 2022, Pandjaitan inaugurated a hydrometallurgical nickel laterite production facility at IMIP that could produce fifty thousand tons of nickel a year.

When I visited, shortly after Pandjaitan's inauguration of the plant, Bahodopi was still a place on the cusp of being finished, in the process of being ripped from one way of life and thrust into another. Boutiques selling streetwear that would not have looked out of place in any trendy neighborhood in any international city stood next to wooden shacks that appeared to be barely supporting their own weight. On paydays, long lines of men formed early at ATMs that were soon out of cash.

Because of Indonesia's proximity to the Asian mainland, Chinese companies had invested heavily in Indonesia's mines. In 2009, the year before the Volt was released, Indonesia tied with Canada, behind Russia, as the second-largest producer of nickel. By the early 2020s, Indonesia had become the largest global producer of the metal, with 292 nickel-mining permits and 1.54 million tons of mined nickel exported that year. Nickel is the fifth most abundant element on Earth, and is nowhere near as concentrated in the earth's crust as cobalt: Russia, Australia, Canada, and the Philippines all mine large quantities of the metal's ore.

I arrived at Bahodopi in the middle of the night after a nine-hour drive through tiny villages and jungle. Five hours in, our driver had become spooked while passing through a "haunted forest" and proceeded to drive at a jogging pace for the last three hours. At around 4:00 a.m., I noticed some bright lights through the tinted car windows, and then we turned off onto a lane and stopped at a hostel. Inside, itinerant workers snored away in the rooms next to mine.

———

THE ISLAND OF SULAWESI, IF YOU LOOK AT IT ON A MAP, IS FORMED OF FOUR arms, a shape that resembles a monkey, albeit a headless one, midswing. Sulawesi sits astride a shifting, buckling mess of geological fault lines that result from the Eurasian Plate moving southeast, the Philippine Plate moving westward, and the Indo-Australian Plate moving northward. Sometime between one hundred million and twenty-three million years ago, a series of collisions between the plates started to give the island the shape it has today. Beneath the island are four distinct fault lines, and earthquakes abound: In the center is a dramatic spine of mountains. In the north, plate subduction has created a series of active volcanoes that regularly belch smoke and occasionally dramatically explode.

Tens of thousands of workers and I were sleeping in Bahodopi's concrete rooms that night because of events that took place around about the time that dogs, deer, and camels began to roam the earth—around twenty-three million years ago, give or take ten million years. In the Sturm und Drang of the period's plate collisions, a chunk of oceanic crust a little bit larger than Connecticut was grafted onto the Eurasian Plate as the Indo-Australian Plate submerged beneath it. This chunk, or ophiolite, as it is known to geologists, was partially formed from igneous rock—dried magma—which was rich in iron and magnesium. Over thousands of years, this rock was weathered into a red laterite soil a little like that of Katanga and formed the gentle slope of southeast Sulawesi's coastline.

As Indonesia's tropical rains fell on this soil, minerals, dissolved in the rainwater, percolated downward, slowly forming limonites, orange rocks streaked with black. Deeper, they formed garnierites, rocks that ranged in color from creamy yellow to grass green.

As in Congo, geological flukes have made Indonesia rich, at least in terms of mineral resources. In the last decade, particularly, Indonesia's mining industry has boomed, and investment has poured in. The limonite can be worked for cobalt and nickel, and the creamy-yellow and grass-green garnierite is rich in nickel. In such deposits, the metal is dispersed in the rock and nickel extracted from such laterites was long deemed to be uneconomical for use in batteries, as it contained excessive impurities.

In the 2010s and '20s, Chinese firms began to use various energy-intensive processes in Indonesia to produce the kinds of high-grade nickel and cobalt needed for batteries: In one, known as high-pressure acid leaching (HPAL), the ore was fed into a purpose-made vessel called an autoclave, superheated, and mixed with sulfuric acid to create a substance called mixed hydroxide precipitate, and in another, nickel ores were baked in a rotary kiln to produce a nickel-rich matte. These are both highly energy-intensive processes, and produce an acidic slurry of waste that needs to be disposed of properly, but thanks to the development of processes like these and deposits like the ones under Bahodopi, in 2022, politicians in Jakarta could boast that their country was now the world's largest nickel miner, and in 2023, Indonesia overtook Australia to become the world's second-largest producer of cobalt after Congo.

———

ALTHOUGH THERE ARE CLEARLY MANY DIFFERENCES BETWEEN INDONESIA and Congo, the colonial histories of both countries exhibit some striking similarities, even as both countries are once again the subjects of contemporary jockeying for resources among wealthy and powerful outsiders. On Sulawesi, there's evidence that the island's metal deposits, like Congo's ore, have been exploited for at least a thousand years. From the seventeenth century onward, Indonesia was colonized by the Dutch, who, like the Belgians in Congo, were searching for a resource—this time spices—

to sell on European and international markets. Colonialism in Indonesia was often marked by significant brutality, as in Congo, a fact that the Netherlands, like Belgium, has only recently come to acknowledge.

Dutch geologists realized that Sulawesi had rich nickel deposits in the early twentieth century, and Holland began mining the metal in 1934. When the Japanese invaded in 1942, they used the metal in their war effort. Indonesia became independent in 1949. In the 1960s, the U.S. helped extremist groups exterminate between five hundred thousand and a million suspected Communists.

In 1967, the dictator Suharto came to power, and he made mining a centerpiece of his industrial policy, offering a 6.6-million-hectare plot to a consortium headed by INCO, which included a Japanese company, Sumitomo Metal Mining. In 1968, the company began working on the area that included Bahodopi's nickel laterite deposit, which researchers later estimated held some 180 million tons of nickel. By the 2020s, according to its website, Sumitomo Metal Mining had become fully involved in the supply chain of lithium-ion batteries, "a top producer of cathode material for secondary batteries used in products such as electric vehicles, for which an increase in demand is anticipated going forward." (Sumitomo representatives did not respond to my queries.)

INCO's work in Sulawesi would be problematic from the very start. In the Sorowako mine area, around a hundred kilometers inland from Bahodopi, the company displaced locals, paying them two cents per square meter for their land, and mining activity began to interfere with agriculture. "They have had a very long saga with INCO," said Richard Kent, a human-rights researcher who investigated displacement and other abuses on Sulawesi in early 2024. When INCO built a dam, the rice fields around Sorowako were flooded. "People lost their livelihoods, and rice farming became untenable," Kent told me. Local and national officials complained as well. What's more, the contract that Suharto's government had struck with the miners didn't seem to be generating much revenue: The royalty payments on the ore that INCO was exporting were very low. Such arrangements weren't unusual at the time. As a popular protest slogan went, Suharto's government was rife with "corruption, cronyism, and nepotism."

In Bahodopi, the company continued to conduct sampling of the minerals beneath the soil, and there were concerns that the land INCO had won in its concession overlapped with farmland. Suharto revived an old Dutch tactic to populate remote parts of Indonesia with people who lived on overpopulated islands like Java. Known as *transmigrasi*, or transmigration, Suharto's policy moved millions, perhaps tens of millions, of Indonesians to remote places like Sulawesi, where the government allotted them farmland.

Bahodopi was one of the destinations for the transmigrants. Among them was Sadip, the farmer, who came from an area of East Java that the government thought overcrowded. He was allocated two acres of land for a rice field and a residence. "At first it was difficult in those days," he told me when we met on Sulawesi in 2022. "Wild pigs would come out of the forest and destroy the fields." His face was worn, etched with the kind of wrinkles that come from being out in the sun all day. We spoke on the mats of his front room, after darkness had fallen in Bahodopi. Above us hung a framed sheet of Arabic calligraphy proclaiming that there was "no

Industry alongside shanties in Bahodopi, 2022

god but God." Sadip's wife hurried in and out of the kitchen, bringing us boxed iced tea and fried bananas with sweet, chocolaty flesh that she had grown in her garden. "That is the advantage of being a farmer," Sadip said. "You can grow all your own fruits and vegetables."

The serenity he felt on his farm, however, was not to last. "I get nostalgic," he told me, his small dark eyes shining in the buzz of the electric light overhead. "Farming was just really, really peaceful. I don't care about the income, or what I made. I was just at peace farming. But after IMIP came, everything changed."

CHAPTER 27

CROSSING THE RIVER BY FEELING FOR THE STONES

P eter Chao Zhou came of age in the 2000s, a time China was consuming more and more raw materials—from Congo, from Indonesia, from all over the world. A graduate of the University of
British Columbia, in Canada, Zhou had maintained a laserlike focus on
finance from an early age. At college, he took part in competitions related
to investment banking and venture capital, and he was a member of his
university's finance club. He got his degree, naturally, in finance and
mathematics. He had grown up in China. "I was born and raised in China,
so I am one hundred percent made of China, put it that way," he told me
when we spoke in 2020.

During Zhou's youth, China had fully embraced the dizzying economic growth that had sprung from its opening to global financial markets. The country began to officially implement a policy dubbed "Going
Out," which was announced by President Jiang Zemin in 1999. Under the
policy, Chinese businesses were encouraged to make overseas investments, and Chinese citizens were prompted to travel and set up shop
abroad. Zhou's generation was keen to prosper in the new environment
of banking and transnational finance. They wanted to learn how Capitalism with a capital C worked.

Zhou, who was bespectacled and had sideswept dark hair, looked up to Ray Dalio, the swashbuckling U.S. hedge-fund billionaire who was famous for being able to turn a profit in even the rockiest of market downturns. Dalio has written admiringly of how China achieved growth abroad through schemes like the Belt and Road Initiative. "China," he wrote in 2021, "has become a rival power to the United States in most ways and is becoming stronger in the most important ways that an empire becomes dominant." And it was only inevitable that industrial power would lead to political power.

CHINA'S GREAT RUSH FOR PROFIT—AND FOR MINERALS—IN THE 2000S CREated what was known as a commodities "supercycle," an economic term that meant demand for minerals was outstripping supply. Zhou had been assigned to grease the wheels of this cycle, which was largely fueled by a construction boom and newfound prosperity in China. Prices for metals shot sky-high.

In 2009, China hit a major industrial milestone: It surpassed the United States as the world's largest automaker. And, much like U.S. cities in the 1970s, the country's major urban hubs had become polluted with thick, acrid smog.

Politicians were under pressure to reverse the environmental consequences of the country's unchecked growth spurt. During the 2008 Olympics, Beijing had experimented with curbs on cars, allowing only vehicles with odd- or even-numbered license plates to be on the road on certain days, and air quality improved. In the year after the Games, the city's Traffic Management Bureau announced that only 20 percent of the city's 3.6 million privately owned vehicles, and only a third of official cars, would be allowed to drive on the city's roads on any given weekday. Drivers in the city, who saw this as a massive inconvenience, grumbled that the government ought to find another way to limit pollution.

It was time for China to look for a new technology, and it didn't have to look far: Wan Gang, who had served as the minister of science and technology since 2007, was a former auto engineer who had a fascination for electric vehicles, especially the lithium-ion-powered Tesla. He decided to go all in on electric vehicles. Luckily for Wan, Chinese companies like BYD were already ramping up production of electric cars.

THOUGH BYD HAD BECOME A MAJOR BATTERY SUPPLIER IN THE 2000S, THE company's car division was still not doing particularly well. In 2009, fewer than five hundred electric vehicles were purchased across China. The Chinese government had already decided to use the economic powers it had accrued in the past decade and a half to spur the making of electric cars, but at the end of the aughts, under Wan, it went into overdrive. Starting in 2009, the government began to provide individual subsidies for electric cars and buses. "They had the foresight to know that this will be a factor, that they can gain leverage, and they made the investments needed to build those facilities," said Ryan Melsert, the former Tesla Gigafactory design team member. "I guess I would call it a strategic decision."

It wasn't as if Washington didn't see what was happening. At a D.C. launch party for the Tesla Model S in 2009, Senator Maria Cantwell told *The New Yorker*'s Tad Friend how concerned she was about the state of the U.S. auto industry, which was flailing in the wake of the 2008 economic crisis. She then noted that China had 250 companies working on building batteries and electric vehicles. "Getting denser, more affordable batteries built here is the whole game," Cantwell said. "Otherwise, we're soon going to be as dependent on Chinese batteries as we are now on Middle East oil."

Cantwell also mentioned that there was no political will in the U.S. to prod consumers toward electric vehicles using high gas taxes, as was being done in Europe at the time. Earlier that year, the Obama administration's American Recovery and Reinvestment Act had earmarked $2.4 billion for electric vehicles and the creation of a battery infrastructure, but the policy seemed scattershot and unbuilt upon. Highly publicized failures in the States had led to increased skepticism.

While further funding for innovative electric-vehicle projects became tighter in the U.S., Europe's battery ecosystem began to thrive. In Britain, the government encouraged the construction of charging points for electric vehicles and, in 2009, set aside £25 million ($40 million in U.S. dollars) for electric-car research and another £20 million ($32 million in U.S. dollars) for public-sector firms to buy low-carbon vehicles. Other European countries provided subsidies worth €5,000 (about $6,600 in U.S. dollars) to buyers of electric vehicles between 2011 and 2015. Norway and Denmark also agreed to exempt electric vehicles from the vehicle purchase tax, which could be as high as €10,000 (some $13,000 in U.S. dollars).

BUT NOWHERE WAS THE BATTERY INDUSTRY THRIVING LIKE IT WAS IN China. Companies such as BYD were remarkably successful at churning out devices to store power. What's more, by the end of the aughts, China already had a developed electric-vehicle market, albeit not an automobile industry. The decade had seen a huge surge in the country's production of electric bicycles and scooters. Two-wheelers had been a symbol of China's industrialization in the years after Chairman Mao's Great Leap Forward. In the 1990s, millions of people had started to migrate to their electric cousins, which were originally powered by lead-acid batteries. When Chinese manufacturers started using the same 18650 cells that were being used to power Tesla's Roadster, they found that they could make bikes go farther and faster.

By 2011, there were up to 140 *million* e-bikes on China's roads, almost double the number of cars in the country. E-bikes were cheap to acquire (popular models ranged from fifteen hundred to three thousand renminbi— from $225 to $300 U.S. dollars at 2011 exchange rates), and they were cheap to charge. According to one estimation, e-bikes cost commuters as little as twenty-one U.S. cents a day to power, whereas commuting by car set drivers back around $4.16.

At the end of the decade, *The New York Times* reported that the e-bike industry was now worth $11 billion worldwide, calling it an "accidental transportation upheaval," with nothing said about how important state support was for the industry in China. "For each mile traveled, electric bikes cause fewer emissions of the gases associated with global warming than do cars," the *Times* reported approvingly, although it noted that lead-acid batteries came with some pollution concerns. (Lithium-ion technology wasn't mentioned.) "Today, electric two-wheelers are so common in China that they account for 80% of all the greenhouse-gas emissions avoided by the use of electric vehicles—in the entire world," the journalist Akshat Rathi wrote in *Quartz*. "China is now applying the lessons learned from its success in electrifying its two-wheeler fleet to building momentum in the electrification of its passenger cars."

What China was doing with electric cars was even more ambitious than what it had done with bikes. Despite the capitalist sheen, the country still operated a centrally planned economy, and government programs were key to setting the direction of the country's industry. "They brought on scale very fast," Melsert said. "I think they focused on the lower-cost end of the market, you know, initially selling more domestically in China." In 2009, the Chinese government launched an initiative called Ten Cities, Thousand Vehicles. True to its name, its target was for ten cities to each have a thousand electric vehicles on the road within three years.

The program was an example of what Chinese revolutionary leaders had called "crossing the river by feeling for the stones," a sentiment endorsed by Deng Xiaoping, the Chinese premier who had opened China to private businesses. Deng's selection of Shenzhen—the city where Wang Chuanfu established BYD—as the first special economic zone had epitomized this kind of experimental policy. In the case of the electric-car program, the stones would be the ten pilot cities, selected by state planners, each of which would launch a thousand electric vehicles by 2012. Elon Musk may have scoffed at BYD's rival electric car in a *Bloomberg* interview in late 2011 ("I don't think they have a great product," he told the anchor Betty Liu), but Wang's company, alongside CATL, was hungry for profit and inextricably woven into the fabric of the Chinese state. It was already crossing the river, and the other side was in sight.

CHAPTER 28

CHINA'S ANSWER TO DR. ZEE

By the end of the aughts, China had already decided to shift its focus from two wheels to four. The country's efforts had barely been noticed in the West, where Teslas and Chevy Volts were getting splashy rollouts; people were more closely watching automakers in Japan, which had developed hybrids like the Toyota Prius.

When BYD showed its F6DM, a ferrous-battery-powered hybrid car, at the 2008 North American International Auto Show, in Detroit, the press "scurried off," bored by what most people saw as a copy of a Honda Accord made by what they imagined to be a "small upstart Chinese automaker." Had they checked the company's filings, they would have seen that not only was BYD hardly "small," but it was also rapidly accelerating at a speed that meant it would soon challenge traditional auto behemoths. BYD had grown its assets more than fourfold since 2004; the firm had a turnover approaching $4 billion and gross profits of almost $800 million. What's more, the F6DM ran on a revolutionary new type of lithium-ion battery called LFP. Made from iron and phosphates, LFP would come to predominate in the auto industry a decade later. BYD said the car was "green tech for tomorrow."

Matt Hardigree, who wrote for the auto site Jalopnik, was the only journalist who decided to stick around after the press conference. Wang

Chuanfu, BYD's chairman, invited Hardigree to hop into his new car. Technically, the journalist would write, it was the first-ever "test drive" of a BYD car in the U.S. Confusion reigned as Wang turned the car on, sped it up to ten miles per hour ("an uncomfortable speed in the middle of a convention center"), almost hit Hardigree's videographer, and gunned it straight through a press conference on a race series for ethanol cars. "Dr. Chuanfu is China's answer to Dr Z [*sic*]," Hardigree would later write. The stunt generated a few headlines and some amused chuckles, but an auto industry that was still addicted to petroleum quickly forgot about the Chinese car zipping around the convention center.

That was soon going to change. By 2011, only two years after the inception of the Ten Cities, Thousand Vehicles initiative, the plan had expanded to encompass twenty-five cities. The government had stated that it wanted five million electric vehicles on the road by 2020. Increased subsidies of fifty thousand to sixty thousand yuan ($8,000 to $9,600 in U.S. dollars) were offered for electric vehicles. The subsidies would be huge in Europe or the U.S., but for ordinary Chinese people who had always dreamed of owning a vehicle, they were gargantuan—around triple the annual disposable income of an urban Chinese family and around ten times that of a rural Chinese family. In 2014, the government went even further and exempted "new energy vehicles" from the country's vehicle purchase tax.

The increase in Chinese electric-vehicle production was astonishing. In 2011, China produced only 8,368 electric vehicles. But that year, the government inaugurated its Twelfth Five-Year Plan. Among the plan's objectives was the development of batteries and alternative fuels; it called for China to produce a million electric vehicles by 2015. Such a figure was only a little optimistic: 340,000 or so new-energy vehicles would be produced in 2015 alone, and year-on-year sales would increase by 223 percent. Three writers at the *Stanford Social Innovation Review* praised China's "unique strategy" and ability "to launch system-level and sector-wide change in multiple cities and regions with no political opposition."

The Beijing bureaucracy didn't get overly involved in the weeds of how the funds would be spent, leaving the decision up to municipal governments. Funds were spread between different cities, which were selected based on their geographic locations and economics. The cities' leaders chose policies that played to their strengths: Beijing, for example, set up three industrial campuses inside the city and used public policy, including license-plate lottery exemptions and tax reductions, to make electric-vehicle ownership more attractive; Shenzhen, the first SEZ, used its industrial base to create leasing models for electric buses that drove down costs for the Shenzhen Bus Group; in Chongqing, planners took advantage of the abundant cheap electricity that came from the nearby Three Gorges Dam and built what is quite possibly the most robust fast-charging infrastructure in the world.

Martin Chorzempa, a senior fellow at the Peterson Institute for International Economics, remembered arriving in Shenzhen in the mid-2010s, just as China was ramping up electric-car production. "They had decided to electrify their entire taxi fleet," he said. When he spoke to drivers, they would complain about long charging times and the fact that the batteries sometimes burst into flames. But all of them felt they had to go electric because, they said, the government had asked them to.

In the short term, each of the pilot schemes met with difficulties. By the end of the Ten Cities, Thousand Vehicles program, only seven of the cities had reached one thousand vehicles. A U.S. political scientist at the Council on Foreign Relations judged the policy to have been "dismal"; Washington, happy with low oil prices and cheap shale gas, took little notice.

In the long term, however, the pilot plans worked better than their architects could have predicted. China soon became the largest electric-vehicle market in the world. In 2015, if you believe the official figures, the country's EV market was already 50 percent larger than that of the U.S. The million-EV mark appears to have been reached sometime during 2017. By the end of the year, the Middle Kingdom was home to two million electric vehicles.

At times, it could feel like the whole country was somehow in lock-step with the government's goals. By 2020, China was *mandating* that its firms produce a certain percentage of electric vehicles or face a penalty. And, of course, direct government stimulus to firms that had decided to make batteries was key. "China sends policy signals when they're deciding what to invest in and what to do," Chorzempa said. "And in areas where there are huge subsidies for grabs, people run to grab those subsidies."

With all the cheap money available, some manufacturers overstepped, going for as much market share as possible. In 2025, it was estimated that China's fifty battery-manufacturing companies produced batteries whose power equated to forty-eight hundred gigawatt-hours a year (one GWh of battery capacity powers at least ten thousand electric vehicles), or four times what the market needed. Chorzempa blamed this on the government getting *too* involved in business decisions: "Often you know when there's a top-down push for something, because you get this huge over-supply of it really fast."

The pilot plans had long-reaching effects in the cities that pursued them. Chongqing, the city that had focused on charging stations, had installed 72,900 electric-vehicle charging points by 2022, and it made plans to increase the number to 240,000 by 2025, 30,000 of which would be rapid chargers. Chongqing is a single city, albeit a very big one with thirty million people. By comparison, the United Kingdom, an entire country (four countries, in fact), had 34,637 charging points that year, and London, England's capital, had 820 rapid or ultra-rapid chargers. As of 2024, California, that hub of the U.S. electric-vehicle dream, had only 37,909 operational charging stations, according to the website PlugShare.

By the end of the 2010s, the writing was on the wall. "China is about to ban the internal combustion engine," the mine financier Robert Friedland said at the 2018 Commodities Global Summit, hosted by the *Financial Times*. "The Chinese can never catch up to Japanese internal combustion engines, but the Chinese have discovered and determined—because there's really only one landlord in China—that they will definitely, one hundred percent, for sure be the world leader in the electric car."

In 2018, 1.26 million electric vehicles were produced in China. In 2024, China produced a million electric vehicles in a single month. "No other country in the world has made anywhere near as big an investment or instituted as significant regulations," Akshat Rathi, of *Quartz*, opined.

"But then again, no country has the same potential payoff as China. If its bet succeeds, China can look forward to cleaner air, lower reliance on imported oil, and being a technology leader in a new high-tech industry."

To get to pole position, however, China would need to develop and scale up two specific industries—battery production and processing for battery materials. It would also need more metals.

CHAPTER 29

THE DEAL OF THE CENTURY

fter college in Canada, Peter Zhou's passion for numbers—the way injections of money can be used to grow and shape businesses, the complexities and vagaries of the world's financial cycles—led him not back to China but to the Bank of Montreal. Canada, a nation with a long history of extractive industries, was a hub for mining start-ups. "That bank is known for resources, fracking, mining, and oil and gas," he told me when we spoke in 2020. "I got into [mining] according to that wave of M&As, the wave of Chinese companies starting to look overseas for resources." He explained that you could divide Chinese investment in foreign resources into two distinct "waves," as he put it. The first wave, he said, was companies and independent businessmen striking out on their own. This began around 2006 and lasted until around 2012. "That was the first wave. Maybe these investments didn't make money at all. Or got into trouble because they're not used to the local regulatory regimes, or they were caught in the middle of the social problems nearby," he said. "But they became more sophisticated starting from 2012."

A chance assignment at the bank plunged Zhou into the world of mining—or, rather, mine finance, where investors line up to acquire projects. "Before I got the assignment, I didn't know much about mining," he said. He learned quickly, though. When he joined the bank in 2008, "I mean, that's the peak of the mining cycle," he told me. "Especially the Chinese mining companies are waking up to buy resources outside."

The peak would yield the most ambitious Chinese project in Congo. It was midway into Joseph Kabila's first elected term in office, and he and his closest adviser, Augustin Katumba, were looking for deals to help the economy grow. (Katumba traveled to China in 2006 as Congo's itinerant ambassador, for example, with the aim of securing Chinese investment into Congolese oil fields.) The president's policy rested on what he called the Five Pillars of Progress: infrastructure, job creation, education, water and power, and health care.

After the European Union–funded election, Western donors had pledged $4 billion to rebuild Congo in the wake of the devastating war. "For me, the Congo is the China of tomorrow: from now until 2011, the example for me will come from the Asian countries, which we call the 'Dragons,'" Kabila told the Belgian journalist Colette Braeckman. "Congo will surprise."

Western aid was slow to materialize in the aftermath of Kabila's victory, so he turned to the "Dragons" of the East. "Why not try with new friends, without abandoning our traditional friends?" the president's logic went, according to the Congolese scholar Jean Mpisi. Kabila had been to China for his military training in 1998. "There he discovered that China was concentrated on itself to escape under-development in every domain, including the economic one, to the great surprise of the rest of the world," Mpisi wrote. "Joseph Kabila liked this Chinese experience, and he wanted the DRC to inspire itself from it." Congo was groaning under the massive debts that Mobutu Sese Seko had aggregated from both the Eastern and Western Blocs during the Cold War. In one of the more unpleasant transactions of the era, this sovereign debt was sold to private investors who would later take Congo to court to ensure repayment— with interest—against shipments of critical metals like cobalt.

Negotiations between Congo and China proceeded quickly. In September 2007, politicians in Kinshasa announced that they were finalizing a deal with two leviathan Chinese state enterprises—Sinohydro and the China Railway Group Ltd., or CREC. The "Convention of Collaboration," which was signed seven months later, in April 2008, initially provided for a $3.2 billion mining investment and a $6 billion infrastructure investment, which was later scaled back to $3 billion. In return for 8 million tons of copper, 200,000 tons of cobalt, and 372 tons of gold, the Chinese companies, it was announced, would build 12 roads, 3 highways, a railway, 32 hospitals, 145 health centers, and 5,000 units of social housing in Congo. According to a report commissioned by Kabila's successor, only a fraction of the $3 billion was actually spent on infrastructure; in 2023, the Congolese government called for another $17 billion in Chinese investment, commensurate, they said, with the value of the minerals China was extracting.

The mining investment provided for a joint venture between Gécamines, which retained 32 percent of the company, and the two Chinese firms. The mining company that arose from the deal would be christened La Sino-Congolaise des Mines, or Sicomines. Barnabé Kikaya Bin Karubi, one of Kabila's closest advisers, heralded the deal when we spoke in 2025—it was, after all, one of the few ways in which Kabila could deliver on his election promise to rebuild the country. But he said the Congolese leadership made a misstep in not okaying it with the U.S. government. "We did not explain it to the master of the world who is the United States of America: We should have come here and do a diplomatic offensive and explain why we are doing this," Kikaya told me while on a lobbying trip to Washington. "Hence, the American government saw it as us abandoning Western countries and relying on China, and the American ambassador in Kinshasa went berserk, started calling us names and so on." He said that it was unfair to criticize Congo for turning to the Chinese, who after all build infrastructure in European countries without being critiqued. "You cannot punish the Congo, Kabila, for being a visionary, saying that we want to build infrastructure, basic infrastructure, with what we have as an asset, which is our mines of copper and cobalt," he continued. "And when China takes over [producing] whatever is the new technology, microchips and so on, they sell it to America and to every single company that can buy it. So it's a holistic thing."

Surprisingly for such a big deal, one that involved primarily state actors on both the Chinese and Congolese sides, a private company called Huayou Cobalt was also listed as a minority shareholder of Sicomines. And though Sinohydro and CREC had evolved from their roots in hydropower and railway construction, Huayou was the only company that was exclusively focused on the mining, refining, and processing of critical metals.

The firm's inclusion was especially surprising because Huayou, by that time, was well known in Congo as a company that had been involved in some of the worst offenses of the artisanal mining trade. "Huayou itself, the company, is very controversial in China as well," Zhou told me. "There are indeed some companies that are chasing for profit without what I call *moral lines*." He even stressed that he had not been to any mines that Huayou operated or bought from.

Moïse Katumbi was too politic to mention any companies by name when we discussed his governorship of Katanga. "There are some good Chinese companies, which are lumped in with the bad Chinese companies," he told me in 2019. "I closed a lot of Chinese companies that were doing things wrong." But he might as well have been talking about Huayou, or at least its subsidiary, Congo Dongfang International Mining (CDM), which bought from just about anyone in its rush to acquire as much cobalt and copper ore as possible.

At the same time, the world was waking up to the types of things that Odilon Kajumba Kilanga was seeing every day at the artisanal mines in Kolwezi: Congolese miners were using dangerous techniques, and the mines were full of children. One article, in *Bloomberg*, focused on Huayou as one of the worst offenders and suggested that it bought ore from mines with terrible conditions. An official with a UN agency told the journalists that a Huayou subsidiary was involved in "one of the worst forms of child labor." Yang Youngjian, a CDM representative in Likasi, spoke about the horrific conditions in the mines but said he had never been to them. "I've seen pictures," he said. "The conditions aren't so good. They are even working with babies on their backs. They are very grueling conditions." But nobody seemed to be asking why a firm involved in this kind of business was being cut in to one of the world's biggest commodity deals for two of its most critical metals.

HUAYOU WAS FOUNDED IN 2002, ONLY SIX YEARS BEFORE THE "DEAL OF THE century." The man behind the company is a former bean-sprout peddler named Chen Xuehua. According to a company history on the website of Huayou Holding, a company related to Huayou Cobalt, Chen, a well-built man with a squared-off chin, "had to sell bean sprout in the bazaar every day before dawn."

Like Wang Chuanfu and Robin Zeng, Chen was born into rural poverty in China's Zhejiang Province, although his rags-to-riches story doesn't involve success at school and a high-flying university education. He started working as a minor, at fifteen, breeding ducks and long-haired rabbits. He then began his bean sprout business. As the company website explains, "This work lasted for ten years."

In 1993, the factory where Chen was working closed down. He decided to set up his own plant, but he had no savings. The next year, with funds that he had managed to scrape together, Chen created a factory of sorts in a one-thousand-foot bungalow, where he extracted nickel oxide from pharmaceutical catalyst wastes using a few large iron pots. In 2002, with $3.19 million in capital and a partner named Xie Weitong, Chen relocated the firm to an economic development zone that the Chinese state had set up in the city of Tongxiang, in the Yangtze River Delta. The city, with its traditional homes connected by canals, was a place that had been China's "capital of silk" for thousands of years, and one of the trailheads of the Silk Road.

Through the help of entrepreneurs like Wang, Zeng, and Chen, and through dealmaking with countries like Congo and Indonesia, China was building a new Silk Road, only this time the things that moved along the route were futuristic products of all stripes, including the kinds of handheld electronics that were becoming ubiquitous—that is, as long as China could secure the raw materials to make them. "Be nothing or be the first" was one of Chen's mottoes.

Chen decided to focus on cobalt—a minor metal but one that he had a hunch would be key to powering the future. He built a hydrometallurgical processing plant to extract cobalt oxide from the sacks of cobalt hydroxide that were flowing into China's ports.

But where was the cobalt coming from? he asked.

Chen's attention turned to Congo. "The source of raw materials was the top priority, so he set his eyes on the distant African continent," a corporate video explained. Another one of his mantras was this: "The intelligent have no doubts and the brave fear nothing." He jumped straight into the country, even though it was in a state of partial civil war and about to go through a violent election.

In 2003, Chen first ventured to Congo to engage in mining cobalt ore and its processing; by 2006, under the aegis of Huayou, he had established his mining firm, CDM. He knew how to keep the local government happy. In 2013, for example, his firm provided low- and no-interest loans to Gécamines and other Congolese companies, and beginning in 2017, it issued a $4 million loan to Lualaba Province for road rehabilitation, which would be paid back through levies on road traffic. It also gave charitable donations—$10,000 to a school here, desks and textbooks there—but these seemed small compared with the growth of the company. In 2024, Huayou made almost $8.7 billion in revenue.

Chen's was a company that managed to go from a small-time prospector in Africa to a big-time player. CDM, Huayou would later boast, overcame "many difficulties," such as "materials shortage, cultural diversity, [and] financial crisis," but was soon well on its way to its chairman's goal: creating a "complete industrial chain for cobalt and copper mineral resources." As Chen would later say, it was all about control. "If we do not control the resources, Huayou will not be competitive, and it will not develop sustainably in the long run."

At the beginning, CDM's workers slept three to a room in cramped conditions in Congo, but the employees "worked day and night." The subsidiary was soon "a leader in the company's profitability," according to Wang Yuchun, Huayou's deputy general manager. Conditions were harsh, but Wang understood that he was not just a Huayou employee—he was a Chinese pioneer. "Taking the flag over there is a responsibility," he said. "To be bestowed with great responsibility," a company video opined, "one must be crucified with ordeal and tribulation." The chairman led by example, Wang noted: He always left the office later than everyone else. In a few years, Huayou employees were able to live as well as their Belgian forebears: Wang helped build "domestic-garden-style residential areas" for the company's employees in Africa.

Huayou expanded to become a worldwide brand, and Chen set up mines and companies across the supply chain for critical minerals. As the corporate history showed, Wang framed the fight to secure the supply chain in bellicose language, calling it a "battle." "At this moment, let us pay tribute to all those who are struggling in Africa, fighting on the islands of Indonesia, fighting in Korea, Europe, et cetera," he said.

Occasionally, the self-depictions of the workers who appeared in a 2023 oral history of the company felt more like the expressions of a religion or of an extreme political ideology. Another employee admiringly talked about how the goal of the *Huayou-ren*—the "Huayou people," as the company called them—was to "conquer cities and territories." The ideology was based on ideas about strength and conceptions about what power meant in the twenty-first century. "If we continue to become stronger," one of them mused, "we will eventually grow and become stronger."

The company dug nickel mines in Indonesia, built recycling plants in Korea, and even inaugurated a lithium project in Zimbabwe. In China, the firm opened a huge industrial park in the vicinity of the city of Quzhou. Under Chen's leadership, Huayou was not just a corporation; it embodied the very essence of its workers' aspirations. "Huayou has grown from the dream of one man to the dream of many," the corporate video noted. As one Huayou official who worked on the Zimbabwe project put it: "With the dream of lithium in mind, we will act with determination. Success comes only through hard work, and hard work will lead to steady progress." In January 2015, after eight and a half years of preparations, Chen, in a red tie, rang a gong to open the Shanghai Stock Exchange, and Huayou listed its stock, proclaiming to be "a global leader in lithium-ion battery materials."

Chen also grew deeply intertwined with the Chinese government. He became a deputy in the National People's Congress, China's supreme legislative body, and in 2024 gave a speech explaining how Huayou was "vigorously advancing the construction of a modern industrial system" and a "safe, stable, and resilient metal supply chain for energy" based on General Secretary Xi Jinping's speeches and edicts from the government. As the company video stressed: "Huayou's success is inseparable from the efforts of our country, our customers, and our leaders." Chen understood that, to get ahead in China, you had to do what the state told you to do.

His corporate ethos was clear: "God rewards the diligent."

IN KINSHASA, THE MOOD AFTER THE ANNOUNCEMENT OF THE SICOMINES deal was one of elation. It was the largest contract ever between China and Africa. Western governments were not so happy. "The almost unanimous sentiment was, 'The Chinese want to steal *our Congo!*'" Mpisi would later remember. George Forrest, the mining fixer, was even more adamant. "If we allow them to do this," he said about the Chinese, "they will kick us out of Africa. The Westerners are talking about good governance, imposing impossible conditions on their development aid. The Chinese are less scrupulous and are trying to put a spanner in the works." Still, Forrest thought, the resumption of great power rivalries in Africa—a new cold war—might be to Congo's advantage, or at least the advantage of its politicians, who could play one side off another.

The Chinese certainly were less scrupulous. Among the institutions that were concerned about the "deal of the century" was the International Monetary Fund, or IMF. The IMF's officials worried that China might saddle Congo with debt it would find difficult to repay. They also thought the terms of the contract made it so that the repayment of Chinese debt would take priority over the repayment of other debts, which would mean that any debt relief for Congo would essentially be a payment to China. This would be difficult to sell in a world where Beijing and the West were increasingly squaring off. What's more, to satisfy the terms of the deal, Congo would have to take mining permits away from the Kamoto Copper Company and give them to Sicomines. (KCC was the result of the merger between Nikanor, the Dan Gertler–founded company, and Katanga Mining, in which Forrest had been deeply involved.) The Congolese gov-

ernment would owe KCC $285 million if it couldn't find replacement re-
serves of four million tons of copper. "Because of these and other debts
tied to the KCC stake, it is unlikely that Gécamines will collect any revenue from KCC in the short- or medium-term," the Carter Center would
note in 2017. "This, apparently, is the price to pay to access US$165 mil-
lion in China-backed loans per year," the amount flowing out of the Sicomines deal.

Debates followed in Congo's parliament, with the opposition calling the
deal "one-sided" and the government using a phrase that had been hand-
picked from the Chinese playbook: The deal was "win-win." The contract
was renegotiated in 2009. Under the new terms, $3 billion was removed
from the infrastructure loans, and the Congolese government would no
longer have to guarantee the mining segment of the deal. The IMF was
satisfied.

But another problem was the lack of sufficient checks and balances on
where large chunks of the money were going. There were early warning
signs: As the Chinese firms began doling out signature bonuses, Gécamines
paid $23.7 million to Caprice Enterprises Ltd., a firm the anticorruption
NGO Global Witness described as "a previously unheard-of British Virgin
Islands–registered company." The owner of Caprice remains a mystery.

Worse was to come. In 2021, the Sentry, another anticorruption NGO,
"found clear evidence of corruption showing that Chinese corporations
colluded with power players in the DRC to secure access to billions of
dollars' worth of natural resources" using a middleman, Du Wei (also
known as David Wei), to funnel money to Kabila and his entourage at key
moments. At times, Du represented himself as an employee of a Huayou
Cobalt subsidiary or as a major player in Sicomines. The Sentry investigators wrote that "the shell company at the center of the scheme—Congo
Construction Company (CCC)—received $55 million from foreign
sources apparently intended for Kabila and his entourage."

Congolese were starting to wake up to the fact that their country had
been bargained away from under their feet. As the Lubumbashi jurist
Marcel Yabili put it: "The 'contract of the century' was actually the blun-
der of the century."

IT WAS PRECISELY THE TYPE OF ATMOSPHERE IN WHICH KATUMBA THRIVED. Described by a U.S. diplomat as "the power behind the throne" in a leaked 2009 cable, Katumba had created the shadow cash chute that enabled Kabila and other members of his government to prosper from Congo's mining wealth. The Sicomines deal represented a ratcheting up of the corruption machine.

Katumba visited China in 2007 and found the customs and protocol overwhelming. In China, he wrote, "the dimensions, the manners, criteria, procedures and absences of procedures, everything is so different. We had to understand the new universe and regain agility and flexibility necessary to adapt to it." But he managed to adapt quickly enough. "The details of the deal are known," Katumba continued. "Its use for funding the President's strategic constructions and its multiple advantages compared to those, more theoretic than effective, of the West, are also known."

The shadow minister didn't get much of a chance to enjoy the proceeds of his wealth. In February 2012, he was aboard Katumbi's Gulfstream when it slid off a runway while landing in the eastern Congolese city of Bukavu. According to an official investigation, Katumba had removed his safety belt too early. He was killed after being catapulted out of his seat into a bulkhead.

Conspiracies erupted: There were several other politicians on the plane, but none of them died, although several were injured. "They say he didn't put the belt on; I can't comment on this," Katumbi told me, noting that the pilot, who was a U.S. citizen, also perished. Melissa Sanderson, who at the time had transitioned from working at the U.S. embassy to working for a Congo-based mining company, also dismissed the rumors. "Augustin Katumba was not murdered by anyone," she said. The airport had recently been renovated by a Chinese firm. "The material that the Chinese used in rebuilding the runway was not adequate to the task." Ironically, Katumba had just championed Chinese infrastructure construction during the Sicomines deal.

One of the architects of Congo's mining state, and one of its key political dealmakers, was now gone. But the system he had implemented remained in place. And now the Chinese were running the show in one of the world's most important repositories of critical minerals. The U.S.—and, to a large extent, Europe—was nowhere to be seen.

Sanderson attributed this state of affairs to a lack of U.S. leadership. She told me that the U.S. had been "stupid" to hand manufacturing over to China. "Twenty-five years ago, we did not look over a horizon to say, 'China's going to become an economic power whose interests are not aligned with ours,'" she said. Greed, and a lust for cheaper goods, had blinded Washington.

In 2023, when we spoke, Sanderson said that Washington continued to be blinkered to Chinese goals in Africa, specifically in Congo. Moreover, despite the "strong pro-U.S. bias" that she believed many people in Congo felt, she did not believe that Washington had ever possessed the foresight to engage positively with the country. It was a case of heads buried in sand. "The stupidity with DRC is thinking that somehow, somebody else is going to stabilize it and make everything fine," she told me.

And, increasingly, that "somebody else" could only mean China.

Katumba's death ushered in an era of even stiffer competition as deal-makers struggled to occupy the (well-remunerated) role of arbiter of Congo's mineral riches. Gertler appeared to grieve the loss of a close friend, though a text message from Moises Gertner to another associate asking about Katumba's death paints a very different picture. "Our friend is happy he is going to save paying a lot of money," Moises Gertner wrote. "Widow he'll giv(e) a few dollars only."

The day before Katumba's funeral, an employee at Och-Ziff, the hedge fund that had invested in Katangese mining, received a text from the agent whose identity corresponds to Gertler's. "I'm fine . . . sad but fine," the agent wrote. "I will have to help [DRC Official 1] much more now . . . tomorrow the funeral will take place." From the evidence provided, it is more than likely that DRC Official 1 was the president, Joseph Kabila Kabange.

Part 4

CONSOLIDATION

"You have all probably heard Mark Twain's definition of
a mine as 'a hole in the ground owned by a liar,' but a
fairer definition, I believe, is 'a hole in the ground sold
by a lying promoter to a stupid investor.'"

—JOHN HAYS HAMMOND, 1911

CHAPTER 30

NO SUCH THING AS DEATH

On the cusp of a mine site called Kilimabunga, on the outskirts of Kolwezi, André, a thirteen-year-old child, was giving out bags of food to a group of children. It was February 2022. In the background, men and boys were descending into deep holes that had been dug into a hill of red earth denuded by miners' shovels; they exited lugging sacks of ore. André was smaller and skinnier than some of the other children, but he had a wide smile and a confident strut. Adults jokingly called him *Monsieur* André, *le Patron des Patrons*, the Boss of Bosses.

The children were crouched on sacks full of copper-rich ore. They were laughing and licking their chops: The food, they said, was very good. They began praising André and saying, "This is so good, there is no such thing as death."

It was a common idiom in Kiswahili, but it was also a piece of bravado. For the children knew death, and they knew death existed in the depths of the hill behind them. Yes, it was the dry season, when the earth held better than when it rained, but they knew that with every descent into the mines, the hand-dug shafts they were plunging down into could collapse, crushing them and trapping them deep inside Congo's earth.

Children had worked in Congo's mines for a long time. In the country, a culture in which children often helped out, taking on agricultural and commercial tasks, had collided with the system of artisanal mining in the crushing poverty of the late Mobutu years. When Laurent-Désiré Kabila had encouraged artisanal mining and his son Joseph Kabila Kabange had enshrined the practice into the mining code, even more children had gone to work in the mines.

André was unusual among the children working at Kilimabunga. A month earlier, his father had tasked him with making money for the family, giving him some cash to go to the mine and trade minerals. "I would stay at the base of the hill," he told me when we spoke a few months later. "I waited for the children so I could buy the minerals." He spoke quietly but with an unchildlike confidence. "I developed a network thanks to my family."

By buying and selling the minerals, André had risen a step above most of his peers in the supply chain, and even women, boys, and older men would sell to him. "Yes, I had benefits from my job," he told me. "But the conditions for some children were not good."

As the children at Kilimabunga were laughing and making their defiant declarations about death, a man in his thirties wearing a black bomber jacket approached. Vital Kamungu was the program director of a school that sought to take children out of the mines and provide them with an education. He had an offer for them: They could go to school, be fed breakfast and lunch, and take food home for their parents. "There is food, there is water, there is electricity, there is everything you need," he said. Many decided to stay, but André considered his options: He decided to take his chances at Kamungu's school.

A FEW YEARS BEFORE, KAMUNGU HAD BEGUN WORKING WITH A CATHOLIC charity called Bon Pasteur, or Good Shepherd, which operated a two-story school in a mining area of Kolwezi; now he headed the school for Still I Rise, an Italian NGO that ran a school called Pamoja, which means "together" in Kiswahili. He was the program manager there, and his job included traveling to mine sites, where he would try to persuade children and their families to come and study.

Still I Rise had begun as an emergency school on the Greek island of

Samos during the mid-2010s refugee crisis in Europe. The organization's thesis was that much of the education provided to refugee and migrant children by relief organizations was rooted in religious institutions, and Still I Rise tried to provide a holistic education that emphasized a child's mental health.

I had first witnessed its work in Samos back in 2018. During the COVID-19 pandemic, I volunteered at Still I Rise in Greece, which was suffering from a shortage of teachers while travel bans were in place. I taught writing to a smart, funny group of teenagers, all of whom were excited for a better life but stuck in the unintelligible limbo of the European refugee system. Some of them had been on the island for more than a year. We created a newspaper at the end of one class, and a boy from Afghanistan wrote a playful observational piece about the Chinese trader who had set up a hardware store next to the school. Another of my students was Congolese and a keen writer of poetry. "I am a Black Man, an African and a Refugee," he wrote. "Give me a chance."

Both times I visited, I was surprised to see many Congolese people in the camp on Samos. A good number of them were refugees, fleeing violence, but some were not from war-torn parts of the country: They told me that they had come to Europe because there were no opportunities for them back home—a sad state of affairs, considering the vast wealth Congo sits atop.

Still I Rise started opening schools in other countries: Syria, Kenya, and Yemen among them. By 2021, the program raised money to rent a low-slung building in Kolwezi for around $1,200 a month. The school would be both a place of learning and somewhere to imagine a better life. The nonprofit began training teachers. Pamoja's stated aim was to provide "high quality education, targeted child protection actions, and holistic nutrition, psychosocial and health support programs to the child and the whole family." In such a poor region, parents often expect their children to supplement the family's income, even if the work is dangerous.

WHEN I VISITED PAMOJA IN MARCH 2022, IT SMELLED OF DETERGENT AND fresh paint. André and his cohort—the school's first class—had been there less than a month. The school would count 120 students at full capacity, and it had an entirely local staff. The building had three classrooms, a refectory, and a psychosocial support office. Outside, a gazebo was being erected so that children could take classes in the open air. I sat in on a French class and watched without understanding as children played a game where they had to clip and unclip pegs from a clothesline. At recess, Kamungu asked if there were any pupils who might be interested in speaking with me.* Six students volunteered, and we sat in Pamoja's breezy main hall as they told me their stories.

Many of them had similar tales: stories of families barely scraping together enough money to feed them, let alone send them to school. The government had promised that schools would be free for all children, but in reality, money was siphoned off in Kinshasa and teachers remained unpaid, so they levied fees from their students' families or refused to teach. Most of the students told me they often didn't eat at home. "There were days that we ate and days that we had nothing to eat," Julie, an eleven-year-old with tight braids, told me.

"I borrowed things. I bought things. I went and took out loans," Norbert, a ten-year-old, said. One day, his mother took him to a mine site. "My mother told me to hold on to the black ore, and to throw away the white stones. The metals were in the black stones. And other stones—white, brown—I would throw them away." Cobalt? "Exactly, cobalt."

Norbert had "always wanted to study," but he went to the mine because he knew that his family would not otherwise be able to afford to clothe him. "My mother bought clothes for me when I worked with her. After, we would sell the ore. She paid me with clothes."

They talked in cold detail about their time in the mines. Jamila, a twelve-year-old, told me that when her mother died, she and her siblings did not have money to pay for school. "The school kicked us out when we did not pay, and our grandmother had to come and collect us. She said we had to go and work," Jamila said, so she and her siblings went to a pit

---

*I checked before I arrived that the school would confirm with the students' parents or guardians if it was okay to interview their children. I also spoke with the psychosocial support counselor to check whether any child had an elevated risk of being retraumatized by my questions, and whether I should avoid particular topics.

where they spent their days hauling bags of rocks and sorting them one from another.

The army and police had been sent into some sites to secure them, and to make sure that children and pregnant women weren't working there, but they had not done much. "Where I worked, there were soldiers before, and then a security company that replaced them," said Dieumerci, a thirteen-year-old in a crisp new Pamoja uniform. He had worked in the mines for a year before Kamungu brought him to the school. "Anyway," he said, "if anyone found a child working there, you would just pay, and they left you alone." He told me that he would make around ten thousand Congolese francs (about $5 in U.S. dollars at the time) a day, and a quarter of that would go to paying off mine guards.

It would have been highly illegal for people to buy products from children this young; miners were technically supposed to belong to a cooperative and be trained by a standards body called SAEMAPE. Almost three years before, the governor of Lualaba Province, home to the capital city of Kolwezi, claimed that there were barely any children in the mines, and that there would soon be none. On the way into town, he had established a buying center called Musompo, ostensibly so as to better control the use of child labor and ensure that ore was coming from registered cooperatives.

The children told me a very different story: They sold to Congolese buyers, and some said they sold to Chinese buyers, no questions asked. Others said they sold ore directly to purchasers at the mines. Two of them told me they would help other children or family members bring the ore to Musompo, the marketplace that the governor had championed, where children were absolutely not supposed to sell minerals. Musompo still only existed as an informal marketplace by the middle of the 2020s, hardly the modern trading spot the governor had proclaimed more than a half decade earlier. Trafigura, the Swiss commodities firm, had been brought in to centralize cobalt trading, and a new governor had made much to-do about the project, claiming it as her own. But a geologist who has covered the region for years told me that she is privately unhappy with the financial arrangements and benefits that accrue to her from Musompo; she apparently also has a bad relationship with the national mining ministry.

Norbert Kalenga, a thirtysomething trader I had met in Kolwezi, explained that his project to build an equitable cobalt trading cooperative had been frustrated. He vented over text in April 2025, "Until now [Musompo] does not exist!!!"

Despite the promises that children would be removed from the mines, all the children I spoke to told me that they had seen other kids working beside them. "I dug with other children at a site called Tshimbala," Dieumerci said. "I felt tired, and I often had bruises and scratches on my body." He had seen other children suffer much worse, he told me. "I saw other children wounded," he said. "I saw other children in a terrible condition."

NOT EVERYONE HAD SUFFERED FROM ARTISANAL MINING. AFTER HE SEIZED power in 1997, M'zee Laurent-Désiré Kabila had permitted the practice to create, he often said, a new class of wealthier Congolese. Norbert Nawiji was one of them; as a young man, in the 1980s, he had studied pedagogy and nursing, but Mobutu Sese Seko's Zaire was crumbling. "This story started suddenly, with the war and the end of Mobutu," Nawiji said. "People couldn't afford to send their children to school, to find things to eat." By the time Kabila came to power, Nawiji was working at mines near Kolwezi. After a few years, he began work as a *négociant*, buying and selling ore. "The middle class was forming itself," he said.

Nawiji realized that he could make even more by establishing his own *dépôt*, a fixed trading house where he could set prices for ore. He applied to the government to put up shop, showing identity cards and commercial documents; finally, he received his *carte de négociant*, which allowed him to create his *dépôt* in the Musompo trading area. When we met in Kolwezi in September 2019, a few days before his fifty-seventh birthday, Nawiji wore a dark zip-up jacket. His two smartphones kept ringing—a sign, in Congo, that he was a successful man.

By the early 2010s, Nawiji explained, the industry was becoming more regulated for official traders like him who wanted to, he said, follow the letter of the law. "People who understood the industry saw that there was a danger to this artisanal work," he told me. He was against children working in mines and thought it a good thing that the law was catching up. But the change in the law, he said, was being used by politicians to sell off mine sites and declare that only cooperatives linked to them could work in specific areas.

The amendment didn't address all the people who were already working in artisanal mining, though. "The problems started when people started to leave the mines that they had known well, because of certain government provisions," Nawiji remembered. "There began to be deaths, and people began to get wounded. There are people who have been handicapped in every way imaginable."

More and more miners were pushed into working illegally, and *creuseurs* like Odilon Kajumba Kilanga had to bounce from site to site, paying off soldiers and policemen who had been sent to enforce the new regulations. In some cases, entirely new, completely unregulated mines sprung up. *Creuseurs* became "geologists all by themselves," Nawiji said. "If there were minerals there, they exploited them."

In theory, the system laid out by the mining code, with its cooperatives and training bodies, could make artisanal mining into a sustainable career. But in practice, it had become a form of bonded labor for many people. Even if artisanal mining supports poor families in the region, it's hard to applaud it in its current state in Congo. The lives of many *creuseurs* are cut short and marked by suffering. Many are not lucky enough to clamber out of poverty like Nawiji. They are instead left with physical and psychological injuries from mine collapses and violent confrontations with the police and the army.

TO SOME, NO TYPES OF ARTISANAL MINING ARE ACCEPTABLE. THEY POINT out that children who work in the mines are often drugged to suppress fear and hunger. When we spoke in 2019, Sister Catherine Mutindi, the founder of Good Shepherd Kolwezi, told me: "If the kids don't make enough money, they have no food for the whole day. Some children we interviewed did not remember the last time they had a meal."

The dangers for children do not stop there. Mutindi inveighed against one superstition that holds that sex with a virgin girl will enhance one's luck in the mines. It is no rarity for children to die while being raped.

While I was in Kolwezi, Mutindi showed me photographs of the bruised corpse of an eight-year-old girl who had been abducted and raped by a *creuseur* the previous week. (The miner was later apprehended; she sent me a video of him in prison.) In one case, Mutindi said, she saw the body of an eighteen-month-old infant who had been raped by a *creuseur*. In a place like Kolwezi, with its high levels of corruption, people could get away with such murders if they were able to pay off the police.

In 2019, at a school run by Good Shepherd, I met Ziki, a serious boy with big dark eyes. He was fifteen, but because he had been malnourished for long periods, he looked much younger. His parents had been killed in a roadside accident when he was three; afterward, he'd been sent to live with his father's sister. "My aunt sent her kids to school but sent me to the mines," he said. For almost a decade, he lived as a drifter, roving between mines with a group of other children. Soldiers beat them frequently if they could not pay bribes. "I was full of bitterness," he said.

Things had become worse in the 2010s, but the world was also starting to prick up its ears and catch whispers of what was happening in Katanga. Ever since Amnesty International and AFREWATCH released a landmark report on artisanal mining in 2016, the issue of child miners in Congo has become better understood and deplored internationally, but conditions have barely changed on the ground. In fact, in times of higher cobalt prices, more children have been sent to the mines. When the price of cobalt drops off, copper prices allow artisanal miners to continue selling the ore—just at a lower price. In the wake of the report, some companies that have been known to use cobalt in their batteries, such as Apple, said they would stop buying from firms with artisanal miners in their supply chains, firms like Huayou. For the time being, at least.

CHAPTER 31

KASULO

By 2014, Odilon Kajumba Kilanga had been in Kolwezi for about six years, bouncing between mine sites with the two other men in his three-man *creuseur* crew: Yannick and Trésor Mputu. That summer, they heard a story that gave them hope they would soon be rich. The tale began in Kolwezi that June, when a man began to dig into the soft red earth in the backyard of the home he was renting in the suburb of Kasulo. By some accounts, this man was a policeman; by others, a Kasaian migrant who had made Kasulo his home.

As the man later told neighbors, he had intended to create a pit for a new toilet. About eight feet into the soil, his shovel struck a slab of gray rock that was streaked with black and punctuated with what looked like blobs of bright-turquoise mold. He had struck a seam of heterogenite—cobalt ore.

The man took some samples to a mineral trader, someone like Nawiji, although various people also told me this trader was Chinese. At the time, Musompo, the mineral market outside Kolwezi, had not yet been completed, and the road into the city was lined with corrugated-iron shacks, known as *comptoirs*, where minerals were bought and sold.

One trader told the man that the cobalt ore he'd unearthed was unusually pure. The man returned to Kasulo determined to keep his find a secret. He stopped digging in his yard. Instead, he cut through the floor of his house, which he was renting, and dug to about thirty feet, until he hit pay dirt. He carted out the ore at night. "All of us, at that time, we knew nothing," said Zanga Muteba, a baker who then lived in Kasulo. But one evening, Muteba and some neighbors heard telltale clanging noises coming from the man's house. Rushing inside, they discovered that the man had carved out a series of underground galleries, following the vein of copper and cobalt ore as it meandered under his neighbors' houses.

When the man's landlord caught wind of these modifications, the two got into an argument, and the man fled. "He had already made a lot of money," Muteba told me when we met in the town of Samukinda in 2019. Judging from the amount of ore the man had dug out, his neighbors estimated he had probably made more than $10,000—in Congo, a small fortune.

The story, which had the dimensions of a fable, was retold in Kolwezi's smoky bars, in its dimly lit living rooms, and around the mouths of the deep, dark pits where *creuseurs* looked for copper and cobalt. Like many other miners who heard the tale, Kajumba and his team rushed to mine in Kasulo. Hundreds of people there "began digging in their own plots," Muteba said. Many of Kasulo's ten thousand residents were day laborers, and they figured they had a shot at making a fortune in their backyards. Murray Hitzman, the former U.S. Geological Survey scientist who spent more than a decade traveling to southern Congo to consult on mining projects there, saw Kasulo in the early days. He told me that residents were "milling about all the time," hoping for word of fresh finds.

The government quickly took notice and tried to stop the digging. "You're going to destroy the neighborhood!" the mayor warned.

But, Muteba said, "it was complicated for people to accept the mayor's request." Muteba had a thriving bakery and didn't have time to dig, but many people ignored the mayor's order: The prospect of a personal cobalt mine was worth any risk.

About a month after the man who discovered the cobalt up and vanished, the local municipality formally restricted digging for minerals in Kasulo. According to Muteba, residents implored the mayor: "We used to mine in the bush, in the forest. You stopped us. You gave all the city to big industrial companies. Now we discovered minerals in our own plots of land, which belonged to our ancestors. And now you want to stop us? No, that is not going to work." As Muteba recalled, "People started to throw rocks at the mayor, and the mayor ran away. And when the mayor fled, the digging *really* started."

After Kasulo's mayor took off, many residents began tearing into the ground beneath them. Some wealthier locals hired *creuseurs* to dig under their houses, with an agreement to split any profits. Two teams of *creuseurs* could each work twelve-hour shifts, chipping away at the rock with hammers and chisels. A pastor and his congregation began digging under their church, stopping only for Sunday services.

By the end of 2014, two thousand *creuseurs* were working in the neighborhood, with little regulation. Kajumba and his cooperative soon joined in the hunt for minerals. Yannick remembered this period as "the good times." He told me, "There was a lot of money, and everybody was able to make some. The minerals were close to the surface, and they could be mined without digging deep holes."

But the conditions quickly became dangerous. Not long after the mayor formally prohibited excavating for minerals, a mine shaft collapsed, killing five miners. Still, people kept digging, and by the time researchers for Amnesty International visited, less than a year after the discovery of cobalt in Kasulo, some of the holes made by *creuseurs* were one hundred feet deep. Once diggers reached seams of ore, they followed the mineral through the soil, often without building supports for their tunnels. As Hitzman pointed out, the heterogenite closer to the surface—the only type of ore people with hand tools could really expect to access—often contains the least amount of cobalt, as it has been subjected to weathering. *Creuseurs* in Kasulo were risking their lives to obtain some of the worst ore.

One of Kajumba's teammates told me that their cooperative of six—during this period, he, Yannick, and Trésor linked up with three other *creuseurs*—used to regularly extract two tons of raw material from a single pit in Kasulo. But most of the best sites were quickly excavated, and the yield from newer pits was less than half as much. The team was also ripped off by unscrupulous traders and corrupt officials. When I first met Kajumba, in 2019, he lamented that he was struggling to pay his rent of twenty-five dollars a month. "Whenever we dig up a few tons," he said, "I send some money to my family."

VIDEOS OF KASULO TAKEN DURING THE HEIGHT OF THE 2014 COBALT RUSH show orange tarpaulins covering fresh pits and bags of minerals littering the streets. After Michael Kavanagh, the journalist, visited the district a year later, he published an article in *The New York Times* observing that the profusion of holes made it look "as if it had been bombed." At one point, *creuseurs* tunneled beneath the main road running west to Angola, and the road subsequently collapsed.

Kajumba and his crew were part of this initial frenzy. They knew that picking at the rock beneath Kasulo's sandy soil was treacherous, especially during the rainy season, but they were just happy not to be risking arrest, as they did when they broke into the big mines. One day in December 2014, Kajumba and other *creuseurs* were working a pit at Kasulo when they felt a rumble. "It was as if something was falling deep underneath us," Kajumba recalled.

They knew that on the previous day, a group of *creuseurs* working in a neighboring hole had asked a local chief to perform a ritual over a new area where they had been digging. The *creuseurs*—who, after all, lived their days in fear of being buried alive—were superstitious. It was common to employ a magic practitioner, known as a *féticheur,* to bless mining work.

In Kasulo, the *féticheur* who had performed the ritual over the neighboring pit had warned the miners not to enter it for three days. If they went in before that, he'd said, they would anger a slumbering dragon that lived at the bottom. Once the three days were up, they'd been told, the pit would be safe—and full of minerals. Rumors of the pit's riches spread, and a day later, some miners decided to disobey the *féticheur.* "*Creuseurs* have curiosity," Yannick Mputu said. "They wanted to see what was down there."

After Kajumba and Mputu felt the ground shudder, they rushed to the neighboring hole. Part of the tunnel had caved in, trapping their neighbors deep below. Some fifty people vaulted into the darkness, desperate to save their friends. Rescuers suffocated in the subterranean passages. Eleven of the trapped miners died, as did four rescuers.

Following another series of *féticheur* rituals, and another period of waiting, all the bodies were pulled from the hole. Some were horrifically burned. "The last person who escaped from the pit said that he saw a huge flame," Mputu told me. The fire's origin was unclear, but artisanal miners can unearth pockets of flammable gas. To Mputu and his colleagues, the accident had supernatural trappings. "The cause of the flame was none other than the dragon," he told me.

Nine months after the cave-in, on September 4, 2015, another group of *creuseurs* in Kasulo burned a tire in an underground gallery to crack open a stubborn rock face. Five people asphyxiated from the fumes; thirteen others were hospitalized. After the incident, Radio Okapi, a media group sponsored by the UN, interviewed Kolwezi's mayor, who said that a year earlier, he had sent a report to his superiors urging the closure of the artisanal pits. According to Radio Okapi, the mayor "expressed regret that no site was closed because of this request." The report noted that more than a thousand holes had already been dug in Kasulo.

Kajumba and his collective regularly snuck into open pits owned by companies like Glencore and the Eurasian Resources Group. "We enter at night, we work, and leave early in the morning," Yannick Mputu told me. He noted that *creuseurs* set something aside for the soldiers and the police, who supposedly prohibited outsiders from entering: "We give them a percentage of our earnings, and they let us in."

IN 2015, AROUND THE TIME OF THE SECOND CAVE-IN, JOSEPH KABILA KAbange decided to reorganize Congo's provinces. He had fallen out with Moïse Katumbi Chapwe, who had wanted to run for president the following year; Kabila had given no indication that he was going to step aside, and by most accounts, he saw the wealthy Katangese political aspirant as a threat to his presidency. When Kabila asked Congo's governors to back a constitutional change that would allow him to stay for a third term, Katumbi stepped down. "President Kabila wanted to have a third mandate," Katumbi told me. "What I paid is the price for this."

Kolwezi became the new capital of a region called Lualaba, named for a tributary of the Congo River that flows through the region. The first governor of Lualaba, Richard Muyej Mangez Mans, promoted himself as "Papa Solution," a Congolese Mr. Fix-It. In 2019, when I first visited Kolwezi, many bus-stop benches were painted with this nickname. In an interview with the magazine *Mining and Business*, Muyej spoke critically of the cobalt "contagion" in Kasulo. Around fifteen thousand *creuseurs* were working on the site. "A plan is needed to avoid hasty movements that could turn into a humanitarian tragedy," he said. "We have made a project proposal that we will submit to the authorities."

The proposal's particulars, which Muyej didn't disclose at the time, involved granting Kasulo's mineral rights to a foreign company: Congo Dongfang International Mining, the subsidiary of Huayou Cobalt, Chen Xuehua's Chinese conglomerate. By that point, Huayou had begun supplying materials for iPhone batteries. In 2015, it had built two cobalt refineries. According to an internal presentation, by 2017, Huayou controlled 21 percent of the global cobalt market.

CHAPTER 32

BARE BRANCHES

It was late on an evening in July 2022, and I was waiting for a call from a man named Antoine Mutumba, a Congolese copper and cobalt trader with whom Odilon Kajumba Kilanga had worked in the past. We had met just before the rains in 2019, the same afternoon I had first met Kajumba and the Mputu brothers over colas. In fact, Mutumba had introduced me to Kajumba, and now he wanted to bring me to meet his sponsor, a Chinese businessman who called himself by a name that recalled the ancestral Luba king, Kalala.

Kalala was among a legion of small-time Chinese traders who had come to Congo and profited after the twin system of industrial and artisanal mining crystallized in the wake of the 2002 Mining Code. Andrew Bell, a mining prospector who first went to Katanga in 2016, remembers being struck by the sheer amount of trading houses he saw in the 2010s. "Up until 2019, if you went along that road from Lubumbashi to Kolwezi, there were lots of little huts with Chinese names on them," Bell told me. "That was Chinese middlemen who were buying from the artisanal miners. And so they basically sent containers full of dust off to China." I saw them as well, before the governor of Katanga decided to centralize them at the Musompo market.

I had spent several years trying to meet someone like Kalala in Congo.

Mutumba had always promised he would put me in contact with the Chinese traders for whom he performed odd jobs: He started out as a car mechanic and had one day fixed a car for a Chinese businessman named Zhou. "Zhou would call him every time he needed a vehicle repairing," Jeef Kazadi, my guide and translator, who knew Mutumba from his trips to Kolwezi, told me. "This is how he came into the world of the *Chinois*. That's how they knew Antoine could work for them."

The businessman took a shine to Mutumba and employed him to sort out various issues around town—disputes over prices, deliveries, hiring. Soon, he introduced him to an associate, a copper and cobalt trader who went by the sobriquet Kalala and had a taste for liquor. Mutumba also liked a drink, and he and Kalala became close. "There is corruption, there is lack of the rule of law, which gives you more autonomy to be entrepreneurial," Peter Zhou, the Chinese mining banker, said of the way that some Chinese think about working in Congo. (Zhou is not related to the Zhou with whom Antoine originally worked.)

The trader welcomed the young Congolese into his compound, a set of bungalows encircled by a low white wall and guarded by dirty, mean-looking dogs. Mutumba became Kalala's factotum, always ready to fix a vehicle, a deal, or a broken relationship.

At the time, Mutumba told me that the Chinese would bring him wealth and prestige. When I interviewed Moïse Katumbi Chapwe, the former governor of Katanga, in Zambia in 2019, he reminisced about his time in office, and about how he had cracked down on illegal traders like Kalala. "I was the only one who went to chase people from the mines who were stealing," he told me. "Behind every thief there was a Chinese. And behind every Chinese, there was a politician."

Mutumba seemed determined to prove the adage: Behind him were Chinese traders, and behind them were powerful political figures. When I asked to meet the traders, Mutumba demurred, making excuses before finally telling me that it was not the right time. Kalala was typical of the Chinese traders who worked with Congolese artisanal miners rather than the large industrial ones—they were secretive and seemed to be striking out alone.

MANY CONGOLESE RESENT CHINESE SMALL-TIMERS LIKE KALALA WHO WORK in Congo's South. They call them the *petits Chinois,* or little Chinese. Racism against them certainly abounds in areas where they have been perceived to be successful. Sometimes, the small-timers will be targeted, their shops ransacked and their homes burned. "Listen, you know, the big companies working in DRC, most of them are not the real problem for artisanal miners," Vidiye Tshimanga Tshipanda, a close adviser to the Congolese president, told me in mid-2022. "The biggest problem for artisanal miners is the thousands of little Chinese and Indians and Lebanese, these companies who are working in the DRC illegally and ruining the cobalt market. The way they work is that they are lending some money to the artisanal people, to the cooperatives. Then they buy the copper and the cobalt, cheating about everything—the purity, the quantity, the humidity, everything."

I had often heard the same complaints from miners like Kajumba. I was told, for instance, that the Chinese had changed the calibration on their Metorex fluorescence analyzers—chunky, gun-shaped machines that indicate how pure a piece of cobalt ore is. "So they don't pay a fair price. They don't pay the right price," Tshimanga told me. "But, also, they don't care if it's a child who is in the mine. They don't care if it's a woman, pregnant woman. They don't care about security. They don't care about anything. And for a little piece of money that they can give the population, the miners, the artisanal miners, are fighting their battle to find the maximum of product."

Tshimanga bemoaned the greed of foreign traders, people like Kalala, but then told me something about the importance of small-scale traders to the global economy. "We can't let that continue, especially when you see the interest that represents," Tshimanga said, speaking of a government plan to centralize cobalt buying and bypass small traders entirely—a plan that, incidentally, resembled one proposed by Richard Muyej Mangez Mans, Lualaba's governor, in 2018. "When today you are talking about twenty percent of the Congolese production is artisanal production.

When you say twenty-plus percent with the actual statistics, that means that you can double it, because officially, they never give the right numbers." He continued: "We can say that thirty to forty percent of the production coming out from DRC is artisanal products. That means that it [the production of artisanally mined cobalt] represents the biggest cobalt company of the world."

ONE NIGHT IN OCTOBER 2019, FUELED BY CURIOSITY AND A COUPLE OF beers, Mutumba suggested we visit a Chinese-run casino with his wife. Kolwezi had several of these establishments, the facades of which were generally adorned with dancing lights. Inside, the place was sparse, its beige walls undecorated and thick with the fug of cigarette smoke. At a metal grille, a grim Chinese man dealt me chips. I tried to strike up a conversation, but no dice. Mutumba appeared to be enjoying himself, even though the Congolese with me were stopped at the door and forced to wait by the bar; they were not allowed to gamble, because, the casino's Chinese staff stated, Black Africans could not be trusted with money. The sentiment was shared by a host of drunk South Africans at the roulette table—the Chinese stuck to blackjack—who addressed a Congolese croupier as "black man."

No one really knows how many Chinese live in Congo; there are no censuses, and many people overstay their visas to work illegally. Some estimates I heard in the South ran as high as one hundred thousand. Fly into Lubumbashi on one of Ethiopian Airlines' daily flights and you will be joined by hundreds of Chinese workers. In an essay from 2018, the Congolese political scientist Germain Ngoie Tshibambe pointed out that for Chinese, the country's South was a place where they imagined they could do "prosperous business."

Though there are no formal "Chinatowns" in Congo's South, there are what the French scholar Emmanuel Ma Mung calls "a parade of Chinese signs" over new restaurants, shops, and medical centers in the region. Few Congolese use these businesses; they are generally relatively expensive and, as locals told me, they don't have a taste for Chinese cuisine. They mainly cater to Chinese workers, who stay in Chinese-run hotels and eat at Chinese restaurants. The Chinese mostly interact with the Congolese in two ways—first, as Ngoie has noted, Chinese clinics have become popular among locals for their cheap treatments, and second, of course, they interact on mine sites, although even this contact is sometimes limited. In 2011, Jean Jolly, a French journalist, reported that one of CDM's directors of external relations had not even visited the mine site that he represented, though it was only four kilometers away from him.

Outside work, most Chinese avoid even the modest mingling with locals that some other expats attempt. "There are cultural differences where, you know, local Congolese tend to gather together, associate together with their own group of people; the Chinese gather together because of language barrier, or culture," Peter Zhou, the Chinese mine financier, told me. Very few of these Chinese residents know French or Kiswahili, the most spoken languages in Congo's South.

Some of the Congolese I talked to spoke of racism directed at them by Chinese workers; Kevin Kumbwa, a Congolese translator who speaks Mandarin, told me that "Chinese people are coming here for business to make money, you get me, so they can never be our friends." He would overhear his Chinese employers say of the Congolese, "These people, they don't really *think*."

Ady Nawezi, the former minister for youth in Lualaba Province, told me that he used to rent a property to Chinese businessmen. Whenever he came to check on them, he noticed that they had a room filled with piles of ready money. "When there was a problem, they would fix it with cash," he told me.

At some mine sites, the punishments visited upon Congolese workers by their Chinese bosses are reminiscent of the colonial days. In a widely shared video, a guard with a Kalashnikov slung across his back is seen beating a man lying seminaked in the mud, his arms bound at his back. Behind the camera, a man speaks Mandarin and then yells the word for *beat* in Kiswahili: "Piga, piga!" In the shot, eight heavy trucks of the type used to transport cobalt ore idle in the background.

Along with the constant complaint I heard from *creuseurs* around Kolwezi—that the Chinese were cheating them on ore prices—people would say that because labor and safety standards were poor in many parts of China, the Chinese who had arrived in Africa were simply acting in ways they were used to at home. "If they work without shoes there, how can they be expected to give us shoes to work here?" goes a common refrain among locals. One Western mining official I spoke to told me of a mine site he had visited that called to mind an internment camp. "The Chinese were barefoot, they were digging with shovels, and they couldn't leave," he said, explaining to me that the mine he had visited was owned by a small Chinese company, not a big multinational. "It's sad to me, because the Chinese are mining in the way Westerners were mining in the '40s and '50s—cowboy mining."

IN MARCH 2022, I HEARD THAT MUTUMBA WAS NOW THE OWNER OF COMpagnie Antoine—an artisanal mining *comptoir*, one of the trading houses that bought hand-mined cobalt—and also ran an artisanal mining collective. I suspected he might have Chinese backers, and I wanted to speak to him about them.

He invited me to talk at a watering hole, a dimly lit sort of place, green with the glow of a beer ad above the bar. The men—for the other patrons were only male—were all smoking, watching a Mexican telenovela that had been translated into French. Mutumba was leaning low over a pool table. A cigarette hung from his lips, and he muttered for me to wait. After loudly winning the game, he strolled over to me and knocked another cheap stick out of a half-empty soft pack. The Mutumba before me had changed—his beard had filled out, and his cheeks were fuller. He was now, he declared, rich.

Tobacco ash kept falling in fiery chunks from the cherry of Mutumba's cigarette onto his fashionably ripped jeans. "You've come here for business," he said, slurring his words. "I am now very strong in business." He had just that week received a permit, he said, brandishing a piece of paper signed by Fifi Masuka Saini, the acting governor of Lualaba Province. "I have Fifi behind me, and my Chinese," he said.

Compagnie Antoine did not technically belong to Kalala, as such businesses were supposed to be owned by Congolese. Kalala, however, was Mutumba's not-so-secret backer. Mutumba told me of recent trouble:

The presidential guard, he said, had tried to seize his mine site on behalf of a member of President Tshisekedi's family. "But we fought them off—the governor stopped it from happening," he told me. I asked him if I could visit the site where the copper and cobalt he traded came from. He said I could, then maneuvered his face close to mine in a fog of fumed alcohol: "But, as I have said, I have become strong in business. Pay me $500."

When I told Mutumba I couldn't pay him to enter his mine site, he became agitated. He finally agreed to let me meet Kalala. But when the appointed day came around, he stopped picking up his phone, and people near his home in Kolwezi told me that they had seen him roaring drunk with a couple of Chinese men.

I CONTACTED MUTUMBA AGAIN A FEW MONTHS LATER, AND HE AGREED TO meet me at my hotel in the evening. "We must do it quickly," he said. "Tomorrow, I am going to the bush to buy cobalt for the Chinese." I asked him if I could tag along, but he laughed: "Where I am going, there is no law. There is no protection for you."

He told me to get into an idling taxi. Waiting inside was a woman with an updo that recalled the hairstyles of the Ronettes. Mutumba nervously laid out conditions: He would sanction questions; the lady with the updo would translate from French to Kiswahili; the interview would last only ten minutes. He was already drunk.

In the darkness, we entered the Chinese compound to the barking of dogs. It was not particularly large or luxurious, though it was in a relatively upscale part of town. A Congolese woman watched a child crawling about on a patio outside Kalala's bungalow. Above them, mosquitoes buzzed around fluorescent bulbs that washed the front of the house with an anemic glow.

Inside, in a bare-walled room, two men in their late thirties were sitting at a table that was empty save for a cluster of condiments and a packet of Oreos. Kalala, who was the fleshier of the two and sported a wispy goatee, welcomed me, then bade me take a seat. The other, in a gray polo, went to an open window and started smoking. Kalala began by apologizing. He said he spoke neither English nor French but did speak Kiswahili, and then he proceeded to tell me why he had come to Congo.

Mutumba occasionally interjected with loud remarks in Kiswahili that sounded like they were meant to be witty. These elicited the kind of wry smile from Kalala that indicated tolerance rather than the tight brotherhood that Mutumba had boasted he shared with the Chinese businessman. "I came here to find a wife," Kalala told me. "And now I am stuck here with my family." He gestured at the woman with the young child who had followed us in from outside.

IT TOOK ME A LITTLE WHILE TO UNDERSTAND WHAT KALALA WAS TALKING about. Why did he need to go to Congo to find a wife? Kalala explained that he was from a rural part of Shaanxi, in Central China. Thirty-five years of China's one-child policy between 1980 and 2015 had led to a high rate of girl babies being abandoned or even killed. In 2005, in Shaanxi, Kalala's home province, 130 boys were born for every 100 girls. The situation left the poorest men unable to marry, especially in rural areas. These unmarried men were known colloquially as *guang gun*, or bare branches.

Kalala had come to Congo at the age of thirty-six to escape the fate of the *guang gun*, he said, fiddling with the packet of Oreos on the table. He had wanted to start a business and get married. When he arrived, he quickly fell into Katanga's mining industry. "The most important thing here is mines," he said. He dreamed of riches. "You can get more than in China. In China, there is nothing special; here there is copper, there is cobalt. Here I am doing *business*. There in China, it was difficult. Here I have money, a house, a child."

Entrepreneurs like Kalala are "all over Africa," according to Deborah Bräutigam, a Johns Hopkins expert in China-Africa relations. There are numerous theories as to where such businessmen come from. Self-interested Western miners might invoke the idea of a "fifth column" of such workers as they point to China's increasing control of battery minerals in Africa; scaremongering politicians in Europe, in the U.S., and in countries where there are large numbers of Chinese expat workers might use the racist rhetoric of the nineteenth-century "yellow peril." In Congo, angry mobs occasionally attack Chinese businesses. Just over the border from Katanga, the firebrand politician Michael Sata won the 2011 presidential election in Zambia, in large part by stirring up Sinophobic sentiment. Such theories are occasionally given a boost when the international press reveals that the Chinese government is operating "a purported network of unsanctioned and illegal Chinese 'police stations' around the world."

But far from being a vast organized mass guided by a central bureau in Beijing, most of these small-time Chinese businessmen are striking out alone, leaving their homes to find riches in the broader world. They are, in many respects, modern-day equivalents of the Western wildcatters celebrated in capitalist lore. A Chinese community website based in southern Congo evokes the mythos of the Wild West's gold rush when it exhorts Chinese workers to come to "uncover the mystery of the Democratic Republic of the Congo . . . come to pan for gold, the Democratic Republic of the Congo is waiting for you!" Brautigam told me that such strivers leave China and "just go wherever they think they can make money."

Kalala seemed not to be fulfilling this second goal. It appeared that he was getting by, but just barely. In the bush outside Kolwezi, I had seen myriad other Chinese traders who were doing much worse, living in conditions of extreme poverty under roofs made of little more than ratty tarpaulins. "I lost a lot of money in the mines," Kalala told me. "Now Antoine takes care of the trading. I can't work in the mines anymore." At this, Mutumba grinned. "Business in this country is a big problem. The Congolese are always stealing." Kalala looked at me levelly, and his colleague offered me a cigarette. "The Congolese have got a lot of resources, but they don't know how to do business."

Chinese workers in Congo face widespread discrimination, attacks, and suspicion by the Congolese. The government has no official integration policy for Chinese migrants, who are often seen as "invaders," Ngoie has noted. On my travels in Katanga, I often saw Chinese residents belittled or yelled at. The Chinese community website has posted an "inventory of recent violent incidents against Chinese in the mining provinces of Haut-Katanga and Lualaba," and it documents ten occurrences of violence against more than a dozen Chinese workers in the two provinces within the space of less than a month. "The public security situation in Lualaba and Haut-Katanga Provinces in the southeastern part of the Democratic Republic of the Congo (DRC) is severe, and vicious cases involving Chinese citizens have occurred many times," the website states. It details, between May 27 and June 19, 2022, a spate of armed robberies, execution-style killings, and attacks by "inhumane" robbers who used scissors to slice the mouths of their Chinese victims.

But Kalala couldn't simply up sticks and go back to China. He was the child of petty traders and had no business back in Shaanxi. He couldn't take his wife and child there, he said. There was nothing back in China for him. "To be honest, things are not good here either. It has become very difficult here," he continued. These days, the government was always sticking its nose into things, and that meant more people to pay off. "My dream is to start a farm here and to teach the Congolese people how to cultivate. They have so much good land here, but they just don't understand how to *cultivate*."

Kalala was tired: The night before, he had been at the casino until 1:00 a.m. "The casino is the only thing to do here apart from work," Kalala told me. "In the casino, Chinese can spend a hundred thousand, four hundred thousand dollars a night." Mutumba cut in. He had to go—he had cobalt and copper to buy in the bush tomorrow. I wondered if Kalala really knew or cared whom Mutumba bought off to find minerals, what he was up to in the bush. Nobody in the room wanted to talk about the specifics of business, but I imagined that no one would object to Mutumba buying low, selling high, and ignoring anything else.

CHAPTER 33

PAPA SOLUTION

In mid-2017, Chinese workers arrived in the village of Samukinda, half an hour northwest of Kasulo, and quickly constructed two dozen houses with corrugated-iron roofs. Kasulo residents were ordered to leave their neighborhood within two weeks and take a payout of $2,500 or move into one of the houses at Samukinda. The Congolese government revealed that a mining permit had been granted to Congo Dongfang, which would remove the topsoil and then wall off what had once been the neighborhood. *Creuseurs* from an approved cooperative would be allowed to mine the site, and Congo Dongfang would become the exclusive buyer of Kasulo's ore.

A consortium of local civil-society organizations wrote to the governor, Richard Muyej Mangez Mans, protesting that the evictions were illegal, but he pressed on. Zanga Muteba, the baker, told me that on a rainy day a few months later, employees of Congo Dongfang "came with huge trucks to crush our houses."

Getting to Samukinda from Kolwezi takes about forty minutes. I made the trip in May 2019, noticing how faith inflected everything in the region: On the road, I passed the Mount Carmel health clinic, the Salon Apocalypse hairdresser, the Light of God tire shop. Eventually, the road became unpaved. Trucks carrying sulfuric acid threw up plumes of dust as they trundled toward factories where raw minerals were processed.

As I walked through the village, children laughed and pointed at me, shouting "Chinese! Chinese!" The customary chief of Samukinda, a wizened lady named Nama Mavu, told me that the smallest ones among them hadn't seen non-Chinese foreigners. Samukinda was seldom visited, even though factories and mines had been built nearby. Mavu assigned two young men to escort me to the houses that Congo Dongfang had built. A row of modern-looking white buildings rose in the distance. As they came into focus, it was clear that their construction was slapdash.

Few of the homes were even occupied—most of the original residents of Kasulo had just taken the payoff, gritted their teeth, and moved to another sector of Kolwezi. Those families who had chosen to take a house had been shown a brochure with beautiful pictures. But the homes Congo Dongfang built for them turned out to have no electricity or bathrooms. The roofs leaked, and the well at the corner of the development was dry. Moving to a small village with no transport into the city made it impossible for people to keep their jobs, if they had them. Most of the families moved away after a few months.

Muteba was one of the few original displaced residents who remained in Samukinda. He did so only because he had nowhere else to go. In his seventies and retired, he wore a soiled lab coat over his emaciated body. He welcomed me into his house, which was stifling hot. The roof was only roughly attached to the walls. There was little decoration save the moldering skin of a small animal that Muteba had hung up. He had also dug himself a lavatory pit, which was covered with a board. "The water here, it's not good," he said. "The smell of acid and pollutants comes out of any hole we try to dig for water."

The old baker was bitter, ill, and suffering from diarrhea. He recalled his home in Kasulo. "It was a big parcel of land," he said. "It had at least fifteen trees—avocado trees, mango trees. All that was mine." He wondered how long he would last out in the bush inside the collapsing home that the Chinese miners had built for him. "We were chased out of our homes like animals, and now we suffer like strangers."

A FEW DAYS LATER, ESCORTED BY COMPANY REPRESENTATIVES, I VISITED THE Congo Dongfang mine in Kasulo. Signs by the gate said that children and pregnant women were forbidden from entering. Inside the compound, the land that had once been a bustling neighborhood was now just a giant red crater. I saw no children during my visit, but Odilon Kajumba Kilanga told me that they still found their way in; responding to written queries in 2021, a Huayou spokesperson said that the company planned to "gradually eradicate all forms of human-rights violation with a responsible supply chain."

My minders cautioned me not to wander too close to the *creuseurs*, as they were liable to be violent. Not long before my arrival, a group of them had set some company trucks on fire. "At CDM there were lots of messes," James Kitangala, an official from the regional mining ministry, told me. "These people are weird. They can burn you alive without batting an eyelid. They killed a Chinese with their mining tools." But there was scope for improvement now that Kasulo was being "formalized," as he put it. "Artisanal miners now have the opportunity to be well led," he said.

Kajumba noted that Congolese had been employed to mediate between the *creuseurs* and company officials. Often, the demands of the *creuseurs* were not met, and they went on strike. "You go in to work and say, 'No, I won't do anything,'" Kajumba said. "The Chinese will feel unsafe and call in the police." The police, he added, do the company's bidding: "They know they will get a gift from the Chinese, so they will threaten you with tear gas and batons." Kajumba said that he had been tear-gassed by police at Kasulo. "Everyone ran to save his life," he remembered. "We felt defenseless."

Upon my arrival at the mine, I had been given a long run-through of safety protocols, but as I approached the *creuseurs*, it was clear that they had only rudimentary equipment. Plastic jerricans, cut roughly in half and tied to ropes, were being used to haul ore. Many *creuseurs* were shoeless, and I saw none wearing helmets or goggles, even though a confidential 2018 audit by the Korean conglomerate LG Chem had criticized the site for a lack of proper safety equipment.

Some *creuseurs* washed ore in dirty ponds by the pits. "The Chinese are cheating us," one of them murmured. "They're telling us the ore is less pure than it is." Kajumba said he had stopped working at Kasulo six months earlier because he felt that he was being treated unfairly. "It's as if you were working to suffer even more," he told me.

In a warehouse at the site, I watched a man in a spiky red hat, his face grim, pulverizing ore on a concrete floor as two Chinese overseers scrutinized *creuseurs* from behind a barrier of chicken wire. No Chinese employee interacted with me, and nobody responded when I waved in greeting.

AS KASULO WAS BEING READIED FOR ITS NEW INCARNATION, JOSEPH KABILA Kabange announced that, after eighteen years in office, he would not run for reelection. In January 2019, Félix Antoine Tshisekedi Tshilombo became president in an election that most observers said was rigged. Once in office, Tshisekedi, a Luba-Kasai by ethnicity, began to install people

Congo Dongfang International Mining overseers watch a
Congolese worker evaluate minerals at the Kasulo mine in 2019.

from his ethnic group throughout the apparatus of power. Augustin Katumba Mwanke's parallel system paled in comparison with the new regime, as Tshisekedi's advisers traveled the length and breadth of the country looking for *vaches laitières*, milking cows.

Scandals followed Tshisekedi's appointments. In 2022, Vidiye Tshimanga Tshipanda, one of the president's top advisers, was caught on tape negotiating with two people posing as mine investors. As captured in two videos posted online by the Organized Crime and Corruption Reporting Project, Tshimanga said he could help them get a mining deal because he had close ties to the president. "I will take my—how you say that? My percentage of the investment," he said. (He was later acquitted of corruption in a Kinshasa court for lack of evidence.) "At least the Kabila people were efficient," Charlotte "Maman Ocean" Cime Jinga, the former mayor of Kolwezi and a national deputy in Moïse Katumbi Chapwe's party, told me. "These people just steal."

In May 2019, I met with Governor Muyej at his fortified compound in the center of Kolwezi. A portrait of Kabila, the former president, hung on the wall. Muyej said that Tshisekedi would likely maintain the course set by Kabila—"a flight that we must take to get out of poverty."

Muyej told me that he hoped to diversify the local economy through tourism and agriculture. Mining, he said, exacerbated inequalities— "enormous mineral wealth beside a population that lives in enormous precarity." A back issue of *Grands Lacs* magazine that someone had left in the waiting room at Muyej's office in the governorate underlined the point well, with a cover story titled "Lualaba, New El Dorado of the Mining World."

In 2018, *Forbes* praised Muyej's governorship as "a model for bringing together economic prosperity, political transparency and social impact." Yet agriculture was scarce in the province; as for tourism, it was hard for me to imagine Kolwezi becoming a travel destination. In 2019, I tried to visit Katebi Beach Lodge, a new lakeside resort. At the entrance, a metal gate topped with barbed wire, I was shooed away by a police officer toting a Kalashnikov. Apparently, the lake was too polluted to allow visitors.

Muyej often cited the construction of a new governorate building—a gaudy structure rising above a sea of ramshackle cinder-block houses—

to show how he had modernized Kolwezi. Renovations of the local soccer stadium and the town's central roundabout, which features a statue of mine workers, were financed by mining companies.

Muyej told me that he hoped to reform the mining sector by, in part, reducing child labor and centralizing the market where traders buy cobalt, thus instilling transparency into the supply chain. Critics in Congolese civil-society organizations and among the Congolese political opposition called such reforms cynical bids to control and tax artisanal production for personal gain. Muyej, his family, and officials close to him profited from the mining boom. The governor's son Yves was the CEO of a logistics company in Kolwezi that had helped manage the Kasulo mine. Muyej's cabinet chief, Joseph Yav Katshung, was a lawyer whose firm did work for Congo Dongfang. (Yav and Yves both declined to speak to me.)

Muyej said that as many as 170,000 *creuseurs* work informally in his province. The Congo Dongfang mine in Kasulo is among the forty or so sites where artisanal miners are employed as day laborers. At the time, only 800 or so *creuseurs* worked there, down from the 15,000 who had worked there before Muyej intervened. Resentment ran high. "Kasulo is a village that is built on mineral deposits, but not enough *creuseurs* can legally work on official artisanal deposits," Jean-Jacques Kayembe, the president of an artisanal mining collective, told me. "And that's a problem."

Whenever Muyej tried to reason with *creuseurs* who had snuck onto industrial concessions, he was attacked with stones, and in 2019, there was so much unrest in Kolwezi that the military was sent in. By the end of the 2010s, it had become common to see soldiers toting machine guns and rocket launchers around the city. When I first visited the area, in 2019, a tollbooth outside town was riddled with bullet holes. A policeman at the booth had recently been murdered by bandits.

In 2020, the Platform to Protect Whistleblowers in Africa announced that two Congolese citizens had leaked documents exposing numerous improprieties at Afriland First Bank, a Cameroon-based institution where Muyej had at least one account. Muyej, it was revealed, had been moving hundreds of thousands of dollars through the bank. He was forced out of government. According to Radio France Internationale, the Congolese authorities have accused Muyej of not being able to justify 40 percent of his cabinet's expenses. (A representative for Muyej said that the governor had done nothing wrong and welcomed an audit of his finances.) For a while, it looked like ethnic tensions might flare up: Muyej rallied his base, and an organization of Ruund activists, the Dynamique de la Jeunesse Ruund (Ruund Youth Dynamic), began agitating for his return to power, threatening that "a good cohabitation between the people of Lualaba" was impossible without Muyej's return.

His replacement was his deputy, Fifi Masuka Saini, also known as Madame Fifi. She was accused of running artisanal mines for her own profit, much like Muyej had done. Masuka made money through the cooperative structure. Antoine Mutumba confirmed to me that the cooperative he ran for Kalala was connected to Madame Fifi. She was also more directly connected to another cooperative called the Coopérative Minière pour le Bien-Être des Exploitants Artisanaux du Katanga (Mining Collective for the Well-Being of Katanga Artisanals), or Comibakat for short. In a legal document, Masuka was listed as the president of the cooperative's leadership council. In 2021, the cooperative was accused of fraudulently mining at ten sites, including at one for which the Kazakh firm Eurasian Resources Group had a mining permit, and ganging up with Lebanese miners to use industrial machinery.

Masuka's work with the businessman-turned-politician Dany Banza Maloba also raised eyebrows across the South of Congo. Banza, a Katangese Kasaian, had been tasked by Tshisekedi with running the president's Katangese affairs. He became known as the "right arm" of Tshisekedi in Katanga, according to the Katangese political analyst Serge Noël Ngoy Mwanabute. Banza was, as various reports had it, involved in collecting revenue for the presidency from the mining provinces. One morning in mid-2022, I met Martin Fayulu, the presidential candidate whom Tshisekedi beat in 2018's election. We had breakfast at a hotel he owns in Kinshasa. Over fruit and coffee, I asked him about corruption in the president's inner circle. "That's all Tshisekedi's government does," Fayulu replied. "He only thinks about tribalism. He only thinks about himself. He only thinks about getting rich for himself, his family, and his friends."

In 2024, the digital newspaper *Africa Intelligence* published a report that connected Banza and Masuka to artisanal mining cooperatives that were operating illegally on industrial concessions around Kolwezi. "She's just a front for Dany," Bradley Barnett, the president of a U.S. firm called Critical Minerals International, said of Masuka. "And Dany's the bridge between the president's family and the road to riches."

By early 2025, Banza had reportedly fallen out with Tshisekedi and left the country, but he seemed to still be living well. In a video posted to TikTok that March, naming him a "billionaire capo," the footage showed him leaving the Hôtel de Paris in Monte Carlo and getting into a pristine vintage Mercedes-Benz. "One of the Felix T guys who has been managing Chinese money," the Congolese analyst who sent it to me wrote on WhatsApp. "Banza is at the heart / center of Chinese money in the DRC."

CHAPTER 34

THE STEVE JOBS OF METALS

Sadip's life changed in 2015. That was the year IMIP, the giant Chinese-backed nickel complex in Indonesia, began operations at Morowali. "People here just farm cassava now; cassava is the only thing that you can sell easily. Rice won't grow," Sadip, the Indonesian farmer, said. When we spoke, IMIP had been open for seven years. "The rice grew until the mining came. It was impacted by the mining, and the water was polluted and killed the plants. We tried to fertilize the rice, but the crop is really, really bad."

IMIP, which Sadip referred to as "the company," was destroying the livelihoods of the people in Bahodopi, he said, even as a huge influx of workers came to take jobs in the mining sector. He said that there had been an almost punitive approach toward the people who had inhabited the area before the industrial park arrived. "A group of farmers, we made a dam on the river, but the company destroyed it," he told me. "The company destroyed our tunnel and our dam, so we don't have any irrigation. The consequence is that we used to grow rice, but now we have to buy rice from elsewhere."

IMIP's development came as Chinese companies ramped up investment in new Indonesian nickel mines during the 2010s. But although nickel has seen an uptick in demand because of its use in NMC cathodes, the metal had become a complex proposition on world markets by the late 2010s and early 2020s. Prices rose and fell dramatically. Much of this stemmed from Indonesia's 2020 decision to ban exports of raw nickel ore. The idea was that Indonesia could capture more value from the metal if it processed it at home. The ban was challenged by the European Union, which successfully sued the Indonesian government to end it, alleging harm to the steel industry.

In response, Indonesia's president at the time, Joko Widodo, who is known as "Jokowi" and ruled as a "polite populist" between 2014 and 2024, claimed that the Europeans were trying to cudgel his country into underdevelopment. "It's okay to lose. I told the minister to file an appeal," Jokowi said. "We want to become a developed country. We want to open more jobs. If we are afraid of the lawsuit and give up, we will never become a developed country." The value of nickel exported from Indonesia had skyrocketed from twenty trillion rupiah to three hundred trillion rupiah (or from $1.3 billion to around $20 billion in U.S. dollars). "Our trade balance, for twenty-nine consecutive months, is positive, after [it] had been showing negative results for decades. We have been receiving [a] positive balance since twenty-nine months ago. This is our goal."

THE SWINGS IN THE NICKEL PRICE HAD ALSO BEEN OCCASIONED BY OTHER forces. One of these was a single magnate: Xiang Guangda, a Chinese businessman known as "Big Shot" and "the Steve Jobs of metals." Outside China, Xiang is most famous for trying to short the nickel market in late 2021 and early 2022. As it turned out, he mistimed his trade. Prices moved higher, partly on the back of Russia's invasion of Ukraine. By the early afternoon of March 8, 2022, Xiang had lost $10 billion. The London Metal Exchange, the place where the highest volumes of nickel are bought and sold on any given day, suspended trading, and Xiang recorded $1 billion in losses. But he didn't seem put out. "The loss has been roughly offset by the profits of his nickel operations over the same period," a *Bloomberg* article explained.

As it transpired, Xiang was substantially involved in Indonesian mining. His company, Tsingshan, was a Chinese success story on par with Chen Xuehua's Huayou, Robin Zeng's CATL, and Wang Chuanfu's BYD. In fact, he and Chen were close, and they invested in the Huayue nickel plant together. Xiang was from the southeastern province of Zhejiang and had started his career as a mechanic at a state-run fishing operation before founding a company that made car doors and windows.

The early 1990s were as pivotal for Xiang as for the other Chinese entrepreneurs in the field of new-energy materials. In 1992, Xiang turned to stainless steel after a visit to Germany convinced him that Chinese automakers would bring car-parts manufacturing in-house. In the early 2000s, Tsingshan patented a new, energy-intensive way to create stainless steel, and the firm went from a small Chinese steel manufacturer to a major player in a booming industry. In the words of a Chinese article praising Xiang, Tsingshan became "a stainless steel aircraft carrier," suggesting it was through soft power, not conventional military might, that China was projecting itself into the world. In 2008, after the global financial crisis hit, Xiang realized that China produced no nickel, a metal that was vital to his industry, so he turned to Indonesia.

China and Indonesia had checkered relations in the decades after Indonesian independence, the strain stemming mostly from Jakarta's rabidly anti-Communist policies under Suharto and by the persecution of the country's ethnic Chinese. After a treaty called the Indonesia-China Strategic Partnership was signed in 2005, however, China began to dominate Indonesian business, and China's Xi Jinping forged a close personal relationship with Jokowi. Chinese companies invested heavily in Indonesia's mining sector and built more coal power plants there than in any other country. "I do think Jokowi is someone who wants to build the domestic nickel industry," William Yuen Yee, a U.S. researcher who focuses on Indonesia-China relations, told me in 2023. He noted that there is probably a limit to how comfortable nationalists like Jokowi will be with China. "If he sees too much of China's Belt and Road Initiative playbook of bringing in Chinese companies, bringing in Chinese workers, if it's not working for Indonesians, I don't think he's going to be completely open to continuing inflows of Chinese investment. That said, China's also the only game in town. It's like the U.S. talks about that a lot, but we haven't really offered up an alternative."

———

XIANG KNEW HOW TO APPEAL TO JOKOWI PARTLY BECAUSE THEY SHARED A similar philosophy on development. Many of Xiang's speeches came from a similar ideological framework as that of the Indonesian premier, who has prioritized the development of his country, sometimes at the cost of cordial relations with foreign countries. Xiang told an audience in Ningde that he often speaks with foreigners who are critical of China's growth, telling them:

> We can't compare with you now. China has only been developing reform and opening up for forty years, and Europe has developed for more than one hundred years now. In fact, we are the first generation, and you are more than ten generations of wealth. Your ancestors were very hardworking and not afraid of death. If you want to compare our modern China with your grandfather's grandfather, then you will understand our current hard work. The struggle now is also for the future, so that our descendants can live the same life as you.

The message that Xiang wanted to impart—"let foreigners slowly understand our diligence in development and reduce the resistance they give in cooperation"—could have been lifted directly from one of Jokowi's own orations.

CHAPTER 35

DIRTY NICKEL

Xiang Guangda backed up his rhetoric with investment in Indonesia. His first moves in the country came in 2009, but it was in 2013 that he signed a contract for the project that would most define him: the Morowali Industrial Park, or IMIP, the plant that Sadip blamed for destroying his livelihood. Over time, a whole range of large Chinese firms—including firms working in Congo, such as CATL, Huayou, and China Molybdenum—created processing facilities in the park using low-interest financing from the China Development Bank and the Bank of China, as well as other large Chinese banks.

Almost overnight, a new city began to emerge. The only difference was that this city, which boasted a four-star hotel, a private airfield, and a series of coal-fired power plants, was focused on one thing alone: mining. Primarily for nickel but also for cobalt. The jungle around Bahodopi was hacked away, and deep pits were sunk into the red earth so that its laterite ore could be extracted. According to Indonesia's Coordinating Ministry for Maritime and Investment Affairs, a total of $29 billion had been invested in Morowali by 2022.

As IMIP grew, locals like Sadip began to complain about rampant pollution. Because nickel is found in laterite deposits in Indonesia, a great deal of energy is needed to separate it from the iron in the soil. Companies like Xiang's built dirty coal power plants to mine and refine the metals and began to displace families and farmers who had worked the land for generations. By 2021, IMIP was burning nine million metric tons of coal a year. Exhaust may have no longer been poisoning the air in Chinese cities. But instead, it was spewing out across the Indonesian rainforest.

Initially, IMIP focused on producing nickel for stainless steel, but with the rise of the lithium-ion battery, the park began to create nickel for firms that sat in the supply chain of large EV manufacturers like Tesla, Volkswagen, and BMW. In 2018, Tsingshan announced that it would build an HPAL facility on Sulawesi, which would allow it to process cobalt and nickel from the ore mined at Morowali.

IMIP officials insisted that the toxic waste from the HPAL process was being disposed of in the right way. But in 2022, when I visited Jakarta's environmental watchdog network, Jaringan Advokasi Tambang, or JATAM, activists there told me that IMIP's activities were far from clean. In August of that year, JATAM had released a report criticizing Tesla's decision to source around $5 billion worth of nickel from Indonesia, stressing how the project had caused "prolonged suffering of the people and environmental damages" and had "also triggered seawater pollution." The report concluded on a depressing note: "The practice of nickel extraction in Indonesia has triggered the loss of access for most people to food and water, as well as the escalation of conflict which is getting higher and wider due to land grabbing and repressive security approaches to people who defend their living space."

ON A SUNNY MORNING IN NOVEMBER 2022, I TOOK A CANOE THAT HAD BEEN fitted with a puttering two-stroke engine to the IMIP mine site. The canoe's captain cautioned me to keep my head down. IMIP's guards didn't like people prying, and even though we were technically in a public area, he warned us that we could be detained by security for snooping around.

As we motored past piles of coal stacked on barges—raw material to be burned in IMIP's furnaces—and watched IMIP's smokestacks puff out fumes, I noticed a handful of canoes like ours making their way to the front of the site. In one, a lady named Neni told us that she used to fish but now just collected scrap with her daughter from the edge of the mine site, as the water had become too befouled for fishing. "Our livelihood has been taken away from us," she said.

Another problem was the toxic slurry that resulted from the HPAL. Initially, IMIP had planned to get rid of tailings at sea, which would have been harmful to wildlife. After environmental groups kicked up a stink, the application for aquatic dispersal was withdrawn. Still, as Winwin, an activist with JATAM, told me, "When we talk about mining, you have people who will have to be displaced from their homeland." She went on to explain the downstream effects: "People are displaced to the shore, and it is not the answer, because when mining operates, waste runs into the river, and that ends up in the sea. And those people who live at the seashore, they have to struggle. And it means their income, everything that comes from the sea, is also affected. The ecosystem is destroyed."

A statement from a coalition of three environmental groups around Morowali complained of "a trail of destruction of marine and land biodiversity in remote areas, marginalizing the lives of local residents and workers," and it highlighted the irony of the battery industry in Indonesia and "the initial goal of electrifying global vehicles with the use of coal-fired power plants."

When we called out to another canoe, it stopped for us. Hudin, its captain, proffered a sea cucumber, a brown tubular scavenger that makes its home on the seafloor. "I caught it near the harbor—this can fetch ninety thousand rupiah," Hudin said. The sum was just over five and a half U.S. dollars. "The Chinese will eat this."

Hudin, who had eleven children and four wives, had continued fishing as a way to sustain his large family. He seemed fairly happy with his catch. "It's a high-grade sea cucumber," he said. "The polluted area doesn't impact sea cucumbers." Hudin was in his early sixties, he told me, and his leathery skin suggested a lifetime spent out in the sun, fishing. Back in the day, he didn't have to scrape by selling sea cucumbers. "I have been fishing here since before IMIP," he said. "I started around 1980. Before IMIP existed, I could catch fish near here, but now the sea is hotter than those years, so it is so hard to catch fish. The sea is hot because of the waste."

The pollution from the mine affected all parts of his life. "It is so hard to get clean water here," he said. "Before IMIP, I got clean water from the river; now the water is dumped with waste and we have to buy gallons of water because the river is closed and polluted by the mining people. It's five thousand rupiah per gallon, and I use ten gallons of water every day, so it's fifty thousand rupiah." This was about $3 in U.S. dollars, more than half of what he would make from selling the sea cucumber. He estimated that he made somewhere between one hundred thousand and two hundred thousand rupiah a day: "Before IMIP, I would make five hundred thousand to a million rupiah. The fish have gone further away because of pollution and the activity of ships here."

———

IMIP'S PROPONENTS SAY THE PROJECT BRINGS MUCH-NEEDED JOBS TO THE local community. By one count, the plant has created 120,000 jobs, 90 percent of which have been given to Indonesians. In 2022, I spoke to Reza, a twenty-four-year-old who had recently graduated from college and was optimistic about the development that IMIP had brought to Morowali. In fact, his dream was to work for the firm. "They have made an agreement to develop the area economically," Reza said, his eyes wide with the prospect of success. "The investment is enough, even too much. They just need to manage their human resources."

Workers at IMIP told a very different story. Late one evening in Bahodopi, I met with four members of a local trade union in the floodlit courtyard of my hostel. The union, called Serikat Buruh Sejahtera Indonesia, had two thousand or so members in Morowali who were employed by IMIP as monthly wage labor. They were drivers and manual laborers. They said that the much-vaunted jobs IMIP had created were not up to scratch.

The head of the union, Sahlun Sahidi, a thirty-seven-year-old father of three, told me that he worked week in, week out, without weekend breaks. He earned only nine dollars a day, which was a pittance even in Indonesia. "My salary is still considered very low," Sahidi said. It wasn't enough to pull him out of poverty.

The Chinese had brought in their own workers, who lived in different dormitories, ate different food, and made different wages. There was a marked distinction in how Chinese and Indonesian workers treated each other at IMIP, Sahlun said. The Chinese were often brusque with him. "They never communicate using our name when giving an order or saying something," he told me. "It's no different to them talking to an animal."

Still, the thing that concerned him most was the safety of the workers in his union. "Indonesian culture is more like European culture: We implement security first and then we concentrate on the goals of the productivity," Sahlun said. "Chinese culture only concentrates on the goals of the productivity lines, but they don't care about security or the safety. It's number two. The first goal is to get the productivity of the factory up." In August 2020, he was fired without warning by IMIP when he and colleagues from three other unions threatened strike action. Sahlun "refused to mediate and provoked colleagues to hold demonstrations," the company said in their note of dismissal. The strikes went ahead, and Sahlun was reinstated in his position with the company promising better working conditions.

Sahlun was still fighting for his fellow union members. He pointed to all the accidents that had happened at IMIP. Nickel smelters there had a habit of exploding, maiming and killing people. "It's a dangerous industry," he said. "We're just living in a cycle that's really, really crushing."

CHAPTER 36

A YOUNG CONTINENT

By 2012, Peter Zhou had been moved to the Bank of Montreal's Beijing office and onto the mergers-and-acquisitions desk. "There was a wave of Chinese companies starting to look overseas, and to look for resources," he said. He was involved in advising Glencore, the Switzerland-based commodities-trading company, on the $7 billion sale of the Las Bambas mine in Peru to MMG, an Australia-based subsidiary of the Beijing-based group China Minmetals Corporation. That transaction was completed in 2014.

Zhou's next stop was Congo with MMG. The company had acquired the Kinsevere copper mine north of Lubumbashi. At one point, Anvil, the Australian firm that had been investigated for the 2004 massacre at Dikulushi, owned Kinsevere, along with Moïse Katumbi Chapwe. Anvil later sold its share to MMG, but the former governor maintained the right to develop and operate the mine. When Katumbi fled the country soon after the Chinese purchase of the mine, MMG stopped using his company to operate Kinsevere. Katumbi's son sued MMG for breach of contract after his father returned to Congo in 2019.

By the time I arrived, in 2019, Kinsevere was a ruddy crater surrounded by a neat encampment of homes for expat and Congolese workers. "The mine is an open pit, and it is formed of steps, so the possibility of falling is very great," Yannick Makola Kasonde, the mine's acting operation manager, told a group of mining professionals who were on a tour of Kinsevere. Around the concession rambled thick bush, and signs alerted workers to the protocols for treating snakebites. Under cover of darkness, local miners equipped with only hand tools, people like Odilon Kajumba Kilanga, breached Kinsevere's perimeter to scrape a little ore to sell to traders like Kalala.

At Kinsevere, I observed the result of Zhou's work for MMG: Copper cathode sheets were dipped into solutions of copper sulfate and acid to create market-grade metal that would be flat-packed onto trucks and shipped out to international markets. Makola told us the mine had produced eighty thousand copper cathodes in the previous year.

At the end of the tour, we visited a primary school that MMG had built for the local community. Since the mine had been built, the promise of an education—and, perhaps, a bit of profit from artisanal mining—had prompted many families to migrate to the area. "At the beginning, no one wanted to come here; it was a small village," Makola said. The Ministry of Education was not paying teachers, so MMG began subsidizing

MMG's Kinsevere mine, 2019

their wages. The mining company now stood in for the state, educating 290 students. "We've built six schools like this," Makola noted. For Chinese firms, such developments were a perfect way to cement their presence in Africa. It reminded me of the old Gécamines days, only this time it was not a state enterprise stepping in where the normal functions of government had disappeared—it was a private company.

———

LIKE KALALA, YOUNG CHINESE PEOPLE SUCH AS ZHOU SAW AFRICA AS A KIND of frontier that promised adventure and profit at the same time. Among the Chinese who came to the continent to practice business, there was an incredibly diverse cast of characters. Even though Zhou and traders like Kalala were from similar places and had both ended up working in Congo, their ultimate circumstances couldn't have been more different. Zhou had made a success of himself in international business, working with large mining firms like China Molybdenum on multibillion-dollar deals. Entrepreneurs like Kalala had left China with nothing and had probably returned with empty wallets unless they'd managed to strike out on their own, as Kalala had.

For Chinese companies and central planners, however, both types of pioneers represent the opportunity to bring the supply chain—for electric vehicles, for batteries, and indeed for all kinds of raw materials—further into their orbit of control. These ambitions had become formalized after President Xi Jinping took power in Beijing in 2013. Xi promoted the Belt and Road Initiative, a massive government infrastructure project aimed at recreating the old Silk Road, connecting Asia to Europe through a series of ports and transit hubs that China effectively controls. "When they talk about BRI, they're really talking development security. They don't want to depend on, say, an American or Australian company for access to critical stuff," Martin Chorzempa, the Peterson Institute economist, told me. The Chinese, he said, "want to make sure that the resources they need for their development are secured." Huayou Cobalt deemed the Belt and Road Initiative an exemplar of "striving for excellence."

During his time working on the continent, Zhou would come to be part of some of the most lucrative copper and cobalt deals in all of Africa.

"What was shocking to me was the population over there, and how young they were," he told me. "I mean, I know that Africa is a young continent, but Congo seems particularly young." Driving through the city, he said, the first thing that stuck out was the "crowds of people, and many young people, many young kids. That was my first impression." Zhou was surprised by the number of children running around their destitute families' ramshackle homes. "In Africa, the poorer you are, the more babies that you're making," he continued. Zhou kept comparing Congo to earlier stages in China's development, a common touchstone among Chinese who worked in the country: "This was probably the same situation in China fifty, sixty years ago."

Some things did not surprise Zhou, however. "When I first landed, I wasn't too surprised about the poverty, because I grew up in Shanxi Province, which is in the interior of China," he said. "People were living in similar shelters to the ones you will see in Congo," he explained, only the shelters of his youth, he noted, had been made of sturdier stuff—cinderblock rather than the mud-brick favored in Congolese villages.

Shanxi, Zhou's home province, was poor, and like the Congolese Copperbelt, its industry depended on what was found below its soil—coal and clay. The region sits athwart an austere wind-carved plateau of loess, a yellow-gray sedimentary rock. The plateau sprawls over an area greater than the size of Metropolitan France. In parts of Shanxi, the loess has, for thousands of years, been carved into cave dwellings known as *yaodong*. Deep underneath the loess is coal, which was mined to power China's march toward industrialization.

The labor conditions in Congo are also not shocking to Chinese people from areas like Shanxi, as labor conditions there were abysmal until the mid-2000s. A few years before Zhou arrived in Lubumbashi, his home province had been rocked by a scandal that had shaken China. The soil of the region had long been used to create cheap black mud-bricks for construction. Many of the province's brick kilns were in remote areas, and in some, the foremen used slave workers who were referred to as "blacks" by overseers. These enslaved workers included children as young as eight who toiled in factories that resembled prison camps. "They eat food for pigs and dogs, and they do cattle and horse work," Fu Zhenzhong, the journalist who first reported the story, said at the time. When the slave labor came to light, twenty-eight people were sent to prison, and the foreman of one of the kilns was condemned to death.

———

FOR MANY CHINESE PEOPLE, BEIJING'S PRESENCE IN CONGO IS NOT SO MUCH colonial as opportunistic. A common sentiment is that the country's involvement in Africa hearkens back to the old ideological dictums laid out by Zhou Enlai in 1964. "This is something that people grew up with, something that we were told since the time of Mao," a Chinese artist in exile explained to me in 2023. "We *help* African people and people in the poorer parts of the world." For Zhou, the mine financier, it wasn't a case of colonization either. "I don't think one side is acting colonially towards another side," he said. What attracts Chinese to Congo, he said, is the "uber-high rich resources." Rather than being a state-directed resource grab in Africa, he said, Chinese government policies "encourage people to go outside to conduct resource-related projects, because China is the one that consumes the most. Mining overseas is one of the activities they want to encourage most."

Another thing had become apparent by the early 2010s: The Chinese were ramping up their investments in Africa to fill a void left by the U.S. and Europe, which had begun to increasingly ignore vast areas of the globe that had been bitterly fought over during the Cold War.

Beijing intensely focused on Africa and winning over African leaders as Washington and Brussels gave the continent the cold shoulder. Five high-level conferences with leaders from Africa were convened in China between 2000 and 2024; during the same time period, the U.S. convened only two. The Chinese strategy in developing countries is not necessarily one thing or another. It continues to be, just as it was in the Mao Zedong days, mutable and pragmatic. And nowhere is it more mutable than in the Democratic Republic of the Congo.

CHAPTER 37

TOKYO DRIFT

In late January 2012, the journalists of Japan's *Asahi Shimbun* newspaper revealed that Sony was planning to relocate its battery division out of Japan. All Sony's production would move to China and Singapore. Since the early 2010s, Sony's battery division had mainly been making batteries used in watches, cell phones, and power stations; the country's strong currency, however, meant that the firm just couldn't match the low price points of its competitors elsewhere in Asia. The division had even lost money since the beginning of the decade. During the relocation, some five hundred workers at the company's battery factory would be shuffled into different roles or asked to take voluntary retirement. It was a sad end to the battery story for the company that had created the first commercial lithium-ion cells.

Investment banks began to circle, swooping in with offers to sell off Sony's battery division. "Severe price competition has resulted in razor thin margins that favor large-scale manufacturers with weak local currencies," a Reuters report from late 2012 noted. It wasn't just Sony that was feeling the China squeeze. Panasonic, Japan's biggest battery manufacturer, remained a force to be reckoned with—it produced cells for Tesla, after all—but much of its production had also shifted to China.

The move to the Middle Kingdom was not without risks. In September 2010, just over a year before Sony announced that it was outsourcing its battery division to China, a dispute over fishing waters had led to Japanese officials arresting a Chinese trawler captain near the Senkaku Islands of the East China Sea. China responded by banning the export of rare earths, a class of metals whose members are used extensively for the production of powerful magnets in hot demand throughout Japan's automobile and defense industries. Most of these metals are produced and refined in China's industrial heartland, and Japan relied on China for 90 percent of its imports.

For more than two months, China blocked the export of rare earths to Japan, sending the Japanese market into a tailspin. Prices for rare earths jumped—in some cases by as much as ten times.

The captain was eventually released, and the exports resumed. But the Japanese seemed to have learned their lesson, at least when it came to rare earths. "It is critically important for Japan to secure sources of rare earths outside of China," Akira Terakawa, deputy director at the Mineral and Natural Resources Division of the Ministry of Economy, Trade and Industry (METI), would tell Reuters in 2014. Since 2010, the country has managed to reduce its dependency on Chinese rare earths by around 30 percent, and in 2018, it discovered a "semi-infinite" supply of the minerals in the seabed near Minamitorishima Island, about 1,150 miles southeast of Tokyo.

But when it came to lithium-ion batteries, the lessons of "the rare metals war," as the French author Guillaume Pitron has christened it, did not seem to have registered. Short-term business thinking trumped geostrategizing, and the move to China was completed without much fuss or public comment. "Batteries were never going to be core to Sony's strategy," an analyst told the *Financial Times* in 2016.

That year, Sony decided to sell the entire 8,500-employee division to Murata, the world's largest supplier of energy-storing ceramic capacitors. Murata was a company that had never made lithium-ion batteries but talked in vague terms about electric vehicles. The company's CEO had decided to focus on gaming and image sensors (in fact, such sensors, not batteries, would be key to a joint EV venture that Sony and Honda announced in 2022). Batteries were quietly forgotten.

WHEN I TRAVELED TO JAPAN IN 2022, IT WAS AS IF A DEEP SHAME STILL permeated the country when it came to conversations about batteries: *We invented this technology and then we lost it.* As he carefully picked through a sushi lunch, one battery executive explained that Japanese corporate culture had stifled innovation over the years, and that the country had handed the keys to China.

This executive told me that although scientists had already optimized the technology, there were places at the very cutting edge of research where Japanese engineers could still best their counterparts across the Sea of Japan. He was secretive and evaded many of my questions, but he did reveal that his company was focusing on creating a battery that could achieve four hundred watt-hours per kilogram, which was about one and a half times the energy density of the previous generation of lithium-ion batteries.

But China seemed to be stronger than even the executive had predicted. In 2023, just under a year after we spoke, the world's biggest producer of lithium-ion batteries, China's CATL, announced that it had leapfrogged the goal of four hundred watt-hours and created a condensed lithium-ion battery with up to five hundred watt-hours per kilogram. "The first time I went to the CATL facility, I was blown away—I have never seen anything like that," Shirley Meng, the Argonne scientist, said of a visit she had made in the pre-COVID days. "There's a couple hundred PhDs working on the engineering side, materials, quality controls, and things. But the actual assembly—there are very few people there." The factory was all robot arms and razor's-edge tech. When it came to batteries, China wasn't moving slowly.

CHAPTER 38

NO GUTS, NO GLORY

The South of Congo has few paved roads, so you have to drive through Fungurume if you want to get from Lubumbashi to Kolwezi. I came to know this drive well. Fungurume had always fascinated me. One time, I saw a troupe of Rastafarians there hold up traffic for miles, dancing and gyrating in a great parade. The two-lane road that bisected the town was crisscrossed with skid marks from trucks that bore copper and cobalt. Jeef Kazadi had shown me where a truck had plowed into a market a few weeks before, killing some twenty people. The town radiated away from the highway, rambling over the Katangese hillscape. In any one panorama, the smoke from dozens of charcoal fires floated into the sky. One of the region's few commercial activities that does not involve mining is making charcoal, which is used for cooking and heating and which people strap to their backs and haul great distances to sell along the roadsides.

Fungurume is also the site of one of the world's largest copper-cobalt mines. Named for the two towns that sit within the concession—Tenke and Fungurume—the Tenke Fungurume mine is a series of massive open pits and mud-filled ponds gouged into an area that is spread across more than 580 square miles of Katangese bush. From the road, drivers can see massive earthworks and concrete drainpipes traversing barren scrubland that is occasionally punctuated by copses of *musamba* and *muputu* trees.

The land around Fungurume can be lush and verdant after the rains, but during the dry season, it is parched and stained with ocher dust that makes your sinuses ache. Driving through Katanga, I often thought about the studies showing that the dust in the region contains high levels of toxic heavy metals.

Humans have mined around Fungurume for thousands, if not tens of thousands, of years, drawn to the land at Tenke Fungurume because it contains an ore body that is unusually rich and thick. Even in places where the mineralization below the ground would by expectation be poor, Tenke Fungurume is rich in copper and cobalt—or, to put it technically, the "high-grade hypogene carrollite mineralization in the SD-1b subunit of the lower Shales Dolomitique (SD) Formation of the Mines Subgroup" wouldn't generally be expected to be rich in anything at all. At Fungurume, however, the rocks below "boast cobalt concentrations as high as 3 wt% [percentage by weight, that is] along with Cu concentrations in excess of 4 wt%"—in other words, high concentrations. In a mining territory like Katanga, it was only a matter of time before they were dug up.

But Fungurume, surprisingly, didn't see large-scale mining for a long time after the country's colonization. By 1918, Belgium's Union Minière knew there was ore there, but it was never tapped by Congo's colonists. During the Mobutu years, the ore was dangled as a tantalizing prospect for investors. In 1970, a Belgian American mining impresario named Maurice Tempelsman led a consortium to develop the mine. Tempelsman hired Larry Devlin, the ex-CIA chief, to smooth things over with Mobutu and then spent $250 million on development, but he had to shut the project down after the dictator reneged on his promises. The truth was that Tenke Fungurume was being held in reserve by Gécamines as a long-term insurance policy against a time when the mines in Kolwezi stopped producing ore.

By the 1990s, Tenke Fungurume's high-quality ore had remained out of reach for years, but it hadn't been forgotten. The U.S. embassy kept an especially close watch on the plot, and the ambassador even visited Tenke Fungurume in 1991. Two years later, Gécamines officials offered it for tender, and there ensued a series of "chaotic" negotiations that involved a consortium of companies, including Anglo American and Phelps Dodge.

———

OVER THE NEXT THREE YEARS ADOLF H. LUNDIN, THE SWEDISH NATURAL-resources investor whose family motto was "No guts, no glory," emerged as a frontrunner. In 1996, he met with Mobutu in his villa on the French Riviera to hash out a deal. Mobutu, who took a liking to Lundin, gave him a tailored Lanvin shirt and told the Swede to address him as "Papa President." What's more, he said that Lundin could develop a dream concession: Tenke Fungurume.

Lundin was a jovial man with the kind of eyebrows that stay black long after a head of hair turns gray. He was also a passionate anti-Communist. "He used to say that instead of blood in his veins he had oil," said Petter Bolme, a Swedish investigative researcher and journalist who has worked on the Lundin group of companies since 2008. "A lot of analysts loved him in the '90s because someone compared the rapid stock-market rise of his company to stocks like Apple."

Lundin's companies were famous for wild fluctuations on the stock market—rapidly appreciating and just as rapidly depreciating, clearing out the savings of investors whom he had lured into purchasing stock. "He who cringes from risk can't see the possibilities, and therefore will never succeed," Lundin once said. "A person with a big ego is totally consumed with his product and will succeed." The hard-charging attitude might have been profitable, but it also led to a landmark 2021 Swedish lawsuit in which two executives at the company were charged with abetting war crimes in Sudan around the turn of the millennium.

Another secret to Lundin's success was big bucks, offered to any official, no matter how corrupt, who could smooth through a deal. Later, Lundin would recall that he had offered to give Mobutu money for his "election campaign," but Mobutu had never followed up, and no money was ever paid. Gécamines and Congolese officials had seemed completely unable to organize the transaction, and one of Lundin's officials later boasted about how he had to pay each of the fifteen negotiators $1,000 to even show up to meetings.

At the start of 1997, the prospects for Mobutu were looking shaky, so Lundin flew in to Lubumbashi to make a deal with M'zee Laurent-Désiré Kabila at the Karavia and began paying the rebels. (He continued to wear the shirt Mobutu had given him—for good luck.) When Kabila took power, Lundin's company took control of the mine.

According to Bolme, in Congo, Lundin "saw the opportunity, which was creating a big mine that everybody knew about, that wouldn't need much development for it to sell off as an asset." Lundin was not looking to develop the huge mine, Bolme believed. "He didn't have the capacity or the capital to do that," Bolme said. From the beginning, Bolme argued, Lundin was looking at how "he could hold on to it, to do some exploration, make some holes in the ground, start a small mining operation, and sell it for a huge profit."

The company declared force majeure, claiming that it was impossible to mine there while a civil war was ongoing. (The war also would have pushed any potential sale price down.) The site was not developed until 2006, and the first of its ore was exported in 2009.

BY THAT POINT, THE LUNDIN GROUP WASN'T THE ONLY OPERATOR AT TENKE Fungurume. The politician Barnabé Kikaya Bin Karubi, who traveled to the U.S. as part of a Congolese delegation, remembers Joseph Kabila inviting U.S. investment into Congo's mines. "President Kabila, because of us being accused of being pro-Chinese, asked President Bush, look I need the Americans to appoint a company, a reputable company in mining to come to Congo and take the Tenke Fungurume copper and cobalt mine," Kikaya told me. In 2005, after protracted negotiations, Phelps Dodge, a U.S. mining firm originally founded in 1834, announced that it was acquiring a 57.5 percent stake in the Tenke Fungurume mine. "We went to Arizona where Phelps Dodge has a mining venture," Kikaya continued. "We visited a beautiful mine, and we could see that these guys knew what they were doing." Lundin would hold on to 24 percent of the company, and Gécamines would keep the rest.

The deal was closed under murky circumstances in 2007, using U.S. government funding from the Overseas Private Investment Corporation.

Much was not made public about the purchase, leading journalists to call it a "backroom deal." Just over a year later, Freeport-McMoRan, the Arizona-based minerals behemoth, bought Phelps Dodge for $25.9 billion. Freeport already owned a huge copper-and-gold mine in Indonesia called the Grasberg. Kikaya told me that Phelps Dodge found working at Tenke difficult. "The Congolese environment, the business environment, was something that they could not deal with." They would call Kikaya's office and say, "Ambassador Kikaya, we cannot function like this." Permits and simple paperwork were slow to arrive. Congo's bureaucracy may have been ravaged by years of war and corruption, but "an American businessman cannot sit waiting for an administration guy to sit on his paper for two days, three days," Kikaya said. Phelps Dodge had to put up with too much bureaucracy, too much "nonsense," as Kikaya put it, so they sold their stake to Freeport.

Melissa Sanderson, the former chargé d'affaires at the U.S. embassy, had taken a job at Phelps Dodge in 2006. The Grasberg, she told me, was the talk of the company. "It turned Freeport into the mouse that ate the elephant," she said, "which was how that merger was classified when it took place, because PD [Phelps Dodge] had numerous mines all over the U.S. and, at that time, holdings in Mexico, Zambia, South Africa, and other places. And at that time, Freeport had one thing and one thing only. They had the Grasberg." After the acquisition, Sanderson would rise to become director of international affairs for Freeport-McMoRan. After it acquired Phelps Dodge, Freeport became the largest copper-and-gold miner in the world.

But Freeport, which was run by James "Jim Bob" Moffett, a charismatic wildcatter from Louisiana, had long courted controversy at the Grasberg. "Jim Bob was very much a cowboy," Sanderson said. "I'd say the common perception in the halls of Freeport is that the Grasberg was obtained with an exchange of suitcases with Suharto [the Indonesian dictator]." Allegations abounded that the company was violating the Foreign Corrupt Practices Act by paying off Indonesia's army.

The Grasberg ended up being toxic in other ways too. In 2005, *The New York Times* noted that a multimillion-dollar environmental study conducted near the mine had found rivers "unsuitable for aquatic life" because of waste dumping. Norway's sovereign wealth fund divested from Freeport.

At Tenke Fungurume, concerns soon arose about the way Phelps Dodge had acquired the mine, and especially about Gécamines's role in the transaction. What's more, in 2008, a *Dan Rather Reports* television segment showed how local residents had been displaced from Tenke Fungurume as mining began. The U.S. camera crew's members were arrested by local police, who were in the pay of Freeport, as they tried to go to a village near the mine to interview local inhabitants. When a Congolese stringer for Rather's team was able to enter a village with a camera in tow, people told him how they had been displaced to poor farmland, and how they had been paid pitifully small sums to move from land that their families had farmed for generations. In the segment, Rather framed the displacement as a "conflict between a powerful American company and Congo's voiceless poor."

In another report, this one jointly put out by the NGO Swedwatch and the International Peace Information Service, villagers who had been displaced from a village near Tenke Fungurume said they had been paid $360 for around a hectare of land. The fields at their resettlements were far from suitable. "This land is not fertile," one of them told the investigators. "It is *bulongo kwa koza*, rotten land."

PARTLY TO MITIGATE THE BAD PRESS, FREEPORT AND LUNDIN BEGAN TO make efforts to develop the local community. By 2013, the mining companies had built ninety-one wells and six schools, invested $50 million in environmental protection, and paid more than $795 million in taxes and royalties.

Problems, however, began to pile up. There were numerous clashes with the locals, as well as with artisanal miners ("illegal artisanal miners," as a company publication would insist) who kept coming onto the concession. Sanderson objected to the company putting up a fence, which, as she saw it, created a headache for locals but didn't really make the mine safer. "The concern was, we knew we had armed militia groups in the area, and some of them were living particularly in Tenke," Sanderson told me. Among these groups, she said, were the Bakata Katanga, the separatists loyal to Gédéon, the "cannibal warlord."

The pile of problems grew. In 2009, the Congolese government reviewed Tenke Fungurume Mining (TFM)'s contract. At the same time, Congolese police arrested three TFM staff members—including a Belgian, Dirk Vanhooymissen—who were accused of running a visa and work-permit scam. They were prosecuted "for misappropriation of public funds and forgery," as Congo's prosecutor-general put it. Freeport appealed to the U.S. embassy, which declined to help because Vanhooymissen was a Belgian subject, not a U.S. citizen. ("The stinkin' government is no stinkin' use to us," Freeport staff grumbled. "Anytime we need 'em, they don't stinkin' do anything.") In the end, Freeport agreed to pay a $16 million fine to the Congolese government.

Freeport officials began to become less and less comfortable with TFM, and with the realities of doing business in Congo. So-called facilitation payments were one thing that caused concern among employees. Under the Foreign Corrupt Practices Act, or FCPA, such payments, "the purpose of which is to expedite or to secure the performance of a routine governmental action by a foreign official, political party, or party official," are technically legal. Officials from other mining companies governed by different laws (the U.K. Bribery Act, for example, considers such payments to be illegal for British companies) were amazed to see Freeport shipping in "huge pallets of cash," as one put it, to make payments to underpaid Congolese officials. "The legal division went over and over and over this, you know, to make sure that we weren't going to get nailed under the law," Sanderson said. "And the legal division reached a level of comfort with the facilitation payments." She did not, she added, "think it was a good idea."

Company culture, however, was changing at Freeport. Under Moffett, the company had developed a freewheeling attitude that was coming under increased scrutiny. Moffett still governed from on high as the company's chairman, but he had stepped aside as CEO in 2003.

By the early 2010s, Kathleen L. Quirk, the company's CFO, was keen on ensuring that Freeport did not become embroiled in a corruption scandal. "Kathleen, in particular, always just had a strong focus on the importance of legal and a strong awareness that FCPA can be fatal," Sanderson told me. Board members had become uneasy about the risks of doing business in Congo.

At around the same time, Moffett put hundreds of millions of dollars into digging oil wells off the coast of Louisiana through a related company, McMoRan Exploration. As oil prices fell, so did Freeport's stock price. And dig as they might, the wells kept coming up dry. Freeport eventually bought out the remainder of Moffett's company.

The firm began to examine what assets it could sell off to pay down $5 billion to $10 billion in debt. Tenke Fungurume fell into the crosshairs, and Freeport began looking for a buyer. Chinese firms, anxious to capitalize on metals that it saw as key to the next stage of global growth, were quick to mobilize. By 2016, the firm had found a buyer in China Molybdenum, which used cheap debt issued by Chinese state banks to fund the deal. "I did not want to do it," the CEO, Richard Adkerson, told attendees at an investors' conference that year. "It breaks my heart to do it in a lot of ways, but in terms of stepping forward with our strategy, it is a good deal for Freeport."

Sanderson was less impressed with Adkerson's theatrics. "The truth is that they sold the Congo [TFM] because as a company, the biased Freeport leadership was too attached to the Grasberg," she told me.

Congolese politicians tried to intervene. Kikaya told me that the government had been unaware that Freeport was selling to the Chinese. "Without telling us—we just woke up in the morning: We were in bed with a Chinese company," Kikaya said. *The New York Times* reported that André Kapanga, the Congolese general manager of the mine, reached out to lawmakers and government officials in the United States to protest the sale.

The Congolese government, however, quickly became comfortable with the deal. China Molybdenum was certainly adept at currying favor with Congo's powerful. In 2018, according to the Sentry, a transparency NGO, the Chinese firm bought Congo Construction Company or CCC, a firm that was closely linked to figures in President Kabila's inner circle. CCC had recently spent $40 million in an opaque deal to buy the rights to mine a phosphate deposit from a company linked to the president's family. (China Molybdenum argued that the acquisition was strategic—it worked with the mineral in Brazil, and as new battery technology came online, phosphates were becoming an important mineral for battery production—but the Chinese firm would be slow to develop the site.)

Tom Perriello, the U.S. special envoy for the African Great Lakes region at the time, understood that Freeport's exit represented U.S. business leaving a key strategic region, and appreciated the worries that Congolese people were expressing: The Chinese buyers didn't come up to Freeport's record on the environment, and they believed that the firm was less committed to training up Congolese workers. "You know, the Obama administration was quite aware of the concerns," Perriello told me in 2023, seven years after the deal. "Freeport was, in particular, was . . . it wasn't a close call for them." He added, "There were a number of dynamics, both in the company and in the global economy, that were drivers of that decision." And, like their predecessors at Phelps Dodge, Freeport's officials found operating Tenke to be unnecessarily bureaucratic and convoluted.

At the time of the TFM sale, Freeport officials were insistent on clinging to control of the Grasberg. "The Grasberg made Freeport," Sanderson said. But Freeport's flagship mine was also a risky asset: Under Joko Widodo, the Indonesians were looking to gain more national benefits from mining. The Indonesian government changed the law in order to expropriate the mine. "From a political perspective, it was crystal clear," Sanderson said. "Freeport was going to lose operational control."

Two years later, one of Indonesia's state-run mining firms took 51 percent of the Grasberg. Sanderson said the sale had been forced in the way she predicted: "Freeport allowed that to happen through its own arrogance."

CHAPTER 39

TYING UP THE SUPPLY

The Kamoto Treatment Center, Kolwezi, 2022

At the door to one of the four control rooms at the heart of the Kamoto Treatment Center (KTC) hung a notice, printed on a piece of A4 paper: "Safety: What am I going to do? What can go wrong? What am I going to do about it?" The room was sliced from a concrete octagon that occupied the heart of the treatment plant. On beige walls, computer screens displayed flowcharts, beaming messages at men who sat behind desks, monitoring them all day.

One morning in March 2022, I stood with Yannick Makola, the general director of operations and technology at Glencore's Kamoto Copper Company mine, or KCC, on the eastern fringe of Kolwezi. He explained how the treatment plant worked. "Everything that happens in the concentrator is managed from here," Makola said. Outside, seven huge mills rumbled away inside a structure that can only be described as a shed, though one that had been magnified to many times its normal size. "The control room operators are in contact with the people who work on the site to modify and regulate the parameters, making sure that the parameters are the same as the parameters in our procedures."

Parameters, procedures: It was a world away from the artisanal mines at Kasulo or the open pits of Tenke Fungurume. Safety in all things—even just by staying within the crosswalk on an empty road in front of headquarters—was essential, because as a stone memorial beneath the shade of a giant conifer tree at the entrance to the mine's offices reminded visitors, mistakes could cost lives. The memorial, dedicated to "our employees who passed away on duty," was engraved with ten names and ten crosses and carried the inscription KCC WILL NEVER FORGET YOU.

Just look at what we have built down in Congo. That had been the refrain of several Glencore employees and former employees. Glencore was, at the time I visited, the world's largest producer of cobalt, which it refined into hydroxide at the KTC plant. The company was keen on showing me the site. Its books were open, and it had nothing to hide. It insisted that the site was an example of mining better and, in terms of environmental and labor practices, could serve as an example for future investments in Congo.

I was shown the massive Mashamba pit and the entrance to the giant Kamoto tunnels; I was given an explainer on how to cut back into the earth in giant steps, following a diagonal body of copper-and-cobalt-rich rock; I was taken to the KTC processing plant; I saw the electrolytic creation of copper sheets and the leaching of cobalt ore using sulfuric acid.

Francis Banze, a metallurgical superintendent and KCC's cobalt manager, gave me the most detailed explanation I had ever received on how heterogenite ore is broken up, milled into powder, mixed with water, floated, concentrated with acid, roasted, lixiviated, separated, then filtered and dried into a blue hydroxide cake that is ready to be shipped through Zambia to Durban or Dar es Salaam, and then to China.

At the end of my tour, Clint Donkin welcomed me into his office at the site. Donkin, a soft-spoken Australian in steel-toed shoes, was the CEO of KCC. He explained that he wished more people would visit and tell the world what they had seen at KCC. "The more people who come here, the more people will understand it," he told me. "This is like any mine site I've been on in North or South America."

Donkin described how, as he saw it, KCC was paying its dues. Collectively, he explained, Glencore paid "five hundred million dollars in taxes and royalties to the Congolese state, annually." These amounts fluctuate with the price of commodities; in 2022, Glencore would in fact pay a total of $1.14 billion in taxes and royalties to Congo's government, almost half of the $2.3 billion it paid between 2021 and 2023. These are significant proportions of the company's earnings. Glencore doesn't publish the revenue or earnings figures at individual mines, but in 2023, for example, the firm mined about $2.23 billion of copper and cobalt at their Kamoto operation. And the benefits to having KCC in the country didn't stop there, Donkin continued. The mining company frequently contributed to charity: Just recently, Donkin told me, it had donated $1 million to local medical needs, and it also took care of its employees' medical costs.

KCC brought technology and leadership skills to its employees, Donkin said, training up a new generation of Congolese who would be able to generate wealth for their country. The plethora of Congolese employees—engineers and technicians, drivers and managers—seemed to confirm that, even though a November report by Rights and Accountability in Development (RAID), an NGO based in the U.K., and the Centre d'Aide Juridico-Judiciaire, a Congolese legal-aid center specializing in labor rights, criticized Glencore's use of workers hired through subcontractors whose contracts were less secure and provided for fewer benefits than if the workers were full-time employees. Glencore did use subcontractors, albeit much less than at other companies. In 2020, at TFM, for example, a whopping 68 percent of the workforce was subcontracted, whereas only 44 percent of Glencore's workforce was hired through such subcontractors.

Environmentally, Donkin said, KCC was more up to scratch than its competitors and also under constant assessment. The mine operates what is known as a "closed loop": Water is recycled and toxic waste carefully managed. There were still accidents, however. In 2021, for instance, KCC maintenance engineers mistakenly released acid into a nearby river. (Glencore quickly admitted to the spill and said, "Our community officers have not registered any complaints nor concerns from their engagement with the surrounding communities.") And in 2024, a report by RAID and AFREWATCH indicated that the firm was contributing to water pollution in the area. In response, Glencore stated that while its mining assets "make a contribution to society, they may also have adverse social and environmental impacts."

Glencore was often accused of buying artisanal cobalt, but company officials I spoke to vigorously disputed this claim. *Look at the size of our mine*, they said. *Look at the amount of equipment we have. Why would we risk it to buy hand-mined cobalt?* It was a compelling argument, and there is no evidence to the contrary.

"This is a world-class asset, and we treat it like that," Donkin proudly told me. "It's easy to lob grenades over the fence. But we want to be part of the solution."

GLENCORE HAD BEEN A TARGET OF CRITICISM SINCE IT FIRST BEGAN OPER-ating in Congo, starting with the purchase of shares in Nikanor in 2007 and when Glencore gained control over the KCC mine through the merger and share purchases of 2008 and 2009. Glencore officials have always maintained that the company acquired the mining interests entirely legally: The acquisitions were purchases of public companies registered in the U.K. and Canada. In the case of Nikanor, for instance, the firm was listed on the London Stock Exchange, and Katanga Mining's shares could be bought in Toronto.

The aura of disrepute that Gertler had cultivated through his dubious dealings with Kabila and Katumba, however, attached itself to Glencore. NGOs like Global Witness have said that compliance departments should have raised genuine concerns about working with Gertler, and an investigation by the Swiss government noted that the company "failed to ensure adequate management" of the bribery risks presented by Gertler's purchase of minority stakes from Gécamines for less than their value.

Another criticism leveled against Glencore was that it had given the Israeli businessman a series of loans and "entered loss-making deals with him from 2007 to 2010." The loans, the company would later insist, were not low-interest, and were driven by financial motives as well as the desire to do business with Gertler: They were set between 3 and 7 percent over the London interbank offered rate (LIBOR), which sometimes pushed them into double-digit territory. These were hardly low-interest loans, the firm argued. "Glencore has made loans to companies associated with Mr Gertler at various times, fully secured on shares and at commercial rates of interest," the company insisted in a letter to Global Witness in 2012. Glencore went public in 2011, and the company began to renew its focus on compliance after it listed its shares on the London Stock Exchange.

Even so, Global Witness kept up pressure on the company with a series of reports detailing how Glencore and Gertler had outmaneuvered other investors—including Forrest (through a process dubbed "de-Forrestation")—to control the mines. The NGO accused Glencore of "paying the gatekeeper" by way of deals that were "nuanced and convoluted. They involve intricate financial arrangements and secret transactions through offshore companies." The outcome, Global Witness wrote, was that "the mining giant gets its assets and the gatekeeper's interests are taken care of." The gatekeeper, of course, was Gertler.

Former Glencore officials with whom I spoke claimed that the image of its cozy partnership with Gertler is overblown. Several described the relationship as anything but cordial. Gertler had even hounded Glencore when he didn't get his way, threatening to sue. In February 2017, in a transaction that involved Glencore paying Gertler-associated companies more than half a billion dollars, the firm bought Gertler out of KCC and Mutanda. It acquired Gertler's remaining stock in Mutanda for $922 million and in Katanga Mining for $38 million, offset against a series of loans worth over half a billion U.S. dollars.

But Gertler wouldn't just go away. Glencore was also obliged to pay Gertler under Congolese law, because two firms he controlled had bought Gécamines's royalty streams from the Glencore-owned mines.

ON DECEMBER 21, 2017, GERTLER WAS SANCTIONED BY THE U.S. DEPARTMENT of the Treasury for amassing a fortune "through hundreds of millions of dollars' worth of opaque and corrupt mining and oil deals in the Democratic Republic of the Congo." The Israeli businessman had weathered bad press, but the sanctions made him radioactive: Anyone who dealt with him now ran the substantial risk of running afoul of U.S. law, a death knell for a large multinational like Glencore.

Because Gertler's mining royalties were paid out in U.S. dollars, Glencore executives feared their company might be targeted by the federal justice system. Glencore initially stopped paying the royalties—estimated to be around $77,000 a day, or $28 million in 2018—that Gertler was entitled to under Congolese law. (The royalties were paid on monthly and quarterly bases.) Gertler began to put pressure on Glencore and initiated legal proceedings, claiming almost $3 billion in past and future royalties. Glencore officials worried that the company could lose the mine in Congo. "And the Chinese would be there, just waiting to snap it up," one Glencore official told me.

In early 2018, Glencore sent a team of negotiators to Washington. They explained that Chinese enterprises and the Chinese state had been buying up copper-and-cobalt mines all over southern Congo, and that if Glencore were to stop paying Gertler, it would lose the mine. With TFM sold to China Molybdenum, Glencore was the last non-Chinese or non-Kazakh major owner of copper-cobalt mines in the greater Katanga region. That year, Glencore's mined output was 42,200 tons of cobalt—the largest share of any one company globally and around 28 percent of total global production. Did the Trump administration want all that to go to the Chinese?

Glencore's lawyers had devised a solution, and they wanted to know if the U.S. government would object. As a company based in Switzerland, not subject to U.S. oversight, could they pay Gertler in a non-U.S.-dollar currency, such as euros?

No objection was raised.

By the end of the 2010s, the fear of China taking over the production of battery metals like cobalt had firmly lodged itself in people's minds; it provided a powerful counternarrative to concerns about corruption. Time and again, mining executives complained to me that businesses from the U.S. and Europe have to contend with anticorruption legislation. Such laws also exist in China, but they seem to be applied only very sparingly to Chinese firms working overseas, especially in places like Congo. What's more, well-publicized concerns about artisanal mining strengthened the case for those running industrial mines, such as KCC.

In June 2018, Glencore, in conjunction with Katanga Mining Limited, publicly announced that their companies, in order to "avoid the material risk of seizure of its assets under DRC court orders" would "pay the relevant royalties as and when they become due to Ventora [a Gertler-controlled firm] in non-U.S. dollars, without involving U.S. persons, in order to discharge their obligations under the terms of the pre-existing contracts."

The same day Glencore released this statement, Global Witness's Peter Jonas blasted the arrangement, saying that Glencore was effectively acting with impunity and that paying Gertler in non-U.S. dollars was "not an acceptable solution to this dispute." He went on: "U.S. authorities must hold Glencore accountable when those payments to Gertler resume."

But when regular payments resumed, U.S. authorities remained silent.

BY 2024, GLENCORE HAD BEEN INVESTIGATED BY THE CONGOLESE GOVERN-ment, the U.K.'s Serious Fraud Office, the U.S. Department of Justice, the Office of the Attorney General of Switzerland, and the Dutch Fiscal Information and Investigation Service. It had paid more than $1.5 billion in aggregate fines and had admitted to doling out hundreds of millions in bribes. (Most of the allegations did not focus on the firm's operations in Congo, but rather its oil business in West Africa.) "Bribery was accepted as part of the West Africa desk's way of doing business," a British judge said of one of Glencore's offices in London. "Bribery is a highly corrosive offence. It quite literally corrupts people and companies, and spreads like a disease."

In May 2022, on the day it was announced that Glencore was being fined $1.1 billion by the U.S. Justice Department and being compelled to install monitors with wide powers of oversight to look for malfeasance, I was at a conference with officials from the company. They seemed unperturbed. They would pay, they said. The company's stock price hardly budged. The market appeared to share the company-wide conviction that it had evolved from the days of deals that were likely to place it in hot water. That year, Glencore's annual revenue was $256 billion.

The firm was released from its monitorship in March 2025, a year earlier than initially agreed upon, with the U.S. government arguing that it was "no longer necessary" to have them present.

But Congo, which Glencore paid $180 million in 2022 to settle disputes, appears to still be trying to negotiate higher taxes. Ronny Jackson, the Republican congressman who sits on the House Foreign Affairs Committee, met with Glencore officials also in March 2025, shortly after visiting Congo. He told congressmen that he witnessed "shocking" levels of corruption and noted that Congo's government had recently tried to charge Glencore $80 billion in taxes. (The figure was apparently inflated.) "This kind of corruption," Jackson noted, "discourages legitimate businesses from investing."

As ever, the Chinese were not discouraged by this state of affairs. During the first years of the 2020s, mining executives such as Glencore's Ivan Glasenberg continued to point to Chinese firms' control of the world's cobalt supply and their domination of the production of lithium-ion batteries. That China Molybdenum had managed to acquire TFM—and hold on to it—illustrated to them perfectly how Beijing has managed to capitalize on the Congolese mining environment to take charge of the supply chain for battery metals. "China Inc. has realized how important cobalt is," Glasenberg, then CEO of Glencore, said in 2021 at a *Financial Times* conference on the future of the car. "They've gone and tied up the supply." (That's not to say Glencore wasn't selling to the Chinese, U.S. officials were quick to point out to me.)

Glasenberg, who would soon step down from the helm of Glencore, spoke on video chat from a conference room in the depths of the company's gunmetal-gray headquarters in the town of Baar. Up until that point, only a few companies had woken up to China's ascendency in the critical-metals supply chain. "The gap is growing every day, and Chinese dominance is growing every day," Brian Menell, the CEO of TechMet, a firm that invests in projects across the critical-metals supply chain, told me. "If Tesla reaches the goal of twenty million electric vehicles by 2030"—

a goal that even Elon Musk himself would back down from—"that is two times the world's current supply of lithium. Where's that going to come from?"

Tesla had already been thinking about where its supply was originating. Musk's company had begun to look into developing alternative supply chains, and it had started early, in the mid-2010s. "There were definitely concerns and discussions about the availability of metals," said Andrew Stevenson, who worked in Tesla's division for special projects from 2015 to 2018. "Elon's the only one who understands this," a Glencore official said when we spoke in 2021. He explained that Tesla had innovated early by buying cobalt for the long term in so-called offtake agreements, which up until that point had mainly been used by Chinese battery firms and cobalt-processing companies.

Vivas Kumar, a Stanford graduate who worked on Tesla's supply chain for batteries from 2016 to 2019, told me that there had been concerns around cobalt even before Amnesty International and AFREWATCH released their landmark report on artisanal mining in 2016. "There had been a direct interaction from people at Tesla to mines in the DRC even before I arrived," Kumar told me. "People came from headquarters with boots on the ground to see with their own eyes exactly where the cobalt was coming from." They didn't want artisanally mined cobalt in the supply chain, and what's more, they wanted a steady supply of critical metals. Tesla began staking claims on raw materials, from cobalt to nickel to lithium. It started signing contracts that locked in supply for three years, five years, sometimes even ten years. "We wanted to move away from one- to three-month contracts to signing much more long-term contracts."

Other companies seemed to be burying their heads in the sand when it came to what was happening in China. Apple, which used cobalt in nearly all of the batteries for its devices, didn't appear to find Chinese control a threat, even as Xi Jinping was taking his country in a more and more nationalist direction. As one analyst told *The New York Times*, "The Communist Party is firmly in control, and both Western companies and Chinese companies in the private sector have been under attack." But, the *Times* noted, firms like Apple had no plan B.

Furthermore, and perhaps most importantly, governments didn't seem to understand the complexities of the supply chain and the risks of over-concentration in China, even as they passed laws and gave out subsidies that pushed people to buy electric cars. On the keynote presentation in 2021, Glasenberg had issued a dire warning. "The Western companies have not done it," he said, referring to investment in mining. "They either don't believe this is an issue or they believe they are definitely going to get the batteries from China." He concluded with a critical question: "But what happens if that doesn't occur and the Chinese say, 'We are not going to export batteries; we are going to export electric vehicles'? Where are the batteries going to come from?"

CHAPTER 40

GREEN-TINTED GLASSES

As the Kabila years came to a close, more and more mines found their way into the hands of Chinese companies. Sometimes they were pried out of the hands of their owners, people like George Arthur Forrest. Government and Gécamines officials would make it hard to do business. For example, at the Luiswishi copper-and-cobalt mine, which Forrest had operated since the mid-1990s, officials forbade the refining of metals on Congolese territory. "We lost a lot," Forrest later wrote. "We continued to exploit the Luiswishi mine, but it just wasn't as profitable anymore." The new head of Gécamines, Albert Yuma Mulimbi, moved to take control of the mine, and Forrest eventually sold his share for peanuts. "As soon as he had got it back, he gave it to the Chinese," Forrest later wrote. As Peter Zhou told me, Chinese firms were experts at playing the games of Congolese officials, whereby huge kickbacks would find their way into the pockets of people willing to do them favors.

But other mines seem to have been just as willingly sold by foreign firms. China Molybdenum had run the Tenke Fungurume mine since 2016, when it bought Freeport's stake. The majority of the company's shares are now held by a Chinese state body and by Yu Yong, a media-shy magnate who was, in 2025, the world's 307th-richest person, holding a net worth of nearly $9 billion, according to *Bloomberg*. Yu also held a large stake in CATL, the battery giant founded by Robin Zeng. Because of its acquisitions in Congo, China Molybdenum became the world's largest cobalt producer in 2023, overtaking Glencore.

China Moly, as the firm is known in the industry, or CMOC Group Limited, as it is now called in corporate filings, went from being a small molybdenum miner in the Chinese interior to a $21.7 billion international-commodities company during the first decades of the twenty-first century. Molybdenum, like cobalt, is a metal that comes into play quite a bit in modern life—it is used to create powerful alloys, as well as anodes used in X-ray mammography; it is also used in combination with cobalt to desulfurize natural gas and refined petroleum—this means that fossil fuels also sit within Congo's supply chain.

In the wake of the acquisition of Tenke Fungurume, China Moly had become a copper-cobalt giant. Zhou, who worked on financing this acquisition by China Moly, praised the mine's operations. "You'll be shocked to see how modern-equipped it is," he told me when we spoke in 2019. "The people working there—I mean, even the local Congolese employees—they're properly protected, and they have good pay, and they have a good social life as well. There's bars, there's games." But other miners were less convinced about Chinese businesses in Congo. "They cut corners—they don't do things correctly," Charles Carron Brown, the consulting mining engineer, told me. "They don't protect the environment as well as they should."

One of the things that has allowed China Moly to thrive is the company's easy access to cheap financing through large, and often state-run, Chinese banks. Banks and big Chinese companies have close relationships that lead to very low-cost financing. Such deals have allowed Chinese firms to carry out their ambitions in the critical-minerals space. "The strength of China is that they can sponsor low-margin businesses," said Brian Menell, the TechMet CEO. "It's not a level playing field." China Moly's consolidation of its hold of Tenke Fungurume shows how far Chinese banks will go to help their compatriots pursue dreams of domination. "China Moly is a big client for Bank of China," Devon Archer—one of the founders of BHR Partners—told me in January 2022. "The original idea of BHR was to be a cross-border outfit. It was a different time. Oligarchs used to be cool. We had a warm relationship with China," Archer said. Hunter Biden, President Biden's son, was a cofounder. "That soured very quickly."

Archer, who would later be sentenced to prison time for defrauding a Native American tribe in an unrelated deal (only to later receive a full pardon from President Donald Trump), was involved with helping China Moly fund the acquisition, beginning in 2016, of another tranche of TFM's stock. This time, China Moly was interested in buying the tranche that Lundin Mining still owned: 24 percent of the mine for around $1.14 billion in cash. Archer understood that Lundin wanted China Moly's money to cover a commodities shortfall in Lundin's portfolio. "The Bank of China brought us this deal," Archer said. "They needed an affiliate to be an equity placeholder. They loaned BHR money to buy Lundin. It went through a BHR account, and BHR got a fee—$300,000 or $400,000."

It was a no-brainer for BHR and its president, Jonathan Li. "You have this great fee opportunity that covers payroll for the last seven months. Of course you take it. It was a foregone conclusion, and there was no vote," Archer explained. During the sale of the Lundin tranche of TFM, Hunter had sat on BHR's board, but Archer, who was close to the president's son, told me he was not involved in the transaction. "None of the U.S. board members sat on the investment committee," he said.

There was only light due diligence on the deal. "It wasn't like buying a company," Archer insisted. "It was nothing like that. It wasn't a principal investment; we had no equity stake, no ownership. We were just a conduit for capital," Archer said. "It was Bank of China trying to get around their banking restrictions." (The Bank of China did not respond to requests for comment.) There has never been a suggestion that BHR did anything illegal or improper at the time of the TFM deal, although the optics of American citizens being involved in the sale of such an important asset to a Chinese firm rankled some. Gécamines did protest China Moly's purchase of the mine, but it swiftly dropped its objections after the firm paid it $100 million in an opaque transaction that went down in early 2017.

On a separate occasion, BHR had invested around $15 million in the Chinese battery behemoth CATL. It was a successful investment, Archer said, and one that he knew well. He even sang karaoke with the firm's CEO when he traveled to China to close the deal, he remembered. "They were there at the right time, and the right place, and they were"—like China Moly and Huayou—"well capitalized."

China Moly also found partners to mine with, demonstrating just how interconnected large Chinese firms had become at the highest level. In 2020, the firm bought 95 percent of the Kisanfu mine, a greenfield cobalt site near Kolwezi, from Freeport-McMoRan for $550 million. Kisanfu has an estimated 6.8 million tons of cobalt and 3.1 million tons of copper. CATL took a stake in the mine in 2021. The company had already invested in a lithium project in Australia and a nickel project in Indonesia; now it owned a mining concern that controlled one of the most important cobalt deposits in Congo. The web of connections that bound these companies to one another was strong and made them stronger, despite outward appearances of rivalry. During the COVID-19 pandemic, for example, Huayou helped bail out CATL when it suffered a dearth of cobalt due to supply-chain disruption. Other Chinese battery and EV firms, such as BYD and Gotion, a firm that was perfecting fast-charging batteries, would soon follow CATL's lead into the critical-minerals space, investing in projects as far-flung as Chile and Indonesia.

As for the TFM deal, Archer told me that looking back on it in light of China's rapid expansion into the battery space, he felt differently about it in 2022 than he had four years before. At the time, he said, "I was thrilled with the investment. I guess I had rose-tinted—or, rather, green-tinted—glasses on."

CHAPTER 41

SEIZING THE SPACE

The highway along which 70 percent of the world's cobalt
is exported, somewhere near Fungurume, Congo, in 2019

The town of Fungurume announces itself not by a gradual accumulation of buildings but by a barrier. Specifically, a metal-spiked barrier tossed by a set of bored guards into the roadway. A sign tells drivers to STOP and declare the "provenance of products." China Moly had seized control of the public road to ensure that no one was stealing minerals. Technically, the checkpoint was overseen by the OPJ, or judicial police, but I never saw an officer from the unit on any of the eight times I traversed the road.

It was one of the most egregious examples of the corporate seizure of public infrastructure in a country that was replete with them. The road was effectively controlled not by the police or by the Congolese army but by armed guards in the employ of a Chinese company with a sizable investment from the Chinese state. Since the Benguela Railway, which ran through Angola, was still not in use and the line to Lubumbashi was operational only once or twice a month, the highway was the export route by which some 70 percent of the world's cobalt left Congo.

Once, in October 2019, during the dry season, Jeef Kazadi and I entered Fungurume, which consists mostly of low-slung buildings with the occasional two-story arcade: a hotel, a set of shops.

Painted across the pink facade of Chez Papa Lofo, one of the larger shops in Fungurume, are the words DIEU DIRIGE MES AFFAIRES (God Directs My Business). Papa Lofo might as well have written "China Moly Directs My Business," for through the process of running the mine in Fungurume, the company has acquired the power of life and death in the region.

The barrier, however, hasn't prevented people from smuggling copper and cobalt out of the concession. Because of its sprawling size, Tenke Fungurume is difficult to police, and huge communities of artisanal miners like Odilon Kajumba Kilanga scrape a living from the land there.

In June 2019, I had planned to visit the mine, but my scheduled trip was canceled by China Moly at the last moment because an "invasion" of artisanal miners threatened the site. That year, China Moly's subsidiary decided to ask the government to do more to help it rid the site of an estimated ten thousand illegal miners. In response, the government sent in the army. (The company's deputy general director later insisted that TFM had not asked for troops to remove the miners.) Eight hundred men were deployed under the command of General John Numbi, the former national police chief who had been sanctioned by the U.S. Justice Department in 2016 and was notorious for not only his repression of protests but also his alleged role in the 2010 killing of a human-rights activist and the activist's driver. (Numbi denies any role in the murders.) At the time, Numbi told *Bloomberg* that the operation had been personally ordered by Félix Antoine Tshisekedi Tshilombo, Congo's president, but in the spring of 2023, Numbi told me that it was he, in fact, who had suggested to Lualaba Province's governor, Richard Muyej Mangez Mans, that troops be used to force the miners out. Numbi showed no remorse: In fact, he remained proud of the operation to expel illegal miners.

The Congolese army arrived in the settlement of Kafwaya on June 13, 2019. Between Tenke and Fungurume and a little to the south of the east-west highway that connects East and West Katanga, Kafwaya was one of several ramshackle mining boomtowns scattered across southern Congo's bush. There was money to be made in the town, which had sprung up on the back of the illegal trade in TFM's minerals, mined by jobbing artisanal miners. But despite the wealth that lay beneath the soil, the town was not much to look at: It consisted mostly of shanties made from sacks, tarpaulin, pipes, branches, and the odd mud-brick.

After arriving in Kafwaya, Numbi's troops warned miners to leave. Their settlements were on TFM's property, after all. The year before, Muyej had railed against them. "Imagine yourself on less than a square kilometer, with one hundred and sixty makeshift houses built from tarpaulins," he told the UN's Radio Okapi. "It looks like hell." But the Congolese troops were asking too much of the people at Kafwaya, many of whom were subsistence farmers and had scant little to do with artisanal mining. A lot of the miners and the people living in the settlement decided to stay put.

A few days after the soldiers arrived, the local civil-society organization reported that the Congolese troops had begun to burn Kafwaya's market stalls. The message was clear: *Leave immediately.* Still, families remained: Their homes, their school, their lives were there. Then, a couple days later, as night fell, the battalion under Numbi charged through the town, burning dozens of homes and ransacking the school. A three-year-old girl and a fourteen-month-old boy were badly burned. (Numbi denied that anyone got hurt.) "They didn't say anything to anyone," an official in the settlement named Fabien Ilunga told Reuters's Aaron Ross. "The army started to burn down the tarpaulin houses."

"If the investors complain," Numbi told Ross, "the government will take measures (to deploy the army) if it decides the police cannot handle it."

But Numbi's operation—and the checkpoints—did little to discourage artisanal miners from working on the site and trading cobalt and copper ore. Each time I crossed the TFM site by car, I overtook scores of motorcycles laden with the type of cheap polypropylene sacks used to transport artisanally mined ore. Minibuses filled with such sacks zipped to and fro down the road. Shantytowns made up of homes cobbled together from tarps and woven assemblages of the same sacks also appeared to rove about the region. I was told that there were occasional police or army raids, and the make-do dwellings would be dismantled in one location, only for them to spring up somewhere else. It was a game of Whac-A-Mole for the army, although in this case, the tunnelers were human beings operating at the squalid bottom of a globalized economy.

AT AROUND 6:00 P.M. ON A GRAY AFTERNOON IN MARCH 2023, I STOPPED with Kazadi at a settlement near Kafwaya to see if I could speak to artisanal miners there. People were coming off their shifts, walking along the paths that cut through the long grass by the road, heaving sacks and metal bars that they used to pry rock loose. The miners were easy to identify: They had cheap plastic flashlights strapped to their heads or swinging from their hands. A few young miners declined to talk before a man in a checkered shirt came up to me and introduced himself as Marcel. Just then, a kerfuffle broke out, and Marcel jumped into our car. We agreed to drop him off in Likasi, where he said he had an appointment.

Marcel was thirty-four. He politely told us that he was a *négociant*, or ore trader, and that he had five children, all in school. He spent much of his time buying minerals at Kafwaya. "The little I make here allows me to put food on my table," he said. He would spend three days at a time collecting minerals to sell. "If God has blessed me, I can make one hundred thousand francs, even one hundred and fifty thousand francs," he told us. This was around $50 or $60 a day.

"How much do the *creuseurs* make?" I asked.

"It depends on the quality of the minerals," he said. "They can make maybe a million francs for three *bombés*. Maybe a million-two, a million-three." That was around $500 to $650 at the time. (A *bombé* is a unit of measurement that equates to around one hundred kilograms.)

Marcel had worked in the mines since 1997, when his father died and Mobutu Sese Seko was ousted. He was nine then. "Things had been tough at home," he said, "so I went to work in a mine called Milele, on the Zambian border." He remembered the names of all the mines in which he had worked—a long list, names I had never heard of, places far off in the bush and accessible only by dirt road. The list underscored what a difficult job would-be mining regulators were facing in Congo. Some copper-and-cobalt mines were just so far away, so small, and so little known that oversight was near impossible and almost anything could happen in them. "After Milele, I went to a mine site called Kabundji. After Kabundji, I went to a mine site called Sandra, then Kipese, Kimanyuki, and Kamfundwa, which is near Kambove," Marcel told me. "Then, after Kambove, I came to Lwambo, to a mine site called Kalabi, then to a site called Kwatebala, and after Kwatebala, that was how I found myself here at Kafwaya."

He began telling me about how children did not mine in Kafwaya: They washed and carried sacks of minerals. At night, these bags would be smuggled to Likasi, where buyers awaited.

"Where do they sell the ore?" I asked Marcel.

"We sell it in Likasi, to the Chinese," he said. "The Chinese of CDM." Marcel was telling me something extraordinary: CDM, which was owned by Huayou, one of China's largest cobalt-mining and -processing concerns, was stealing from China Moly, also one of China's largest mining firms, in broad daylight. In Indonesia, on the island of Sulawesi, the two firms even worked together on the Huayue Nickel Cobalt Project. But in Congo, it wasn't unusual for Chinese operators to steal from one another. "The Chinese, they hate each other most of all," Marcel said.

CHINA MOLY'S SEIZING OF THE ROAD—THROUGH AN OPAQUE DEAL WITH A local governor—was a powerful metaphor for Chinese investment in the region and in the cobalt supply chain. The control of public space by private enterprise was nowhere more obvious in Congo than here. In fact, China Moly could conceivably hold up all the copper and cobalt shipped out of Kolwezi, at least for as long as the company controlled the highway.

Little of the wealth being generated a few miles away appeared to have trickled down to the people living in Fungurume, a town of dirt roads on the verges of the Cobalt Highway.

When I visited, in 2019, Fiston, a motorcycle driver, told me he had worked in the mines at one point but had left when he realized that "it was another form of bondage." He showed me his home—a hot, cramped four-room house made of cinder block—which he shares with his wife and two children.

Kazadi and I walked through Fungurume with Fiston and visited a medical clinic that had barely any equipment or medicine. Suddenly, a man appeared, claiming to be with the Congolese secret police. He tried to arrest us but had no vehicle. He had to hail a moto-taxi, which couldn't catch our pickup. We were soon clear of him.

I asked Kazadi whether all intelligence agents in Congo were so poorly equipped. They had no funding, he explained. Even though the country is wildly rich in natural resources, it cannot afford to spend money on public services. Public funds are plundered by corrupt politicians, and large companies pay off politicians to take control of the country's mines. Big industrial miners liked to draw a clear dividing line between their practices and those of the "illegal" artisanal miners, but in many ways, they were two sides of the same rusty coin, depending on each other to exist.

YEARS AFTER THE TFM SALE, BEGINNING UNDER THE FIRST TRUMP ADMIN-istration and continuing into Biden's term of office, there would be some soul-searching as to *how* the U.S. let China, its main adversary in an increasingly complex world, take control of the world's largest copper mine. Republicans and Democrats blamed each other for being asleep at the wheel. The GOP pointed to Hunter Biden's connection to the second sale of TFM shares, and in 2020, as Joe Biden ran for president, Trump's campaign took to calling him "China's puppet." In 2021, *The New York Times* ran an article headlined HOW HUNTER BIDEN'S FIRM HELPED SECURE CO-BALT FOR THE CHINESE. Therein, Lundin and Freeport executives expressed puzzlement at BHR's participation in the deal, but naturally, they did not go into detail about how *they* had helped secure cobalt for China Moly.

Chinese companies like China Moly, Huayou, and Sicomines had expanded in Congo under administrations from both parties. China Moly had bought assets under Obama and Trump, and Biden was in power when CATL bought its share in Kisanfu. Perhaps culprits could be found among the U.S. foreign policy establishment and the State Department, which had underinvested in Africa; diplomats would often say that Africa was playing sixth string in Washington's continental orchestra. But the general antipathy of most U.S. companies toward working in Africa was also another important reason for the decline. Far better, executives stateside had theorized, to let China do the dirty work and then reap the profits further down the chain.

The 2020s were turning out to be a mirror of the late 1970s, back when Congress had wised up to Soviet stockpiling of critical minerals. "There seemed to be a real desire to suggest there was some way the Obama administration had screwed this up and let their eye off the ball," Tom Perriello, the former African Great Lakes envoy, told me. "The real story was that the U.S. government just doesn't have that many tools to address this." He went on: "The great strength in our system is that our companies are independent of our government. But it also creates a weakness, whereas there is a fairly tight relationship, or at least a *dynamic* relationship, between Chinese corporations and the Chinese government." U.S. corporations in the 2010s and '20s did not operate like they had during the twentieth century. "During the Cold War, U.S. corporate interests typically aligned with expansion of the American middle class at home and with governments not going pro-Soviet abroad." Their focus these days was shareholder interest, not patriotic duty. "U.S. companies not only don't have an obligation," he said, "but often see it as, well, it's *not* their obligation to advance U.S. foreign policy."

Part 5

WAKING UP

"High speed is the critical factor which makes transportation socially destructive. A true choice among political systems and of desirable social relations is possible only where speed is restrained."

—IVAN ILLICH, *ENERGY AND EQUITY*

CHAPTER 42

GOD IS SMILING ON CONGO

Y*ou know what, God is smiling on Congo,* the thought went. *He hasn't smiled on Congo in a long time, but now he has begun to.* The sentiment belonged to Amos Hochstein, an energetic forty-nine-year-old. It was September 12, 2022, and for just over a year, he had been the senior adviser for energy security at the U.S. State Department, as well as a special presidential coordinator for global infrastructure and energy security under President Biden. Hochstein, who had a mop of dark hair that he combed across his head when on official business, looked slightly out of place among the stuffed suits of Congolese officialdom. As he sped through the streets of Kinshasa, past gray tower blocks whose only development since the days of Belgian colonialism had been in the realm of rot, past heaving slums filled with those who had fled the country's lawlessness and violence, he hoped he could convey this message—that God had indeed smiled on the Democratic Republic of the Congo. In return, he hoped Congo would have the good sense to smile back. And, come to think of it, he hoped Congo would especially smile on the United States.

Hochstein's life had been dominated by questions surrounding the world's energy supply: He was born in Israel, in 1973, just as the Arab-Israeli crisis was working its way to a bloody nadir, and in his adult life, he had worked on energy issues for three Democratic administrations, starting in the 1990s. In Kinshasa, a Congolese newscaster hailed the arrival of Biden's "Mister Energy."

In 2022, the world's energy map was looking more uncertain than it had for years. In Ukraine, Russian forces were retreating against an onslaught of Ukrainian troops. Vladimir Putin was becoming more and more aggressive in his threats to use natural gas as a weapon. Pipelines had been bombed, coal mines reopened, climate goals ignored. European politicians braced for a long, cold winter ahead. In Paris, the mayor had just announced that the Eiffel Tower would go dark after 11:45 p.m. to emphasize the need to save energy. Russia, along with the Organization of the Petroleum Exporting Countries, or OPEC, was just about to announce a cut in production of one hundred thousand barrels that October. Oil prices would go up. Inflation in a post-pandemic world was spiraling out of control.

Hochstein had wanted to visited Congo ever since he'd been appointed to his position. An earlier trip had been canceled when he contracted COVID. He had become particularly concerned with the fate of the Central African nation, even in the midst of rolling Middle East crises and revanchist mayhem in Ukraine. He was worried that the country was falling into Beijing's increasingly totalitarian orbit. The U.S. had for years neglected a swath of countries that had become integral to the battery supply chain, even as Xi Jinping's China continued to break ground at construction sites and woo leaders around the developing world. Now the U.S. was reviewing its critical minerals and where they came from in the wake of Biden's Inflation Reduction Act, as well as Executive Order 14017, which directed the Department of Defense to scrutinize its supply chains.

Hochstein wanted the Congolese to say to themselves, *We have a natural resource that is as critical in the twenty-first century as oil was in the twentieth century*. Batteries and renewable energy more broadly were, to him, "a national security imperative. It is a national, a global, economic imperative." A supply-chain review by the DOD predicted that, with electrification, "reliance on China will grow and China's relative cell dominance is projected to remain stable." The review also urged the U.S. to work toward making sure that "reliance on China's cells and material does not inadvertently grow."

Hochstein feared that the U.S. had ignored these issues for too long, and that they would very soon come back to haunt the country. His trip to Congo was an attempt to stanch the bleeding before it was too late. He wanted to get one thing across: The U.S. was getting serious about critical minerals.

ONE OF THE INITIATIVES THAT HOCHSTEIN WAS TRYING TO PIONEER WAS haunted by the ghosts of the 1970s. The Biden administration wanted to use the Benguela Railway, which had been destroyed by Simba rebels during Angola's civil war, to ship Congolese ore across a tract of central Angola and western Congo that it was calling the "Lobito Corridor."

In 2023, the U.S., Angolan, and Congolese governments signed a seven-way memorandum of understanding "to accelerate growth in domestic and cross-border trade along the Lobito Corridor." The prime way that they would do this would be by reopening the Benguela Railway, which had originally been built in 1903, during the colonization of Katanga. The Americans and the Europeans now thought they could extend the Lobito Corridor to Zambia, further connecting the region.

Envoys from Washington spoke about investment, but little came. Michael R. Hollomon II, the commercial director of a mining firm called U.S. Strategic Metals, told me that he had tried to convince the Biden administration to allow his firm to invest in Congo. "There were payments that had to be made in Congo," he said, not specifying exactly what payments these were. "The State Department said no." A State Department official lamented to me that the U.S. government was "playing catch-up" in African countries, because the perception is that it is difficult to do business with Washington. There are no U.S. businesses in Congo, the official said, "apart from a Coca-Cola bottling plant," because U.S. firms are scared to invest. A Washington lobbyist told me that the U.S. government does not help investors who want to buy stakes in Congolese mines, despite overtures from the government about "critical minerals."

What little U.S. investment there was barely seemed serious. In 2021, *The New York Times* ran a piece titled "Who Are Congo's Cobalt Entrepreneurs?," naming figures who were famous in the U.S. business and celebrity arena: everyone from Akon, the Senegalese American R&B singer, to Jide Zeitlin, a disgraced fashion-industry CEO, whom the report claimed had a "laser focus" on Congo's mining sector, to Dikembe Mutombo, a Congolese American basketball player. Mutombo told the *Times* that "his investment plan could give him and his partners access to one of the country's largest cobalt and copper mines by pressing the Swiss-based mining giant Glencore to sell it." He even showed the paper a letter, signed by seven Democratic senators, saying that Glencore was thinking about selling its mines in Congo. But anyone who knew anything about Congolese mining had never heard of Mutombo as a mining investor. Glencore certainly hadn't, and since the *Times* had published their piece his name had not surfaced again.

People involved in the Congolese mining world found the names more than a little unbelievable. By 2024, of the "entrepreneurs" mentioned by the *Times*, only Erik Prince, the mercenary CEO of Blackwater infamy, had anything that appeared to be a sustainable business in Congo, and that was to train guards at Chinese mines and support the government in its fight against rebels in the East.

Akon had invested in a mine alongside Red Rock Resources, Andrew Bell's firm, which Glencore had already squeezed out in Kolwezi. Akon's project looked to turn a little-known site previously exploited by a George Forrest–linked company into a full-fledged industrial cobalt mine. After more geological work was done, the mine turned out to be "exhausted," and Congo's finance directorate announced that it was opening an investigation into the company's finances.

The Lobito Atlantic Railway was an attempt to get serious. The development had the potential, as a report by the Cobalt Institute (an industry body that tracks the cobalt market) put it, to "counter Chinese involvement in the African Copperbelt." It was popular in Washington.

But it ignored one key detail: It was China, not the United States, that had spent $1.83 billion rebuilding the railway in Angola. By 2019, twenty Chinese workers had died building the line. The first shipment of copper and cobalt had taken place under Chinese auspices. Government officials in Luanda were keen to work with the Chinese again; by 2024, Angola was the second-largest country for Chinese overseas investment, and Chinese firms were discussing the possibilities of investing in Angola's huge offshore oil reserves.

THE LOBITO ATLANTIC RAILWAY WOULD CERTAINLY BE A GOOD THING FOR Congolese people who lived on the Cobalt Highway out of the country. Exporting cobalt by road had been an improvised measure, devised in the 1970s to get around the problems in Angola. Since the time of the rebels and the Angolan Civil War, however, politicians and powerful local interests in Congo had come to control trucking routes. Why would they destroy their lucrative business moving material south?

In 2019, I visited Kasumbalesa, the border through which most of Congo's copper and cobalt is shipped. The lines of trucks running their engines stretched for miles, and the towns around the border posts were thick with acrid fumes. The road is also dangerous. Trucks careen off it and kill people with alarming frequency. Adelle Hollmer, a student who wrote about artisanal mining in Congo in 2024, saw a man crushed in an accident: "There were no ambulances, no police, and most shockingly no one seemed alarmed or even unnerved." A representative of Ivanhoe, a firm that has developed the huge Kamoa-Kakula mine, told me that the mine prohibits foreign employees from driving on the road, as it is deemed unsafe.

Congolese on the Cobalt Highway complain that the drivers of the cobalt trucks, who stack long-distance round trips of forty to fifty days one on top of the other, are not in full control of their faculties. One evening in 2022, a truck smashed into the side of my vehicle at a checkpoint outside Likasi, about midway between Kolwezi and Lubumbashi.

The driver of the truck, which was laden with sacks of cobalt hydroxide on their way from a Chinese-owned mine near Mutanda to the port of Durban in South Africa, was Zambian. His firm had been subcontracted by the French logistics giant Bolloré, which ships much of the copper and cobalt out of Congo's mines. He was stoned. After the crash, he stumbled from the cab of his truck. I pointed to the mangled fender and door of my rental car. Slurring his words, he asked if I wanted to go and smoke a joint.

Would the Lobito Corridor end up like so many other well-intentioned plans for Congo and Central Africa, confined to memory and a series of documents gathering dust in an archive? In the summer of 2024, the *Financial Times* heralded the project as a soft power blow against China, calling it "the U.S. railway that could set off a copper war."

Later that year, President Biden visited Benguela, the terminus of the corridor, with Angola's president and Congo's president, Félix Antoine Tshisekedi Tshilombo. (U.S. officials had privately expressed frustration that Congolese officials working on the project had constantly asked them for money as a precondition to meetings.) At a meeting, Hochstein pointed out that Biden had "set the vision" for the corridor at the U.S.-Africa Leaders Summit in 2022. "Critical minerals our world needs for electric vehicles and semiconductors can be found here," the president would say in December 2024, on the only trip to Africa of his presidency. "Clean energy we need to power artificial-intelligence data centers and economic growth can be built here." He continued: "Nations across the Lobito Corridor have solutions to some of the world's toughest problems." Tshisekedi hailed the project as "a driving force for economic and social transformation for millions of our people."

All this didn't change the fact that the project hailed by Washington still didn't really exist, not beyond a few printed pages, memoranda of understanding, signed by diplomats and politicians. A new, vastly more unpredictable president was coming to Washington: When Trump took power again in 2025, the new government said it was focusing on more concrete dealmaking with Kinshasa. In policies that hearkened back to a former era of U.S. policy, Trump wanted the extraction of critical minerals, not just their means of export. He was agitating for them in Ukraine and Greenland as well as Congo.

Massad Boulos, the father-in-law of Trump's daughter Tiffany, was appointed senior adviser for Africa. In his first trip to Kinshasa in his new role in April 2025, Boulos said he envisioned "fostering U.S. private sector investment in the DRC, particularly in the mining sector." The railroad wasn't mentioned, and some thought Trump would move away from a project around which Biden had centered his critical minerals policy in Africa. ("I mean, the Lobito Corridor, what was *that?*" mused one businessman connected to Congolese mining early in Trump's term.) Boulos was more focused on dealmaking. He shuttled to Central Africa in order to broker peace between Kinshasa and Kigali, after Rwanda had supported rebels in snatching a chunk of Congo's mineral-rich east. After an agreement to end hostilities was signed, in mid-2025, *Bloomberg* reported that a U.S. consortium that included several former intelligence and special forces officials was looking at buying Chemaf, a firm that owned several mining permits in Congo and three projects with processing plants. When a Chinese company had tried to buy Chemaf earlier in the year, the Congolese government had held back necessary approvals for the deal, presumably in a show of good faith toward the China-phobic Trump administration.

The Lobito Corridor's future was unclear. But James Story, the U.S. ambassador to Angola, told Reuters that the Trump administration was still committed to developing the Lobito railroad and even that the government, which was cutting spending elsewhere, was continuing with the $550 million loan it had pledged to the project. When KoBold, a U.S. mining firm backed by Bill Gates and Jeff Bezos, began looking at developing a lithium mine in the southern Congolese town of Manono, officials there imagined the means of egress for the lithium ore would include an extension of the Lobito Corridor. But, still, by the time Trump's envoy was flying to Congo, little copper and cobalt had left Congo and passed through Angola by rail on its way to the U.S. For the moment, road transportation still ruled.

CHAPTER 43

THE SPICE OF LIFE

A wall of earth curls through over fifteen hundred miles of howling dust and scrub on Africa's left ear. Around it, the vast desert plains of the Western Sahara are empty and flat, though it is occasionally marked by horseshoe dunes that move with the wind like a cast of crabs traversing an ancient seabed. The Berm, as the wall is known, is no natural phenomenon. Built by the Kingdom of Morocco, it is the longest defensive fortification in use today and second only to the Great Wall of China in terms of absolute length. It stretches from an Atlantic peninsula at its southern tip to well inside Morocco's border at its north end. It is visible from space. Over one hundred thousand Moroccan troops man its barricades, and it is surrounded by land mines and barbed wire.

Amos Hochstein's trips overseas had concentrated attention on critical minerals in countries like Congo; places like the Western Sahara hardly got a look-in. But inches beneath the contested desert, and sometimes even at its very surface, is phosphate, a resource that the world would soon be clamoring for as it sought cheaper, cobalt-free batteries.

The United Nations considers the Western Sahara to technically be "a non-self-governing territory." Those who don't mince their words call it "Africa's last colony." On the desert side, a movement called the Polisario Front is fighting a low-intensity war on behalf, it says, of the people who inhabit the region—the Sahrawis. The rebels speak Arabic and Spanish, because their land was originally colonized by Spain. In Spanish, their group's name stands for the Popular Front for the Liberation of Saguia el-Hamra and Río de Oro.

The parameters of the war were effectively frozen in 1991, when a ceasefire was agreed to, pending a referendum that never came. Like any war, this conflict affects civilians most of all.

Perhaps the best-known Sahrawi internationally is Aminatou Haidar. She is from Laayoune, the capital city of the occupied part of Western Sahara. Laayoune is a jumble of pinkish concrete that is often racked with demonstrations against Moroccan rule. A slight, birdlike woman in her fifties, Haidar, who has long been advocating for nonviolent resistance, has been beaten, imprisoned, detained, and interrogated by the Moroccan security services since she was a young woman. She was nominated for the Nobel Peace Prize in 2007, and she has been nicknamed the "Sahrawi Gandhi."

Haidar is physically frail: After a hunger strike, she acquired a severe calcium deficiency that left her bones brittle and the vertebrae in her back warped. She wears powerful spectacles because her eyesight has been dimmed by the four years she spent blindfolded in a Moroccan jail. When I met her in 2017, Sahrawi activists in Laayoune had just been attacked by the Moroccan police during a demonstration. This was not unusual, according to Haidar. "During the last six months," she said, "we have registered, I think, eighty-six demonstrations that were stopped or repressed."

Phosphate was at the center of the lithium-ion battery boom, even though few people seemed to realize it. The essential element itself was also a resource whose scarcity portended an apocalypse: Its supply on our planet is finite. And when it goes, we go too. The science-fiction writer Isaac Asimov put it best back in 1975:

> Life can multiply until all the phosphorus is gone and then there is an inexorable halt which nothing can prevent. . . . We may be able to substitute nuclear power for coal power, and plastics for

wood, and yeast for meat, and friendliness for isolation—but for phosphorus there is neither substitute nor replacement.

JOHN B. GOODENOUGH'S LAB AT OXFORD HAD BEEN THE FIRST TO CREATE A battery cathode that contained phosphate. In 1987, Goodenough and another scientist, Arumugam Manthiram, wrote a paper showing that lithium ions could be intercalated at room temperature into a type of compound called a polyanion oxide. The oxide contained iron with tungsten or manganese. No cobalt or nickel was used in the structure.

The two scientists' attention shifted, and in the early 1990s, they assigned a PhD student named Geeta Ahuja to continue their work. She expanded their investigation to look at other polyanions. Some of them contained the mineral that is found in such abundance beneath the desert at the Western Sahara's Bou Craa mine—phosphates.

Ahuja's project was eventually shelved because the scientists at the lab believed that the materials she was investigating weren't going to be able to produce as high a voltage as the cobalt cathodes Sony had recently brought into production in Japan. But later in the decade, they decided to revisit the subject.

In 1997, Goodenough, along with two other scientists, announced in a paper that they had discovered a material from which could be made "a cathode of good capacity, and it contains inexpensive, environmentally benign elements." They described this safe and cheap cathode material as "olivine $LiFePO_4$," but it would soon become known as LFP, or lithium iron phosphate.

Their experimentation, the trio of scientists wrote, "shows this material to be an excellent candidate for the cathode of a low-power, rechargeable lithium battery." In electric-vehicle terms, this meant that LFP batteries would probably not power a high-spec, top-of-the-line Tesla sports car, but they would be suitable for a city runabout. Cost rather than mileage would give such a car an edge, as LFP batteries were cheap. Automakers—especially those in China's fledgling car industry—would soon take note.

A couple of decades would pass before such batteries began to be adopted, but in the second half of the 2010s, more and more companies, especially in the electric-vehicle arena, started switching to LFP. Significantly, an LFP patent had not been filed by its inventors in China in 1997, and firms in the country were able to mass-produce the batteries quickly. The batteries would be engineered to become more powerful by companies like BYD and CATL, and China would dominate the manufacturing of the technology.

In 2024, according to the International Energy Agency, LFP cathodes represented around 40 percent of the world's market share. NMC cathodes, the high-powered ones used in the Chevy Volt and in many Teslas, still retained top billing, at 60 percent. But phosphate batteries were on the rise: In the West, they were touted as safer, cheaper, and less geostrategically complex than their cobalt-containing cousins. In China, their low cost and high safety specs were praised by CEOs like BYD's Wang Chuanfu.

The scientists in Goodenough's labs had successfully read the tea leaves. LFP cathodes produced less power than their high-powered cobalt- and nickel-containing counterparts, but they had serious cost benefits. As records show, they anticipated some of the debates around the scarcity of lithium battery materials that only really started to preoccupy battery pioneers, miners, and device manufacturers some two decades later. Their primary motivation, they admitted, had been the cost of materials like cobalt and nickel, which were already expensive to acquire in the 1990s.

———

IN 1975, THE WESTERN SAHARA WAS TAKEN FROM THE SPANISH BY THE MOroccans after an invasion and a series of backroom deals with Francisco Franco, then Spain's dictator. A deadly war was fought on the territory's sands, which were mined and strewn with unexploded bomblets. The territory is the last place in Africa that is subject to the UN's Decolonization Committee, and it remains divided in two by the Berm, which Morocco built to effectively corral the Polisario Front into the most arid and inhospitable corner of the territory. Some two-thirds of the Western Sahara are occupied by Morocco, while a third is controlled by the separatists of the Polisario Front.

The partition's logic is partially economic, thanks to the region's store of phosphates. In the Moroccan sector is the Bou Craa mine, which accounts for, by some estimates, just under 10 percent of Morocco's phosphate production. The mineral is shipped out to sea on a sixty-one-mile conveyor belt that has held the record as the world's longest since the Spanish completed it in 1972.

The Western Sahara's drab plateaus are stacked with phosphate because, like Katanga, the region was once the bed of an ancient sea. Over millions of years, dead animals and plants, shark teeth and skeletons, sank to the bottom of the ocean. Through a series of complex interactions with bacteria, this seafloor litter was stripped of its phosphates, which coalesced into rocks that were pushed to the surface of the desert by the shifting of tectonic plates and the collision of continents. A 1970 CIA report on the region pointed out that the phosphate of the Western Sahara is of "a quality comparable to the best grades of other producers," and that it is "exposed on the surface and can be worked by open-pit mining methods. Thus, although the initial investment requirements are large, operating costs should be low." The report's authors suggested that the phosphate exports of the Western Sahara could "nearly equal" those of Morocco within five years.

But the mere presence of phosphates, still a relatively low-cost commodity, doesn't justify Moroccan expenditures in the territory. For every dirham of profit that Morocco makes from the Western Sahara, it puts in seven. An official at the state-owned phosphate mining concern told me that the kingdom already has huge quantities of phosphate, and that the Western Saharan phosphate makes up less than 10 percent of Morocco's annual product. In a 2014 speech, the country's king, Mohammed VI, said that the Western Sahara is seen as a limb without which Morocco, or at least its monarchy, could not survive. "Morocco will remain in its Sahara, and the Sahara will remain part of Morocco," he said, "until the end of time."

Still, the Western Sahara has its uses. Thanks to Bou Craa, for example, Morocco has been able to set prices on the phosphate market, and its phosphates are used to procure international trade agreements. The kingdom produced forty million metric tons of phosphate in 2022, second only to China. As our global population grows and grows, creating ever more mouths to feed, demand for phosphate is booming. *World Fertilizer* magazine calculated that the market for phosphate fertilizer was growing at a compound rate of 5.5 percent every year. *Morocco World News*, a pro-Moroccan publication, called phosphate "the spice of life."

CHAPTER 44

LITHIUM IS SEXY

In 1987, when she was twenty, Aminatou Haidar organized a demonstration in advance of an official United Nations visit to the Western Sahara. "We wanted to transmit a clear message to the UN that we wanted independence," Haidar told me. "Morocco underwent a huge campaign of arrests that targeted both sexes. Women and men, old and young—nobody was spared. They started by arresting over five hundred people."

One night, three or four days before the UN was slated to arrive, the Moroccans came to Haidar's house. "I was arrested," she said. "They put me in a car, and they started to quickly drive around the street." She was taken to a police barracks near her home. The car circled the streets of Laayoune to give Haidar the impression that they had traveled farther than they had. She was worried that she, like some of her relatives, had been taken to a secret jail inside Morocco, and she feared she would never return.

For most of Haidar's imprisonment, her family didn't know where she was. She had been disappeared by the Moroccans. (Under international humanitarian law, enforced disappearances are considered crimes against humanity.) She was held in solitary confinement for a year. "When it was very cold, it was freezing; on the ground, where I was lying, we had no covers," she told me. "In the first year, I contracted rheumatism because I was thrown into a corridor where it was really cold. And in the summer, it was boiling." Some of her fellow prisoners were bitten by dogs set on them by policemen.

"We didn't know if we would be able to leave from this hell," she said. She thought she would be locked up forever. "Our faith in God helped us through. And we were also determined, because we were convinced that we had all the justice of our cause and international law on our side. We were convinced that the Sahara is not Moroccan and that tomorrow or after tomorrow, it will gain its independence."

In 1991, Haidar was released into an atmosphere of new hope for the region. Morocco and the Polisario Front had agreed to stop fighting and to hold their referendum. But, as earlier noted, through the 1990s the promised referendum never arrived. In the early 2000s, James Baker, the former secretary of state who was then the UN secretary-general's personal envoy for the Western Sahara, probably got closest to negotiating a long-awaited deal, but Morocco's king managed to scupper it. "Mr. Baker's proposals endanger the very founding principles of the Kingdom," Mohammed wrote in a letter to President George W. Bush, raising the threat of the "redeployment of terrorist groups in the region." In the margin of his copy of the king's letter, now at the Princeton University library, Baker wrote "WRONG!"

Morocco sought to control the narrative after that point. Plans for a referendum were continuously delayed and undermined. When journalists attempted to report on the Western Sahara, they were followed, harassed, and removed. In the late spring of 2017, I tried to visit Laayoune, the capital of the territory. I was stopped before getting off a Royal Air Maroc flight and then sent to Las Palmas, in the Canary Islands. By chance, Haidar, who had become an internationally renowned activist after her imprisonment, was also traveling to Las Palmas, where she would attend a conference; she told me about her long history of activism, all her hunger strikes and protests to bring awareness to the plight of the Sahrawi. "I have lived the suffering in my own flesh," she said.

In December 2020, as hostilities flared up between Morocco and the Polisario Front, President Trump announced recognition of Moroccan sovereignty over the Western Sahara through a tweet ("Morocco's serious, credible, and realistic autonomy proposal is the ONLY basis for a just and lasting solution for enduring peace and prosperity!"). The United States set up a consulate in Dakhla, but it stopped short of actually recognizing Moroccan sovereignty. Since then, both sides have fired heavy weapons over the Berm and used drones to attack each other's positions. (The Bou Craa phosphate mine has not been affected.) The problem of the Western Sahara, Baker said, "has not been handled well, and that's why it continues to persist."

———

THE SUPPLY-CHAIN ISSUES SURROUNDING BOTH A TERRITORY THAT IS QUITE literally at war and a commodity whose very existence the world relies upon to feed itself should be concerning. But nobody seemed to be considering it, neither among State Department officials I interviewed in 2022 and 2023, nor at battery conferences in Michigan and Stuttgart. No one professed to have considered supply-chain constraints on phosphates.

One of the arguments in favor of adopting LFP batteries is that the materials needed to make them are much easier to access than nickel and cobalt. Iron, after all, is abundant. "We're definitely not going to run out of iron. There's so much iron it's insane," Mujeeb Ijaz, the U.S. EV start-up Our Next Energy founder, told the *Financial Times* in March 2023.

Only a few people had concerns about a phosphate supply crunch. On a 2021 investor call held by Mosaic, one of the largest producers of phosphate fertilizer in the world, company officials noted that three hundred thousand tons of purified phosphoric acid had been redeployed from use in fertilizer production to use in the making of LFP batteries. That year, India's stock of phosphates plummeted by around a third. Farmers committed suicide because they could not secure enough phosphate for their fields. "It's going to become a battle, and let's face it, fertilizer manufacturing isn't exactly sexy," one of Mosaic's officials said. "Lithium is."

Awareness of the limited supply of phosphates only really began to hit home in 2023. "LFP batteries also contain phosphorus, which is used in food production," a report from the International Energy Agency said that year. "If all batteries today were LFP, they would account for nearly 1% of current agricultural phosphorus use by mass, suggesting that conflicting demands for phosphorus may arise in the future as battery demand increases."

By 2024, the auto industry was using 5 percent of the world's purified phosphoric acid; Benchmark Mineral Intelligence, a consulting firm that tracks minerals used in batteries, predicted that this figure would increase to 24 percent by 2030. Morocco, on the other hand, had become a key piece of the Chinese electric-vehicle supply chain—Chinese firms used the "Made in Morocco" designation to circumvent U.S. tariffs on goods made in China. In 2024, Morocco announced plans to electrify 60 percent of its vehicle production over the next half decade.

As for LFP, it was on the rise everywhere. Jose W. Fernandez, the Biden administration's undersecretary of state for economic growth, energy, and the environment, told me that "LFP battery chemistries are

Aminatou Haidar on Gran Canaria in 2017

expected to comprise a forty-two percent share of cell demand by 2030....
Phosphate refining is expected to be a bottleneck in the future, as only
three percent of total phosphate product supply is currently suitable for
lithium-ion battery applications, given the refinement qualifications."

And, of course, most of that phosphate was refined in one place: China.

CHAPTER 45

NEXT ENERGY

LFP batteries were a cheaper, safer alternative to their LCO and NMC cousins, but they were less powerful. This was a tricky problem that, in the early 2020s, was being solved by China. Historically, advancements in battery technology had been made in Europe, in the U.S., and in Japan, but suddenly it was Chinese scientists who were leaps and bounds ahead of their colleagues elsewhere. The massive research departments at firms like CATL and BYD were the size of entire battery-research communities in Western nations. CATL's R&D department was said to have eighteen thousand people in its employ, a number that equaled or even surpassed the total number of battery scientists in Germany, a country endeavoring to lead the European electric-car market; BYD, which was also a car company, employed a staggering 110,000 researchers in 2024. "We are not leading in materials innovation," Philipp Wunderlich, a battery technology specialist with the consulting firm Accenture, lamented at a 2023 industry gathering in Stuttgart.

By 2024, the most prominent leap in LFP battery technology had been made by Wang Chuanfu's giant car-and-battery company, BYD, which had created a new form of battery. When BYD announced its innovation, which it called Blade, on its website in 2023, it explained that it had invested €1.3 billion (just shy of $1.4 billion in U.S. dollars) to develop the factory that made Blade. A slickly produced video showed a huge facility with fully automated production lines and a plethora of robot arms.

The video claimed that BYD was "setting a new era for the EV industry." The battery, which had been designed to maximize space in the electric car, featured very thin cells placed in a flat, bladelike fashion.

The website claimed that the battery was safe and that even when punctured with a nail, it didn't emit smoke, heat, or flame. Clearly, such batteries would have definite advantages in electric vehicles, and they would be safer in the event of a crash. They were also lighter and allowed more cells to be aligned in an aluminum honeycomb structure. It wasn't just car batteries either: By the 2020s, BYD had over ten thousand engineers and some one hundred thousand employees working on what it called "the fruit chain." Someone's sly sense of humor masked a profitable "side hustle" for BYD, as *The Wall Street Journal* put it. The company was making around 30 percent of Apple's signature tablets: iPads. In November 2024, Tim Cook, Apple's CEO, told reporters that the company "could not do what we do" without Chinese suppliers like BYD.

By 2024, too, BYD had built not only its own manufactory town but also its own ship, the *BYD Explorer*, a bulk carrier that allowed it to transport cars. In the previous two years, BYD had increased its car sales by one million vehicles. The vehicles were selling in Europe, but Trump-era tariffs on Chinese cars, which the Biden administration had retained, meant that it was uneconomical to sell them in the U.S. *The New York Times* called BYD a "Tesla killer," and even Elon Musk admitted that without tariffs, BYD would outcompete his firm. According to *Bloomberg*, Wang was worth almost $19 billion that year, making him the 110th-richest man in the world. Long gone was the workshop in Buji, when BYD's initials stood for nothing and human beings were putting batteries together by hand.

Early in 2024, BYD announced that it would be releasing an even more powerful version of its Blade battery that year. The latest Blade could give its vehicles a range of one thousand kilometers. It was even lighter and took up even less space. That year, *Bloomberg* mused that the CEO might be "China's version of Henry Ford."

IN JUNE 2023, I VISITED NOVI, MICHIGAN, A SUBURB OF DETROIT, AND MU-jeeb Ijaz's firm, Our Next Energy. ONE, as everyone called it, was focusing on an LFP-powered future. Detroit has long been the headquarters of the American auto industry, and the U.S. government was betting that companies in the area would be able to spearhead a new electric-vehicle revolution. ONE, which was producing several different types of flat batteries, with cells arranged in aluminum casing, had some of the most interesting new products to show in Michigan and elsewhere in the U.S.

ONE had been coming up with impressive results, despite not having the heft of a BYD or a CATL. One of its batteries, code-named Gemini, was incredibly powerful, so much so that Ijaz was claiming it could power a vehicle over six hundred miles without recharging, finally putting an end to questions about "range anxiety." It paired LFP technology with an anode-free range-extender cell; the lithium coating on the battery's current collector formed an anode without a negative electrode, the graphite part of the battery that Akira Yoshino discovered back in the 1980s.

In 2024, the firm brought a BMW iX, fitted with one of its Gemini batteries, to Los Angeles for what is known as a Worldwide Harmonized Light Vehicles Test Procedure (WLTP) evaluation. The wheels on the car spun arms on a machine that simulated the resistance of everyday driving. Engineers nervously watched the computers as the distance that the car drove slowly ticked up. After a few hours, the counter ticked past six hundred miles. "That SUV traveled six hundred and eight miles on a single charge," Ijaz said proudly at the end. "I'm so excited about that."

Ijaz, a mustachioed auto-industry veteran with smiling eyes and a penchant for sleeveless fleeces, is the son of two Virginia Tech physics professors who came to the U.S. from Pakistan. ONE was not his first electric-car venture: At college in the 1980s, he built the power train for a solar-powered car project. He went on to work for Ford until he joined A123, the ill-fated battery start-up, where he ran the company's automotive business. He spent some time in China when the firm was sold to a Chinese auto-parts company, and he saw firsthand how A123's technology was being openly copied by its Chinese suppliers, including one that had been making bedsheets prior to acquiring the U.S. company's battery blueprints.

When he returned to the U.S., Ijaz was hired by Apple to work on a secretive project that entailed designing a car, which never came to fruition (he doesn't like to speak about it). He came up with the idea for ONE during the COVID pandemic, while he was in lockdown with his family. He chose LFP as his cathode material, he explained, after studying the supply chains for cobalt and nickel and hearing about Tesla's battery fires, which he thought were only going to become more common as more and more people bought the vehicles. Ijaz wanted to "push limits," he said. "There are three key things for our company: double the range of EVs, avoid nickel and cobalt, and establish a North American supply chain."

In the long term, new technologies like LFP could have a dramatic effect on demand for cobalt, because a large percentage of global cobalt demand comes from electric vehicles. But Ijaz told me there was no immediate worry on the part of cobalt producers. Cobalt would remain key for things like small electronics—cell phones, computers, and portable music players. "You know, the world market of consumer electronic devices, they must use cobalt," he said. Batteries like LFP were just not powerful enough at a small size to warrant ditching cobalt. "There's no voltage option to get away from cobalt," he said.

Shirley Meng, the materials scientist who now heads a division at the Argonne National Laboratory, told me that it was a matter of science. "Cobalt matter has really nice packing density," she explained. "With a smaller volume, we can deliver more energy. For mobile devices, lithium cobalt oxide is still desired." The issue with cobalt was the way it was extracted and shipped around the world, she said. "I think cobalt will still play a role, but hopefully with a much more stable supply chain."

Despite the bad press that the mineral has been getting, it remains popular for energy storage. In 2023, 93 percent of cobalt's demand growth came from batteries, which were in demand for electric vehicles, especially in the U.S. and Europe, where nickel-manganese-cobalt batteries were the standard.

And even if cobalt were to be removed from all batteries, Congo still has huge reserves of other metals that are used to store power. In fact, experts said that beneath the town of Manono, in northern Katanga, Congo has one of the largest reserves of lithium in the entire world.

In 2016, AVZ, an Australian mining firm, had acquired the exploration permits for Manono, but under Félix Antoine Tshisekedi Tshilombo, the Congolese government had blocked them in a series of backroom deals that seemed to benefit only a handful of figures who were close to the president, a powerful Chinese businessman, and the Chinese mining behemoth Zijin Mining. When I visited Manono, in 2022, the project was stymied, and the town was suffering economically. In 2024, Graeme Johnston, AVZ's technical director, told me that the Chinese firms were "a bunch of gangsters" trying to wrest control of the Manono lithium concession. "We are facing a concerted and well-funded disinformation campaign in the DRC paid for by our Chinese mates," he said. The fate of the mine should be of global concern, he added. "Since this deposit could supply twenty percent of the global lithium supply, it is logical to assume that those who end up with Manono will be able to control the world's lithium price."

Later that year, KoBold, the Gates-and-Bezos-funded U.S. mining firm, visited Manono. Someone captured photographs of KoBold executives in the town's airport and posted them to the social media site X, and the community of AVZ watchers were intrigued. It turned out they were in exploratory talks to buy AVZ's share in the mine; the Australian miner had effectively been shut out of the country. "It is an amazing deposit and it would be an amazing opportunity for KoBold," Jennifer Fendrick, the director of public policy at KoBold, told me. She said that the company did not want to repeat the extractive mentality of prior generations of foreign companies. "The immorality of working in Congo only comes into play if you are part of adding to the cycle of killing and theft. It's not immoral if you are part of electrifying a village, if you are part of doing positive things, then that is a different story." In May 2025, KoBold would announce that it was buying the license to mine lithium in Manono from AVZ, although the actual exportation of the metal would have to wait until a Chinese company finished a proposed highway to the town.

—

AT ONE'S HEADQUARTERS, SCREENS FLASHED A NIGHTTIME CAPTURE OF planet Earth as seen from space. Some places were lit up with electric light, while others were shrouded in darkness. ONE's logo was emblazoned across the photograph. "You see the dark spaces?" Ijaz had said to one of his employees. "That's ONE's mission: We want to bring light to those dark spaces."

ONE had been a beneficiary of the Biden era. Among the administration's aims was the goal of growing the EV market: Biden had spoken about wanting half of new U.S. vehicles to be electric by 2030. And he wanted to spur battery production and innovation across the country. The way he planned to do this was by massively increasing spending and by ring-fencing China's supremacy in the supply chain. "We are now seeing the ability for one nation to use its dominance in critical minerals to exert leverage over other nations play out," Undersecretary Jose W. Fernandez told me. "We want to ensure that the United States and its partners and allies don't face a similar vulnerability."

In August 2022, the president signed the Inflation Reduction Act (IRA), a hefty piece of legislation that stretched over 274 pages. The act earmarked just shy of $370 billion for investment in renewables. It extended tax credits for electric vehicles and provided "grants for domestic production of efficient hybrid, plug-in electric hybrid, plug-in electric drive, and hydrogen fuel cell electric vehicles."

The legislation was also designed to protect the U.S. supply chain. It only provided credits for vehicles that were produced in the U.S. or in countries with which the U.S. had a free-trade agreement; vehicles that originated in a "foreign entity of concern" were excluded. It went without saying that the foreign entity of most concern was China.

Administration officials including Amos Hochstein also lobbied to get sanctions on Dan Gertler lifted. The logic for this was explained to me one evening in Washington by a lawyer who was involved in advising the U.S. government on critical minerals policy: "People think that you need someone like Gertler in there, to counter the Chinese." I asked what special skills Gertler had. "He's got the experience, and he's tough. Americans aren't so tough anymore." Sanctions on Gertler, however, were not lifted during Biden's presidency.

The authors of the IRA were keenly aware that China controlled the refining of cobalt and other critical metals. Companies like Chen Xuehua's Huayou, the parent company of CDM, had poured money into refineries, where raw material from the mines is purified into chemicals that are suitable for batteries. Often, this is achieved through dirty, pollutive processes, and since the era of Deng Xiaoping, the U.S. and Europe have been happy for Chinese firms to do this work.

The year after the IRA was signed, as if to underline the concerns that had led to the bill, the Chinese government announced export controls on gallium and germanium, two rare earths used in solar cells and fiber optics. China controls the market for these metals (60 percent of the world's gallium, and almost 90 percent of the world's germanium is produced there). It argued that the materials could be used for military applications and thus had to be exported under a different set of regulations than they had been before. Then Beijing announced new restrictions on graphite, the material used in the anodes of lithium-ion batteries. As an analysis by the Center for Strategic and International Studies, a Washington think tank, put it: "China is leveraging its dominance of the global critical metals and raw materials supply chain to respond to expanded economic security policies in the West."

In 2022, China was the largest producer of refined cobalt, creating 76.1 percent of the world's supply. And despite Europe and the U.S.'s "expanded economic security policies," China was growing its refining capacity: In 2023, it would create 78.5 percent of the world's refined cobalt. When I asked about what the U.S. needed to do to catch up, Fernandez, the undersecretary, listed the development of "processing capabilities for critical minerals" first. They are, he insisted, "particularly concentrated and need to be more diversified."

In 2024, when Trump was reelected, he promised to do away with the IRA. (By the time of Trump's reascendance, many IRA-funded projects were hampered or paused due to higher-than-expected costs, even as Washington was abuzz with a sense of urgency to do something about the critical-metals gap with Beijing.) In his 2025 One Big Beautiful Bill

Act, Trump killed electric vehicle credits, leading to a very public spat with an erstwhile ally, Elon Musk. Biden-style encouragement-through-investment seemed to be dead in the water.

ALL THIS POLITICAL FLIP-FLOPPING IS BAD FOR THE DEVELOPMENT OF THE U.S. battery industry. By 2023, ONE was three years old and had raised $325.6 million from investors. It was also a recipient of IRA funding, because its goal was to create batteries that were compliant with the act. That year, too, the firm had announced it would put $1.6 billion into building a factory in Michigan. ONE looked like it might be able to compete with Chinese firms on some levels. But two years later, under Trump, the company's future would be anything but certain.

For U.S. firms, especially start-ups like ONE, it was notable that the IRA had created credits that were good for ten years. A problem with European Union legislation, one I witnessed when I visited automakers like Tesla and Stellantis (the parent company of GM, Citroën, and Fiat), was that it was often unclear, shifting, and mutable, unlike legislation in China, which had clearly set and defined goals. In 2024, officials at Stellantis

Mick Brown, of AVZ, surveys the Roche Dure mine at Manono in 2022.

seemed unsure about which credits applied to their vehicles. The U.S. act, Fernandez told me, "provides ten-plus years of policy certainty through clean-energy tax credits." The point of this was "so companies can think big and bold and pursue long-term projects." But of course, Trump's new administration would reinject massive levels of uncertainty into the system.

Though Ijaz thought the renewed focus on the U.S. battery industry was important, he also believed that China was already several steps, if not several thousand paces, ahead of the U.S. "They have innovated and invested in energy storage for twenty years in a way like no other country has," he told me. "And their leaders now, I would say top scientists, top capability, they invest in them. And there is also a lot of clear technical know-how and leadership. It's absolutely clear in my mind that we have a very long and difficult journey ahead to try to out-innovate and industrialize."

CHAPTER 46

RED SEAS

Pertamina is a state-run oil-and-gas firm that has its origins in Royal Dutch Shell. The company was born when Shell was nationalized in Indonesia in 1957. During the Suharto dictatorship, Pertamina was infamous for corruption, a little like Gécamines under Mobutu. According to the historian Adrian Vickers, "through Pertamina, Suharto, and other members of the power group, had ready access to an ongoing source of funding, which meant that they were not accountable. They ran this cash cow into the ground."

By 2022, Pertamina was trying to turn a corner. The firm was focusing on growth through nickel-based battery chemistries. "We want to use the nickel resources in Indonesia as much as possible, and as soon as possible, we want to push NMC," said Adriel Simorangkir, who at the time worked as a corporate strategist at Pertamina's arm for new and renewable energy. It would play into two of Indonesia's strong suits—nickel was the country's cash cow, but Indonesia was also boosting its production of cobalt. In 2023, according to the Cobalt Institute, Indonesia was the second-largest producer of cobalt in the world: That year, the country's output of the metal rose 86 percent and contributed to a quarter of the global supply growth. By 2030, the institute estimated, Indonesia would produce 16 percent of the world's cobalt supply (a distant second to Congo, predicted to produce 67 percent of the world's supply by 2030).

Indonesia's GDP was benefiting from its nickel. So was Xiang Guangda, the nickel magnate, who had become a billionaire many times over. His firm, Tsingshan, had started looking at other new-energy metals in other places, like lithium in Chile. In late 2023, the Chilean government announced that a Tsingshan subsidiary would join BYD in investing in the country's lithium production and refining infrastructure. Chile's president, Gabriel Boric, said that, together, the two firms would be investing more than half a billion dollars in his homeland.

Indonesia's investment in nickel paid off. When I visited in November 2022, Indonesia was aboil with the news that Elon Musk had worn a Batik Bomba shirt from a Central Sulawesi village, which environmental activists took as a bad omen for that island's forests. In August that year, Tesla had signed a $5 billion contract for Indonesian nickel. BYD would follow suit in early 2025, when it announced that it would be building an electric vehicle plant in West Java.

―――

IN JAKARTA, MANY PEOPLE WERE OPTIMISTIC ABOUT THE DEVELOPMENT prospects that nickel had ushered in. Simorangkir, the Pertamina strategist, was a recent graduate of Imperial College London. In his twenties, he wore round spectacles and sported shorn hair. When he was not working on energy, he helped out as "chief munchies officer" at a trendy bar in South Central Jakarta. Simorangkir told me that he had decided to come back to Indonesia rather than stay in Europe because he was particularly excited about the creation of Indonesian electric vehicles. His focus was not on electric cars but on scooters, which are abundant in Indonesia. "We're trying to introduce nickel-based batteries into the two-wheeler market," Simorangkir said. China, with its enormous economies of scale, was an all-powerful competitor. "We're directly competing with Chinese batteries, which are much more affordable."

In Jakarta and on the island of Bali, Pertamina had already teamed up with Gojek and Grab, two of Indonesia's "super-apps"—through which one can buy groceries, hail taxis, make deliveries, and even pay shops—to provide battery-swapping stations, or "swap shops." At these depots, appended to already-existing Pertamina gas stations, depleted lithium-ion batteries could be swapped out for fully charged ones, making the prospect of driving an electric delivery bike seamless and cutting out long charging times. (The concept had already caught on in China, and when I spoke with Akira Yoshino, he predicted it would soon be a global trend.) "We have ten locations in Jakarta, fourteen battery-swapping stations in total," he said. "And we just built another six in Bali."

Pertamina had seen rapid growth in demand for the battery-swapping service since it was launched earlier that year. Competition between the apps was fierce, Simorangkir said. "Indonesia has the potential to become the leader in electrification," he insisted, but he acknowledged just how difficult the transition to a battery-focused economy would be. "Most of our battery cell production now is just very dirty."

The Harita Group's site on Obi Island. Nickel and cobalt are being loaded onto ore carriers. In the 2020s, Harita mostly shipped its products from Obi to China.

ACCORDING TO ACTIVISTS IN JAKARTA, SULAWESI WAS, IN FACT, AN EXAM-
ple of "best practices" in the Indonesian mining industry. Obi—the largest of the Obi Islands in the province of North Maluku, part of the vast
Maluku archipelago—was an example of real degradation, they told me.
In 2021, the isle had received a huge investment from Chinese firms to
create a nickel- and cobalt-processing plant. But it had also been criticized for destroying a remote part of Indonesia where there had been little development for hundreds of years.

In December 2023, a few days after meeting Simorangkir, I boarded a
rusted ferry called the *Queen Mary* to make the nineteen-hour journey to
Obi from the island of Ternate, which I had reached on a series of mean-
dering flights. I woke the next morning to the sound of vendors hawking
nasi kuning—warm turmeric rice rolled into cone-shaped plastic bags—
and jungly hills rose from bright-blue water. We rounded a headland and
Obi came into view, its hills higher than I imagined, swathed in fog and
thundering cloud.

Upi, a local environmental activist, told me that as legend has it, they
are home to ghosts who watch over town from the woods. Endemic species of butterfly, flying foxes, and paradise-crows all live among the trees,
despite the encroachment of loggers in recent years. Under the sea, in
areas away from the mine, shallow coral reefs alive with clownfish and
Moorish idols stretch for miles.

The trees had once made the islands—or at least those who controlled
them—fabulously wealthy. For hundreds of years, Ternate and the Maluku archipelago, known by Europeans as the Spice Islands, were the only
places on earth where the small aromatic clove flower grew. In the fifteenth century, whoever controlled the Spice Islands was bound to make
a fortune selling spices like cloves and nutmeg at highly inflated prices
in Europe. When Columbus set out on his journey west from Spain, he
was searching for a quicker way to get to the Malukus. When the Dutch
colonized the islands, they exchanged the tiny islet of Run with the British for another one of their overseas properties: Manhattan.

But the forests and the seas were now being ignored and destroyed in favor of a far greater stock of wealth in a world obsessed with energy. And that stock of wealth lay beneath the soil.

On Ternate, I had met a burly man in a blue polo with SAVE THE CORAL emblazoned across the back. Muhammad Aris was in his forties and had a penchant for slim cigarettes, which he smoked constantly. Aris had studied fish all his life. For twenty years, he had been teaching aquaculture at Ternate's Khairun University. "I want to make sure that the ocean is clean for the long-term life of the marine ecosystem," he explained. There, Aris ran a project to plant seaweed near the village of Jailolo. "It's a technology transfer to the locals so they can have new ways of making a livelihood," he told me.

Through his obsession with fish and aquatic life, Aris had been inadvertently propelled onto the front lines of the resource rush for battery metals. His wife was from Obi, and by the late 2010s, he had started to hear reports that on the island's southeast coast—and on Malamala, one of the islets adjacent to Obi—a huge nickel mining operation was stripping the hills of lush tropical forest and had gouged deep scars into the ocher earth. The sea, elsewhere in the islands a shimmering palette of turquoise, teal, and aquamarine, had turned a dull brown color that the locals call *kemerahan*, or reddish. At least eleven different species of fish there had been poisoned by heavy metals, which had destroyed their organs and ovaries and could potentially cause health problems for humans who ate them.

The main town on Obi, Laiwui, which is perhaps a three-hour ride from the mine site by way of motorized canoe, was a laid-back place of cinder-block bungalows that occasionally reverberated with bassy music coming from passing vehicles. "Obi is a unique place," Abdul Kahfi, the thirty-eight-year-old chief of Laiwui, told me. "There are no Indigenous people here, so many people have come over the years from all over Indonesia—from Java, from Sulawesi. It looks like a miniature Indonesia."

But on Obi, Kahfi told me, many people complained—as they did in Bahodopi, in mineral-rich parts of Congo, in the phosphate-rich Western Sahara, and indeed in many, if not most, places where new critical minerals were being dug up—about serious changes to their world. They complained about forced displacement, low reimbursement for land that had been grabbed by mining companies, pollution of their farmlands. And like in Congo and in the Western Sahara, the concerns had coalesced into a separatist movement of sorts, or at least one that sought a federal status for Obi and the surrounding area that gave those who lived there more control over their own destinies. Their concerns echoed those expressed almost a decade before in a report by the Australian Non-Judicial Human Rights Redress Mechanisms Project at PT Weda Bay Nickel, a mine site on the nearby island of Halmahera. The report found that a "paramilitary arm" of the Indonesian police had pressured residents to sell their land to the mining company for obscenely low compensation. On Obi, the people I spoke to insisted that such practices were alive and well.

Chinese companies like Huayou were forging ahead with development. In 2024, Huayou would boast that its plant at Weda Bay, the world's largest HPAL project, would have an "epoch-making impact" on the global nickel market.

———

THE MAJOR MINING COMPANIES ON OBI AND ON SULAWESI ARE EITHER majority-owned or significantly invested in by Chinese groups, and local islanders to whom I spoke complained about the employment of Chinese workers over Indonesians. A worker for the Harita Group, the conglomerate that operates the mine on Obi, told me that there were over a thousand Chinese workers at the facility. Almost without exception, people mentioned that the mining operations had turned the sea "reddish" and that it had become almost impossible to fish in the vicinity. "The fish don't come near here anymore," said Lisman Musahati, a twenty-seven-year-old fisherman in the village of Soligi, which is around four miles from the main mine site on Obi. His grievances echoed what I had heard in Sulawesi. "The waste from the mines is destroying the coral reefs, and we can't fish in this area anymore. We have to go farther out to catch anything—twenty miles out and further in small boats."

Shortly after the rapprochement between China and Indonesia in 2005, the Harita Group applied for a permit to begin mining there. Harita started out in 1915 as a grocery store founded by a Chinese immigrant to Indonesia. By the first decade of this century, it had expanded into timber, palm oil, and gold and nickel mining, making its owner a billionaire.

By 2008, on Obi, the beginnings of a mine site that would come to look like a miniature city had been erected, and operations had begun. Four years later, Harita planned an initial public offering of its nickel division in Singapore to raise $250 million. It advertised in the city's Central Business District with the slogan "Mum was wrong. Slow and steady *doesn't* win the race."

Perhaps the only entity on the island who seemed to be in any kind of rush was Harita, true to its advertising. In 2015, only seven years after it had first broken ground, Harita announced that it was once again expanding on Obi, this time teaming up with subsidiaries of Xinxing Ductile Iron Pipes Co., a company based in China's Hebei Province, to build a $320 million smelter for the refinement of nickel ore. But this was nothing when compared with the $1.05 billion investment in 2019. Financed by Harita and Lygend, another Chinese company, and fired by a huge coal power plant, this new project, a nickel-and-cobalt smelter, opened in 2021. Two Chinese EV battery firms, GEM and Easpring, agreed to buy nickel and cobalt byproducts from the new venture for eight years.

On a sunny day in May 2021, Jiang Xinfang, the president of Lygend Resources, gave a rousing speech at a ceremony celebrating the first batch of products from the new smelter. For the occasion, he was dressed in turquoise factory workwear and a red hard hat. "We have every reason to believe," he said, "that this project will achieve the record of the fastest production reaching the standard in the industry and create more miracles!" He went on to laud the "professional planning scheme" he envisioned for Obi, an island that, outside the mine site, is still remote, even by Indonesian standards. He imagined an airport, supermarkets, and shopping malls, as well as a "new town with complete supporting facilities, livable environment, and humanistic characteristics."

One of the problems that Harita encountered, however, was where to put the waste, or tailings, that resulted from the production of nickel. Two years before, according to the conservation news site Mongabay, Harita had backed down, in the face of protests, from a plan to pump six million tons of waste into the deep sea, but the damage had already been done. In early 2022, the Indonesian magazine *Tempo* reported that waste was being pumped out of a black pipe and that mountains of waste were being washed into the sea. When I was there, the waste pipe had apparently been moved, but the sea around the mine site was still "reddish," and fishermen complained that it was almost impossible to fish. "The thing I love most about Obi is the fish, but now after the mining, the fishermen can't catch fish," Kahfi, the chief, told me. (In response to such charges, Anie Rahmi, Harita's corporate communications manager, wrote to me to stress that "Harita Nickel is fully committed to meeting both Indonesian and global ESG standards.")

Aris, the professor whom I met in Ternate, said that he had approached Harita in 2010 about doing an environmental study near the Obi mine site. After nine years, Harita finally allowed him to start taking water samples from the area, although his access to many sites, including the main mine, was severely limited and, in some cases, prohibited.

Nevertheless, Aris was able to examine water, fish, and fluted giant clams, and he prepared three academic studies to present his findings. In one paper, he wrote that his research had found "degeneration and cell necrosis" in the fluted giant clams, which are listed as a "conservation-dependent" species because their natural habitat has been impacted by humans. "This change is thought to be influenced by heavy metals," he continued. "Heavy metal content in liquids exceeds the quality standard threshold." In another paper, he noted the degradation of fish tissues. His sampling detected levels of ammonia, nitrate, dissolved oxygen, nickel, and iron that exceeded water-quality standards set by the government. In the conclusion to his paper on fish, he described it as "a reference for warning of heavy metal pollution in Obi Island waters."

When it came time for Aris to publish his results, Harita contacted him. "They asked me to change the data," he told me. When he refused, the company tried to prevent the study from appearing. "They sent a letter to the university," he said, "and the rector spoke to me directly and told me not to publish that the water pollution level was over the limit." (Khairun University did not respond to multiple emails seeking comment.) Aris decided to publish his work despite the pressure, and he has not received funding to conduct another study in the Obi area since. A separate water sample tested in 2022 by *The Guardian*, the British newspaper, revealed that the cancer-causing chemical hexavalent chromium, of Erin Brockovich fame, was present in worryingly high levels in water near the mine site.

At Kawasi, the village nearest to the Harita mine, I saw just how the landscape had been wrenched from wilderness to dystopia by Harita's nickel operations. Tall smokestacks belched a reddish-gray fug, and a river gushed reddish-gray water into the blue sea. Kahfi would later tell me that he was arrested and jailed for two months after protesting at the mine. Local journalists, he said, had also been detained after they approached the mine. A pair of Indonesian warships loomed menacingly on the horizon.

Our boat landed on a drab beach pocked with litter. Chickens pecked at plastic bottles and packaging. Despite the state-of-the-art nickel refinery yards behind the town, there were neither drains nor trash collection services, so the sea was used as a dump by the local population. Through the sheet of rain, we saw the glow of a fisherman's clove cigarette. He beckoned us onto the porch of his home. We watched the rain for hours and talked about how the company had plans to displace the people of Kawasi. The minerals their land held were so powerful that no one could live there.

CHAPTER 47

BURIED UNDERGROUND

On the morning of March 28, 2019, Nitunga Ilunga woke early at his home in the Kanina neighborhood of Kolwezi. The area was pocked with deep holes in the ground where people had dug pits to look for cobalt and copper veins. The government was conducting a "sensibilization" campaign, and now people in the neighborhood did not go down into the holes, or at least they said they didn't—many were still gaping open. Sometimes, at night, small children would tumble into them, falling to their deaths in the darkness below.

Ilunga, who worked as an artisanal miner, told his wife, Françoise, that he was heading out. They spent a few minutes calmly chatting about the education of their seven children and how things were going at school for each of them. Then, another familiar topic came up: the piece of property they hoped to buy using the money that Ilunga earned as a miner. "We discussed buying a small plot of land to build a home," Françoise said. "But since we didn't have enough money yet, we were renters."

Artisanal mining was a profession that Ilunga had chosen out of necessity, not out of any particular passion for clambering into the earth and digging up minerals. "He liked this work only because there was no other work for him," Françoise said. "He had hope for his children. He worked for his children. Through the work, we thought that our children were growing up well, that they were able to go to school, that they ate well, and we were excited to buy a parcel of land."

Ilunga worked on a team that illegally entered industrial mine sites and dug up ore to feed the insatiable global appetite for cobalt. "He would go depending on the opportunities: If there was a chance to go in during the day, he would enter; if he had the chance to go in at night, he would go in then," Françoise said. His team would get into the mines by "fraud," she noted, by paying off the guards.

That morning, he arrived at his job site, illegally entering Glencore's KCC concession around 7:00 a.m. (Glencore stressed that miners like Ilunga trespass on land that has been permitted for industrial mining and that the company has taken active measures to try and prevent them entering.) There were no safety briefings or registrations: Ilunga's team paid off any guards who asked questions and headed for a shaft dug into a secluded area near the mine. Hundreds of other miners were doing the same, starting their day at the bottom of the global battery supply chain.

Toward 9:00 a.m., the ground rumbled, and the pit where Ilunga was working collapsed. He was quickly crushed, suffocating under the soft soil.

Eight hours later, Françoise was awaiting her husband's return when her nephew came to her with the news. "There was a cave-in this morning, and your husband is dead," he said. She burst into tears.

The family rushed to the mine to try and recover the body, but when they arrived, the rescue teams had come and gone. There was no one to help them remove Ilunga's remains from the ground. It took them two days to find his body; when they finally got it out, they took it to the Cimetière Mwangeji, where they held a hastily arranged funeral.

"The government didn't help us during the death of my husband," Françoise said. "There were lots of deaths that day. I don't have the exact number."

"At least a hundred and fifty people died there," Françoise's nephew told me when we were sitting with her. "The government said only four people died, so they only bought four coffins." The rest were wrapped in sheets and lowered into the ground.

When I met Françoise, she was struggling to make ends meet: With-

Françoise Ilunga in Kolwezi, 2022

out any income from her husband, it was hard to bring up her seven children. She had grown to hate mining. "It's a bad work," she told me. "It only brings us death and destroys the mentality of people." But every day she was reminded of her husband's death, because every day she returned to sell manioc roots at the edge of a nearby mine site. She had taken up the work in the wake of Ilunga's death, and hated it. It was squalid, and she barely made enough to feed her children. But she would never allow her children to go down into the mines. She wanted them to go to school. "We suffer a lot," she said. "But I still have hope."

SUCH CAVE-INS WERE REGULAR OCCURRENCES IN KOLWEZI. "THEY JUST follow the veins, and the way they do is extremely dangerous," Charles Carron Brown, the consulting mining engineer, told me. "The accident rate in artisanal mining is appallingly high." When I visited Kolwezi two months after the landslide that killed Ilunga, I was ushered into the office of Erick Tshisola, the chief of staff to the minister of mines for Lualaba. He was on the phone and gestured for me to sit. "Yes, *Excellence*," he said, speaking to the governor, confirming that there had been a cave-in. "Two deaths and nine wounded," he continued. "No, we haven't told the families yet. . . . Maybe they can give some money to the families?"

Tshisola hung up and looked a little embarrassed—we had caught him at a bad time. "At Tenke, there were uncontrolled *creuseurs*," he admitted. "At three a.m. they snuck onto the concession, and there was a landslide."

A month later, in June, as international investors flew in for a mining conference in Lubumbashi, more than forty *creuseurs* were killed in a landslide after breaking into KCC. Odilon Kajumba Kilanga and his friends were also at the site that night, but they were working a different seam. "The worst thing I've seen as a miner is the sheer number of dead bodies when there were cave-ins," Kajumba told me when we spoke later that year. The KCC cave-in was one of the worst. Glencore shares briefly plummeted on the London Stock Exchange but rebounded when it became clear the cave-in wouldn't affect the firm's operations.

The night after the landslide that Kajumba witnessed, a Glencore employee told me, "People snuck back in and continued digging."

A WIDELY PROPOSED SOLUTION TO THE ARTISANAL MINING CONUNDRUM IS traceability, which has advanced in fits and starts since the practice was enshrined into the country's 2002 Mining Code. This would involve tracking of the supply chain, which some companies have said can be done using blockchain technologies. The European Union has even planned "battery passports" to be implemented from February 1, 2027, which will show the origin of each of the minerals mined for use in particular batteries.

The problem with such schemes is that they rely on sometimes extremely shaky information; people input that data, and people, especially in a poor country like Congo, are susceptible to all kinds of external influences. A supply chain is only as clean as its dirtiest link, and in Congo, a country with high levels of poverty and corruption, links can be very dirty indeed. Traceability "is not intrinsically driving the supply chain to become more sustainable," said Ilka von Dalwigk, the director general of Recharge, a battery industry trade body in Europe. She said that the EU should be focusing on new schemes to promote social sustainability in poorer countries and poorer member states.

Companies have also been trying to address the issue of data insecurity. Benjamin Clair, the founder of Datastake, a start-up that seeks to, as he put it, "develop and deploy software so that local stakeholders can digitize their available information and provide data," said that there are ways of remunerating local communities for providing data along the supply chain. "There's a ton of data available locally from local actors, and there's a ton of demand for data, for impact monitoring, for risk management, for opportunity assessment, for market research," Clair said. "So let's just find ways to connect this supply and demand of information."

Another solution is the formalization of artisanal mining, whereby people who do it receive safety training and equipment. Kasulo was one such project, although when I visited, the safety standards appeared to be limited. Carron Brown argued that it was in the government's best interest not to allow artisanal mining, because *creuseurs* make sites less economically viable. "The artisanal miners go down following the high-grade veins," Carron Brown said. "They can take a deposit that would have been extremely valuable to a commercial mining company, and in so doing they ruin the deposit, because they take out the high-grade portions of it. So that means that the low-grade portions can't be mined." But Carron Brown acknowledged it would be hard to stop the practice. "It's very difficult to sort of put the cat back in the bag once you let it out."

At the Mutoshi mine in Kolwezi, another formalization scheme was instituted by Pact, a human-development NGO based in Washington and jointly funded by Chemaf and Trafigura. The Mutoshi formalization project ended in March 2020.

It was hard to tell if Mutoshi was a success: I was not permitted to visit during my trips to Kolwezi. But Dorothée Baumann-Pauly, a professor and human rights expert at the University of Geneva, managed to interview sixty miners and local residents at the mine in 2023. Her research showed how safety improvements and the integration of women—who were previously thought to bring bad luck to male miners—had tangible positive effects on the local community. Child labor was eliminated, and school attendance rose as mothers were able to support their children with extra income.

It appeared that formalization increased salaries and quality of life. "There were over 3 million hours worked without a lost-time injury and no artisanal mine fatalities at the fenced Mutoshi pilot site during the roughly two years of the pilot project," Baumann-Pauly wrote. "Since the end of formalization in March 2020, there have been seven work-related deaths, most related to tunnel collapses." She noted that local businesses had suffered, and that an estimated three hundred children now mine on the site every day because many of the women who were supporting their families during the formalization period are earning less. Formalization had its critics: Monopsonies, in which single buyers of minerals are able to dictate prices, don't usually work out so well for the people selling minerals at the bottom of the supply chain. But, Baumann-Pauly argued, at least they allowed for the "setting and enforcing [of] basic human rights standards for the extraction process."

———

KAJUMBA THOUGHT FORMALIZATION WAS JUST ANOTHER SCAM TO PAY MINers less. Since Kasulo had undergone its formalization process, the prices that Congo Dongfang paid for ore had cratered. By 2019, he wanted out. The Glencore cave-in that year convinced him that risking his life in an artisanal pit was simply not worth it.

He continued mining until 2021 but resolved to save up and find another job. When he was hired at a restaurant, he decided to quit artisanal mining. After he lost that job, he occasionally went back down into the pits to make ends meet. "I was able to help my family, thanks to what I earned from being a *creuseur*," he said. "But there were lots of problems. We did this job without security."

Yannick and Trésor Mputu, Kajumba's colleagues and roommates, continued mining and working with Antoine Mutumba's cooperative.

The last time I saw Trésor was with Mutumba in 2022. I was outside church with Kajumba when we bumped into them. They were going to a party, they said, with Kalala. Did we want to join? Kajumba shook his head. He was less and less interested in Mutumba's world and the world of the *creuseurs*. He saw it as deadly.

Kajumba's resolve was impressive. Some people would stay in the mines no matter how dangerous they were. When last I met with Kajumba, in 2022, he told me he was hoping to get a subcontracted job with Ivanhoe, but it never materialized. We stayed in touch, and he told me that shortly after, he had managed to find work at an engineering firm. "I am trying to fight and to organize myself little by little," he said in 2024. "And to find solutions to go forward a bit."

An artisanal miner named Jolie, Manono, 2022

CHAPTER 48

DETAINED

On the morning of July 13, 2022, I had planned to meet some people who said they could introduce me to Gédéon, the "cannibal" warlord whose group was blamed for most militia activity in Katanga. In 2009, he and six of his followers had been sentenced to death by a court in Likasi, but he had escaped from prison and fled into the hills around Upemba National Park, where he began to reconstitute his militia with the tacit help of some figures in the military. Following a speech made by a Yeke princess, he started to say that he wanted to *kata Katanga*, or "cut Katanga," and began agitating for a separate state. Occasionally, his Bakata Katanga—as his militia came to be called—attacked mine sites around the region and raided the road upon which minerals left Congo. In 2016, Gédéon decided to turn himself in and was kept under house arrest, but then, in 2020, under Félix Antoine Tshisekedi Tshilombo, his forces attacked town, and he went on the run again.

I had been told that Gédéon funded himself through artisanal cobalt mines, although the allegations seemed far-fetched. Nevertheless, I wanted to ask him myself what involvement he had with the mining world. One of the people I was supposed to meet on July 13—Willy Nkuwimba, the demobilized militia member who said he still supported Gédéon—had told me that the warlord had been getting mining licenses for an artisanal cooperative outside Lubumbashi. By my reasoning, if some of the cobalt that powered the world's cars and cell phones was benefiting a cannibal warlord (or a purported one, at least), the outside world ought to know.

By 2022, Gédéon had been on the lam for two years. He was accused of planning several attacks against military barracks in Lubumbashi. Nkuwimba insisted that he could connect me with the warlord and urged me to send an interview request to Gédéon by way of a personal video greeting.

Our meeting had been set to take place at the bar of Hotel Ouagadougou, on the outskirts of Lubumbashi. I knew some of the people who crowded into the dingy dive that morning (Nkuwimba and Mura Mutomb, who described himself as Nkuwimba's assistant), but I did not know the rest—grim-faced men who spoke neither French nor Kiswahili, only Kiluba, the language of the Katanga Luba. They were introduced as Gédéon's "emissaries."

The negotiations were winding. Nkuwimba and Mutomb had been introduced to us by the head of a local civil-society organization. This person had told Jeef Kazadi Kamwanga that they were demobilized members of the Bakata Katanga militia, and he had even said they had taken part in peace negotiations in 2016. We didn't have any reason to distrust them.

DURING THE MEETING, I NOTICED A SERIES OF MEN IN TIGHT-FITTING football shirts hawking cheap Chinese radios at the bar. I went outside and was just sitting down when a large, fleshy man placed his hand on my shoulder. I was impelled toward the door of the hotel restaurant. "Come with me," he said, firmly. "Routine immigration check."

My immediate thought was that I had fallen into a scam. The fact that the radio sellers had also drawn up behind the man with his hand on my shoulder reinforced the feeling that something untoward was happening, and that the large man was not an immigration officer.

The man, who would later introduce himself to me as "Joseph," kept insisting that he was from the Migration Directorate of the Congolese government. "We won't keep you long—it's routine," he insisted, hustling

me through the dimly lit lobby. He pushed me outside into the bright afternoon, and as my eyes adjusted, I saw soldiers, Kalashnikovs drawn,
jumping from the back of seven pickup trucks.

"My passport," I said, feebly. I had left the passport inside, tucked into
a satchel with my cameras; inside, too, was Kazadi, who had been helping me set up the interview. If I could only get to Jeef, I thought, he would
be able to explain to this Joseph guy that it was all a mistake, that we were
journalists, that we were fully accredited to be working in Congo.

I CONVINCED JOSEPH TO LET ME RETURN TO THE RESTAURANT: INSIDE,
pandemonium had broken loose. Kazadi was on the floor, the barrels of
several rifles aimed right at him. A man in a gray *abacost*—a type of sa-
fari suit pioneered by Mobutu Sese Seko—was barking orders. "Tie him
up!" he shouted, and a plainclothes officer began to tightly bind Kazadi's
arms using his sweater. Wincing, he looked at me and slowly shook his
head.

I was being detained, taken to an unmarked two-story building in the
center of town. Joseph, who had originally packed me into the car outside
the hotel, turned out to be an officer in the Agence Nationale de Rensei-
gnements, or ANR, Congo's secret police. Kazadi and I, along with the
six purported members of the Bakata Katanga, were taken to different
areas of the agency's headquarters. The building was almost farcically
unkempt: Wires dangled out of walls where lights should have been, even
in the main conference room, and a man with a can of black paint was
painting the windows. I sat watching for hours as he painted the grilles
without bothering to cover the glass, flecking the glass with paint.

I was guarded in a series of conference rooms by Mechaque, a fresh-
faced member of the special forces who, after a few hours of chatting and
smoking cigarettes with me, took me through the mechanics of his RPK
light machine gun. He told me he had used it against Bakata Katanga rebels, whom he barely distinguished from bandits. "This is a weapon for
killing people," he said with a smirk. "Much of the time, they only have
bows and arrows and machetes."

My phone, my books, all my effects, were confiscated. I asked—politely at first, then more and more forcefully—to speak to the U.S. embassy. Each person I talked to seemed to reply from the same script: "Don't worry. You're safe. *Treat this as your own home.*" At first, I was annoyed at being held up—precious, expensive hours of reporting time were being wasted by bureaucratic nonsense—but I soon became more and more worried, especially when I heard what sounded like shouts and screams from the cellblock where Kazadi had been taken.

Finally, late at night, I was moved to a brightly lit conference room with exposed wiring, as elsewhere in the building. The man in the *abacost* was among my interrogators. The questions they asked became increasingly nonsensical. I was apprised that someone had told them I was planning to smuggle a separatist leader out of the country to supply him with weapons.

"That's absurd," I replied. "I'm just a journalist. How could I smuggle rebels out of the country?"

The interrogator who was writing up the transcript of our interview—in longhand—looked up at me. "Well, you might not have the means, but your employers are powerful and rich," he said, smiling. I realized I was caught up in a story where everyone believed the most extreme conspiracy theory offered to them. My working hypothesis about Congo, formed over some four years of writing about the country, was now being proved in the most terrifying way—namely, that people had been gaslit, lied to, and repressed for so long that their senses of reality had been irretrievably warped.

At another point in the interrogation, the agents asked me whether I believed in God, and whose fault it would be if "the inevitable" were to happen to me in the course of my reporting. When I asked what they meant by that, the man in the gray *abacost* looked up from his notes and replied: "The inevitable? The inevitable is *death*."

———

I WAS GIVEN A THIN MATTRESS AND TOLD TO SLEEP ON THE FLOOR OF THE conference room. A guard sat and smoked nearby, keeping a close watch on me as I swatted mosquitoes and tried to rest. I worried for the people at home: When I travel on reporting trips, I try to check in each evening to let people know I'm okay, and I hadn't checked in. The next morning, someone asked me for fifty dollars so they could buy me cornflakes and milk. I hadn't eaten since the arrest, so I gave it to them.

Kazadi and I were reunited, and we spent a listless day planning what we would do when we were released, which an agent named Thierry assured us would be that afternoon. Through Thierry and some of the other agents, I also started to get a picture of the ANR and its attitude toward freedom of the press. "You may have been authorized by the Ministry of Communication," one said to me, "but you should have told *us* of all your activities. Ministers come and go, but *we* are the government."

Morning turned to afternoon, and Joseph suddenly burst into the room. "And now you are leaving!" he shouted. "Now you are going to *Kin*." He pronounced that last syllable, the abbreviated name of the capital, with a flourish, a dramatic exhalation that caused his colleagues to begin whooping with excitement.

BEFORE WE KNEW IT, WE WERE BEING HUSTLED INTO CARS WITH OUR SUITcases and shuttled toward Lubumbashi's airport, most famous as the place where, in early 1961, President Lumumba was deposited before he was killed.

All of us were bundled onto the evening flight to Kinshasa, Congo's capital. The stewardesses, clearly not apprised that we would be on the flight, shouted at our escorts. "Are those political prisoners?" one asked. "Prisoners cannot be transported on commercial flights!" The agents ignored her.

When I had been taken by the secret police, anything I could have used to record my detention—pens, pencils, electronics—had been confiscated. On the plane, I fished around in my pockets, and among a jumble of tissues and paper, I felt a pen.

I had something to write on: the back pages of the paperback edition of Ryszard Kapuściński's *The Emperor*, which I had managed to retrieve from my luggage. I made two trips to the bathroom and scribbled down notes that I slipped to two strangers on the plane. One was Congolese and wore a cowboy hat; the other spoke French with a heavy Belgian accent. Later, I learned that both had dropped the notes off at the U.S. embassy the next morning.

At the bottom of the stairs, a bunch of shouting men with Kalashnikovs herded us across the tarmac. On the plane, one of the ANR agents— referred to by his colleagues as simply "*Monsieur* Joe"—had given me an assurance that I was still trying to process as we got off: "Don't worry, they won't torture you. The ANR doesn't torture journalists." (I later learned that the ANR has been accused of all kinds of aggression against members of the media. Only two months before, three local journalists were tortured by the security service at Boende.)

In the hot night outside was Kinshasa's N'djili Airport. It had been newly built by the Chinese. I wondered if something else had percolated into the loam of Congolese society with China's famously no-strings-attached, no-questions-asked investments. Was the entire notion of personal freedom— admittedly never particularly strong in a place that had endured seventy-five years of colonialism, almost four decades of dictatorship, and three rigged elections—being forgotten altogether?

I WAS TAKEN TO THE ANR'S HEADQUARTERS IN KINSHASA, A HULKING BRU-talist structure in the center of town. I was searched and made to declare all my effects in a floodlit courtyard in front of more shouting men with guns. This was a theme that would play out over the next couple of days. These declarations showed the agents' bosses that they had not stolen anything from me, even as they pocketed objects and cash I had been in-structed to leave off the lists.

After the courtyard, I was forced to make a second declaration in a shabby second-floor office. My last pen was discovered and confiscated. In the background, an old documentary about Mobutu was airing. When he came on-screen, the guards cheered: "Papa Mobutu!"

Inside the headquarters, I was confined to a room with barred win-dows on the seventh floor. I was told that my cell had been prepared for François Beya Kasonga, a former Congolese security chief who was detained in 2022 under shadowy circumstances. In August of that year, after seven months of incarceration in Kinshasa, the previously healthy Beya had to be sent to France for emergency medical treatment.

The room was grimy and filled with mosquitoes, and it smelled of smog from traffic down below. Outside, a nightclub thudded out music late into the night. But at least I had my books, running water for some of the day, a view of Kinshasa, and a bed with sheets. I fashioned writing instruments from burned matches and wrote a diary of sorts, which I hid in my clothes whenever I heard the guards enter. Kazadi and the other men were confined inside a grimy cell in another building and had to defecate in plastic bags.

ON MY FIRST DAY IN THE KINSHASA CELL, I WAS ALLOWED TO PURCHASE A baguette, some water, some spoons, and some smokes from a guard for another fifty bucks. I had to rewrite my declaration. More cash was siphoned off. I would mostly survive on these supplies, three tins of sardines, a packet of muesli, and a packet of cookies that I kept stowed in my bag for the rest of my imprisonment. (When I had traveled to Manono a few months beforehand, I had often ended up in villages that had no shops and no food, so I was at least a little prepared.) I was constantly asked if I had enough food by officials who said I was not eating enough. I quickly realized that their purported concern was something of a sick joke—the mentioned food almost never came.

That night, a Friday, I began to panic. I sat up for hours, unable to sleep. One of the guards told me, cryptically: "At home, they know you are here." When I pressed him, he clarified: "Your people, the *embassy*." I wasn't sure whether I trusted him or believed him when he said: "I will try and explain that it is not good that you have been taken. You were doing your job."

AT SIX THIRTY THE NEXT MORNING, ONE OF THE GUARDS ENTERED MY room, joked with me a bit, and smoked one of my cigarettes. He then shuttled me out to another interrogation, a long one this time, with Kazadi, Nkuwimba, and one of the Bakata Katanga emissaries. It was led by a sturdy agent in a red plaid shirt who took notes in longhand on printer paper. We were forced to rehash the entire story again.

At one point, the agent began harrying Kazadi, condemning him as a traitor. Kazadi turned to me and started pleading: "Nick, if you're hiding anything, please tell them. I'm here in prison with you."

I replied that I was telling the truth, that I wasn't a spy. "We've known each other for three years," I told him.

When the interrogator finished, he handed the notes to another man, bald, with snakelike features, who typed up the notes on a laptop using two fingers. Every so often he would flick out his tongue and wet his lips. Occasionally, he would erupt in fury and scream, "You're a spy!" By midafternoon, the interrogation had been over for a good while, but he was still typing.

Everyone in the room lapsed into slumber. Sometime later, I woke to general confusion as the agents huddled for hours around the computer, furiously whispering. "The boss wants it as a PDF file!" I heard one say. At another point, the officer in the checkered shirt went through my passport and asked me why I had been in each and every country stamped there. He was especially interested in a Rwandan stamp from a trip I took transiting the country in 2019. Were we Rwandan spies?

A few hours later, Jean-Hervé Mbelu Biosha, the head of the security service, called me into his office on the third floor of the headquarters building. In a waiting room with red wall-to-wall carpeting, an assistant and several figures in military uniform watched a thriller on television. Everyone occasionally glanced over at a padded door with a fingerprint scanner through which people entered and exited reverently.

Mbelu was, as Radio France Internationale put it, "a man of the [president's] household." He had roots in Kasai, and under Mbelu, the ANR had become a tool for Tshisekedi and the people around him to go after perceived enemies of the president. During his time as head of the ANR (he was dismissed in early 2023), Mbelu presided over an intelligence state that was taking more and more of a sinister turn: The ANR carried out audits of various Congolese adversaries to the president and delayed the departures of opposition figures who tried to leave the country by private jet. Mbelu even implicated himself in the detention of Beya, the former intelligence chief, according to *Jeune Afrique.*

———

I DID NOT UNDERSTAND WHY WE WERE BEING HELD. WE WERE TOLD THAT we had been trying to interview a warlord and that somehow, improbably, the service thought we might be attempting to smuggle him out of the country.

But there was another possibility. One of the accusations that kept getting lobbed at us was that we were prying into the presidential family's business in Kolwezi and had no reason to. I had heard several people complain about the presidential family being linked to artisanal mine cooperatives, but they had always been a bit vague. Antoine Mutumba, the fixer in Lubumbashi, had tried to tell us a story about how the president's family was involved in mining with Vice-Governor Fifi Masuka Saini, but he became annoyed when I told him I could not pay him for information. "You'll see," he had said. Had he mentioned my reporting to the ANR?

Mbelu's office was stacked full of pictures of him and Tshisekedi: Mbelu bowing to the president, Mbelu closely following Tshisekedi at a crowded event, Mbelu in various other supplicant poses with the head of state. Sitting with Mbelu on gilded chairs were two officials from the U.S. consulate. One of the officials, the vice-consul, gave me a list of lawyers in Kinshasa, some granola bars, and some peanut butter. Mbelu insisted on having an ANR translator explain everything we said. I was told by one of the consular officials that I couldn't transfer personal messages, and that they couldn't say what would happen to me.

I returned to my cell even more convinced that it would be my home for a very long stretch of time. I wrote in my improvised diary ("It seems I will be here longer than I thought") and grimly consulted the list of lawyers provided by the consular staff. The latter came with a reference roster on the last page titled "Types of Cases Handled by Attorney," which ran from "1. Adoptions" to what I assumed would be my category: "25. Criminal." I envisaged a lengthy trial. Also in my diary: "No dinner."

The next day, my fifth day of detention, I recorded as "a day without anything." I was locked in my room, and I didn't leave. The only remarkable thing was that I was fed: A scrawny chicken leg and gelatinous rice in a blue plastic bowl were passed through my door that afternoon.

ON MONDAY, I WAS WOKEN EARLY. ON THE THIRD FLOOR, U.S. CONSULAR OFficials were there with two medical orderlies. "You're being COVID tested," one said. I felt a pang of hope: COVID testing was required to leave the country. (Congo's deputy health minister had at the beginning of the pandemic accused a "mafia network" of political figures of embezzling government funds for COVID services.)

I was taken back to the interrogation room and asked to unlock my phones. I sat there and watched the agents delete my photos and files: my early drafts of this book, months of research and interviews on mining, history, sustainable agriculture, and so on—all gone. "It's all not allowed, everything you have here," one said.

Several more hours of waiting followed, and I chatted to Papy, a genial ANR agent who wore chinos and a blazer. "I am happy to report that I have *never* seen people tortured here," he told me. We talked about his work, about how he made around $300 a month (which mainly went to schooling fees for his girls and to rent), about his fear that his neighbors would find out he was an agent of the ANR, and about the relentless swarms of mosquitoes that plagued the headquarters. "When we moved into this building," he said, "the mosquitoes were so bad that we wore socks on our hands and hats on our heads to stop being bitten."

Later, I was made to sign a declaration stating that I would never return to Congo, that I would not write about Congo, and that I would forget everything that had happened there. An agent scolded me, saying that I was poorly raised, and then proceeded to steal one of my rechargeable power banks.

I was informed that Mbelu wanted to see me in the late afternoon. He was once again there with the consular officials, who looked hopeful this time. Mbelu reprimanded me again, telling me that I had been detained for my own safety, and then instructed me to re-sign the declaration. He ordered my equipment, my camera footage, and so on to be impounded.

My equipment would stay, but I would be released. (I would never get my laptop, footage, and cell phones back.)

An embassy car sped me to the airport, and one of my jailers accompanied me—to get me through customs. I relished the textures, the people, the unfamiliar faces of the crowded Kinshasa streets. The agent rushed me through security onto the Air France flight to Paris. Before I climbed the steps, he embraced me and wished me good luck. "We are together," he said, flashing a sad smile. I thought of the threat of violence that hung over his life.

Kazadi was still in jail, and it would be two more weeks before he was released. (He later told me he had not been tortured.) Nkuwimba and Mutomb would be freed in several months. They told me they had been tortured, and Mutomb sent me a picture of Nkuwimba in the hospital. Later, I found out that Mutomb had helped the ANR set up the sting. But the reason why they wanted to arrest me in the first place was never made clear.

I never learned what happened to the other people I was interviewing: Some suggested they were freed in an amnesty. I have heard since, from various Congolese figures associated with the intelligence services and the Katangese opposition, that Gédéon is not believed to have any direct involvement in mining. But separatism continues to be an important topic in Katanga. In the days after my arrest, several Southern Katangese political figures, people like Patrick Masengo Kalasa, president of the Alliance de Forces Populaires de Katanga, or AFPK, a separatist-federalist group that urged peaceful disobedience, urged me to write "the reality, how we have no freedom in Katanga, how even to speak of separatism is to be arrested by the authorities in Kinshasa."

Masengo himself would be disappeared by Congo's secret state on September 20, 2024.

CHAPTER 49

AN AFRICAN ELECTRIC CAR

The Kiira EV factory, near Jinja, Uganda

I could not go back to Congo, but in 2023, I heard that the neighboring country of Uganda had built an electric-car factory. Uganda's chameleonic president, Yoweri Kaguta Museveni, has been in power since 1986 and has pushed a revolutionary "doctrine of self-sufficiency" for all Africans and for Ugandans in particular. To this end, the president's wizened countenance, invariably sheltered by a wide-brimmed hat, can be seen at a seemingly endless procession of ribbon cuttings and openings for industrial parks, factories, farms, data centers, oil fields, and—a personal favorite of the president, who has said that he comes from a seven-thousand-year lineage of cattle herders—centers for animal husbandry.

Uganda is trying to become that rarest of African states—one where finished products are *actually made*, rather than imported from China. At a 2023 conference in Kampala, the country's capital, Monica Musenero, the country's science minister, said that the government was "propelling technological innovations aimed at optimizing transport and mobility" for Uganda's overwhelmingly young population.

These are lofty aims for a country that is, by most metrics, almost as poor as Congo. The World Bank has argued that Uganda should focus on building human capital and reducing inequality. But such humble goals are not what Museveni has in mind for his country.

In a green field just next to the headwaters of the Nile, near the city of Jinja, sits Kiira Motors' headquarters, a factory that was designed to look like a bus but in fact looks like a spaceship. Kiira is one of Museveni's favorite projects: In 2011, Kiira's founders created one of Africa's first electric vehicles. When I visited Kiira, Ugandan engineers were striving toward an even more lofty goal. They wanted to refocus the battery supply chain in Africa.

THE STORY OF THE UGANDAN ELECTRIC CAR BEGAN AT MAKERERE UNIVERsity, one of sub-Saharan Africa's most venerable institutions of higher learning. The Kiira project was the brainchild of Paul Isaac Musasizi, a Makerere lecturer, and Sandy Stevens Tickodri-Togboa, a professor who, in his early teens, had dedicated his life to studying electricity after receiving a hefty electric shock from the power main at his parents' house. In 2008, Musasizi and Stevens Tickodri-Togboa accompanied a handful of their students to an event called the Vehicle Design Summit in Turin, Italy.

The summit, which gathered university teams from twenty-five institutions around the world, sought "to catalyze an Energy Space Race." Its ambitious goal—in 2007, a year before Tesla launched the Roadster—was to produce a plug-in electric hybrid for the Indian market. The students open-sourced the best ways to build such a machine. When the MIT students who started the project met with *The New York Times*' Thomas L. Friedman in late 2007, he wrote enthusiastically of the project, suggesting that they were making "the Linux of cars." The students, Friedman said, "blew me away."

By 2008, the students at the summit had created a prototype, and it was time to go home. But the Ugandan team, mugging for pictures inside the skeletal frame of the VDS prototype, had only just begun their work on electric vehicles. When they returned to Kampala, they applied for funding from Museveni's office to make, this time, a fully electric car. The president began to take a special interest in Musasizi and his team.

Over the next two years, the team assembled a car, the Kiira EV, a buggy hatchback that was painted in Day-Glo green (in case the electric vehicle's green credentials weren't clear enough).

Museveni arrived at Makerere for a test-drive in late November 2011. Musasizi piloted the car, and the president rode shotgun. "This is very good," Museveni said over and again, above the signature hum of the electric vehicle's engine (the car was powered by sixty-four lithium-ion batteries connected in series). "What is happening at Makerere is a renaissance; you fellows are waking up from a long slumber," he told the deans after the test-drive. Kiira's green EV could represent the first step toward having a lithium-ion battery supply chain that began in Africa and stayed there.

IT HAS BECOME FASHIONABLE ACROSS AFRICA TO INSIST THAT, SOMEDAY, batteries and even electric cars will be made on the continent. It is hard to overemphasize the propaganda value of such stories. In early 2025, for example, Burkina Faso's military dictatorship said that they had produced an "indigenous" electric vehicle, but the story was pure fantasy. It turned out that the cars were actually old Chinese cars that had been delivered to the country by a businessman connected to the Chinese state. Once they reached the country, they had been re-badged and used as a propaganda tool by a government that claimed it was casting off the yoke of French imperialism but had in fact pivoted to new masters in Russia and China.

In Congo, at the 2021 DRC-Africa Business Forum, President Félix Antoine Tshisekedi Tshilombo declared that the country should start building a battery industry "right after this forum." The rhetoric has come thick and fast, even as actual progress has been slow. "Faced with the challenge of a carbon-free industry at the global level, it is common knowledge that the Democratic Republic of Congo has the ambition to become one of the major producers of electric batteries by 2030–2040," Vuko Ndondo Kakule, deputy executive director of the Congolese Battery Council, told Germany's DW news service in 2023.

This ambition, which has gained some traction in certain corners of social media, presumes that complex, power-intensive industries can be made to work in a country with an unreliable infrastructure and shaky electricity generation. At the moment, 80 percent of Africa's raw materials are exported unprocessed, and African countries lose out by not refining materials. The concept of "value addition" (adding to value by moving down the supply chain and creating more and more complex products) only works when the conditions to create the products—the batteries and their precursor materials—already exist in the country.

Wracked by years of conflict and theft, Congo is no such place. A working paper published by the Harvard Kennedy School in 2023 looked at how Congo could improve its refining of copper and cobalt and produce battery precursors. The report's authors cited various obstacles to Congo's plans to move up the value chain, including corruption, spotty power generation, and a terrible transportation infrastructure. "Precursor production is not simple," the report noted. "Australia—a country with an effective grid, functional logistics, a skilled workforce, strong governance, and dynamic capital markets—has had to prioritize a national multi-year plan to produce precursors at commercial scale." Most of Katanga doesn't even have electricity. In 2022, on my final trip to Samukinda, for example, the new chief, Rikomeno Samukinda, gestured to a power transformer and wires running off into the distance: "We don't get any power." How could a country that cannot power a village next to one of its richest mine sites expect to produce batteries with any efficiency anytime soon?

The report's authors urged Congolese policymakers to prioritize better facilities for making copper wire and to "reframe precursor production as a long-term vision rather than an immediate strategic priority."

U.S. State Department officials pointed out that the Biden administration's Minerals Security Partnership might provide opportunities for countries outside China's orbit to move down the value chain—that is, to capture more money by producing more finished products.

But there were always more immediate priorities in Congo. Education, electricity, roads—so many other critical things were screaming out to be given pride of place before investments in battery precursors.

———

WHEN I VISITED JINJA IN SEPTEMBER 2023, THE KIIRA FACTORY WAS STILL under construction and looked like it had a long way to go, despite the company's promises that it would be ready by Christmas. An electric bus was parked in the sunshine outside.

Inside, Richard Madanda, a Kiira engineer, showed me some promotional videos, including one in which a new Kiira invention, a tricycle, helped small farmers water their fields and carry produce. The factory floor was empty save for a few prototypes, and as I was walked around, I was asked to imagine machinery, conveyor belts, robot arms, and scurrying workers. That was all in the future, Madanda promised me. The buses are currently built in an army-owned factory a few hours north of Jinja.

We went over to one of the cars, shrouded under a gray cover. Madanda pulled the covering back, and I realized it was the same vehicle that Museveni had been so impressed with all the way back in 2011. Kiira claimed it could reach 150 kilometers an hour and run for 80 kilometers on a single charge. Had more not been built? Not yet. By 2024, only around twenty thousand electric vehicles had been sold across Africa, and the continent had fewer than a thousand EV chargers, though at the end of that year, several African countries pledged to build batteries, electric cars, and electric bikes. Kiira would finally open its factory in late 2024.

Cina Vazir, one of the authors of the Harvard study, cautioned against seeing African EV factories as a panacea. First off, he wondered, who would buy them? "Africa's EV projected demand is not too high," he said. "Africa is also logistically kind of hard to shift things from one part to another? So, do you have the demand there?"

And at Kiira, it didn't seem like Uganda had completely exited the globalized supply chain. I asked where the batteries that power Kiira's electric vehicles come from. "China," Madanda answered. "We get them from CATL, though we used to have a different supplier."

CHAPTER 50

UNCLE BUNKER

In 2022, when Biden signed the Inflation Reduction Act, with its emphasis on building a resilient North American supply chain, into law, historical mining communities that sat atop critical metals within the country questioned whether they would once again see digging and drilling near their homes.

The town of Kellogg, Idaho, population 2,314, is one of those communities. It is a place that once had some of the richest mining anywhere on earth, but after the mines around it closed in the early 1980s, people moved away, and the area's income dropped off a cliff.

On an evening in June 2023, about two dozen people were enjoying Sam Ash's tri-tip barbecue and sipping beers on the porch of his house in downtown Kellogg. One of the guests wore a T-shirt that read I'M A KENNEDY DEMOCRAT, in support of Robert F. Kennedy Jr., who had recently announced he would be running against Joe Biden for the Democratic nomination. Another guest arrived bearing a sweet, creamy pudding topped with marshmallows. Almost all of them were employees of the Bunker Hill Mining Corporation, which hoped to revive mining in the area.

Part of the Biden administration's plan was to use private businesses to restore a supply chain for critical minerals in the U.S. and allied nations, thereby reducing reliance on countries like Congo. In Toronto, in 2022, Jose W. Fernandez, the energy undersecretary, had helped launch the Minerals Security Partnership, or MSP, an association of North American and European countries, as well as Japan, Korea, and Australia. The aim of the MSP, Fernandez said, was "to promote public and private investment in responsible critical-minerals supply chains." Mines like Bunker Hill could be central to such plans. "The private sector has been instrumental in identifying opportunities and leveraging expertise to promote shared regional goals," Fernandez told me.

In Idaho, the sun began to set over the wooded hills of the Silver Valley. "If you would have come here in the 1980s, in 1981, when the mine closed, there wouldn't have been a single tree growing on any of these hillsides," Ash, who is the CEO and director of the Bunker Hill Mining Corporation, told me. They had been poisoned by smoke from smelters. "From a strategic-metals perspective, this is really the story of, you know, the impact and the outcome of an initially unregulated industrial expansion in the United States, which resulted in the United States recognizing that it needed to enact environmental regulations and also initiate cleanup for the damage that had been done. That resulted in the mining industry shrinking significantly in the United States."

Ash was doling out the final plates, stacked with meat and beans, when Richard Williams, the company's executive chairman, motioned for everyone to be silent. Before entering the mining industry, Williams had served as the commanding officer of the British Army's Special Air Service, or SAS. His affect was one of quiet gravitas. "I'm very proud of everyone here," Williams said. "I'm proud because you have chosen to be part of the revitalization of this community. I look, and I see what we have built, and it really is incredible."

—————

DIGGING THINGS OUT OF THE GROUND STATESIDE IS A COMPLICATED PROPosition. Until the 1970s, mining had been part of the bread and butter of U.S. progress, but then a series of highly publicized disasters, including at the Sunshine Mine, just east of Kellogg, shone a spotlight on an industry that resisted scrutiny. The conflict between activists and miners helped birth the environmental movement. After Nixon's government founded the Environmental Protection Agency, Washington began to more strictly regulate the industry—so strictly, supporters of mining say, that entire communities were degraded and mining ceased to be a viable business on all but a small handful of sites.

The Bunker Hill site's origin story reminded me of Kasulo's, in Congo, for its element of chance. It began in the 1880s, when a carpenter named Noah Kellogg arrived in a remote corner of northern Idaho that the Schitsu'umsh ancestors of the Coeur d'Alene Tribe had long called home. By the time of Kellogg's arrival, a gold rush had brought all kinds of fortune seekers. "His donkey, he ran off while he was camping, and he chased that donkey up to the top of the mountain, and the donkey kicked over a rock," Ash told me. The rock glinted, and Kellogg decided to examine further. "He found one of the biggest silver mines that the world has ever seen."

Mining began in 1886, and over the next hundred years, one and a half billion ounces of silver were produced at Bunker Hill. Over time, lead and zinc began to be extracted there too. "It was instrumental in the industrialization of America," Ash went on. "It was considered absolutely a strategic asset by the U.S. government." Lead from Bunker Hill was used for munitions in World War I, World War II, and the Korean War. The mine, which became known as Uncle Bunker, formed the heart of the community, providing livelihoods.

But in the 1970s, Bunker Hill fell on hard times. In 1973, a fire damaged the mine's filtration systems, which were not replaced, and toxic chemicals were released that poisoned children living near the mine; emissions from a zinc smelter damaged trees. When sulfur dioxide began to come under regulation, Bunker Hill could not meet the standards set by the EPA. "Ultimately, what shut the mine down was its environmental performance," Ash told me. In 1981, the mine closed.

A decade later, the Coeur d'Alene Tribe brought a lawsuit alleging that Bunker Hill and other mines around the Silver Valley were poisoning its water supply. Fish and clean water are part of the Coeur d'Alene people's patrimony; their ancestors survived off the salmon they fished.

Until 1968, the mining company had dumped toxic waste in a local river, killing off aquatic life. Now lead leaching through the soil and dripping through Uncle Bunker's tunnels was killing the fish and poisoning the water with heavy metal.

The U.S. government joined the lawsuit in 1996. Bunker Hill had been declared a Superfund site, which meant that the EPA would take responsibility for cleaning it up. A huge processing plant was built in the valley beneath the mine to clean the water that would gush out of Uncle Bunker's tunnels after the rains.

For a while, the mine's ownership decided to fight the lawsuits; the site had changed hands many times, and the current owners said they were not responsible for the cleanup. "The EPA is probably the worst thing that's ever happened to America," Bob Hopper, the mine's bombastic, chain-smoking owner told a radio journalist in 2003. He had hoped to rekindle mining when he bought Bunker Hill in 1992, but instead, he had assumed the mine's liabilities. Hopper died in January 2011, and a few months later, the mining company that owned Bunker Hill paid $263.4 million to the U.S. government, the Coeur d'Alene Tribe, and the state of Idaho.

Williams had come across the mine as it was limping from crisis to crisis. A series of investors had tried to revitalize it and had been stymied by the toxic legacy and high reinvestment costs. But Williams was up for a challenge. He had found solutions in far harder places to harder problems, after all: When he left the SAS in 2008, he began working in the private sector, building a mining business in Afghanistan; from Afghanistan, he moved to South Africa and from South Africa to Canada, where he worked for Barrick, one of the world's largest gold-mining corporations, as its chief operating officer. There, he met Ash, who then ran one of the company's mines in Zambia. In 2021, Williams raised $3 million to study an old mine that he had found in Idaho's Silver Valley.

Bunker Hill was registered in Canada, because the Great White North has historically been friendlier to mining than the U.S. has been. But Williams told me that he was interested in seeing whether mining could be reintroduced south of the Canadian border. It was partly an interest in history that motivated him; during the time I spent at Bunker Hill, employees excitedly showed me old newspaper articles and clips about the wealth of their mine. "We wanted to have our anchor asset here and then see where it took us," Williams said. "We were big believers in regeneration."

Ash and Williams strove to work with the EPA. Their vision included "building an ecosystem of new and green businesses" for the local community. After the initial $3 million investment, Williams raised another $20 million.

Bunker Hill's proposal was that mining could be done in a sustainable way.

———

SILVER, LEAD, AND ZINC ARE NOT LITHIUM-ION BATTERY METALS. THEY ARE, of course, used in closely related industries: Lead is used in lead-acid batteries, zinc and silver in electronics. But cobalt is found about three hours south of Bunker Hill under the woodland of the Salmon-Challis National Forest. A 2022 *Atlantic* article by Michael Holtz proclaimed that "Idaho's cobalt rush is here" and detailed the struggle by Jervois, an Australian company, to build a clean mine. Idaho could become an important source of cobalt for the U.S., though Holtz noted that the estimated annual production—1,915 tons—was a "drop in the bucket compared with the output of mines in the Democratic Republic of the Congo."

The development of what is known as the Idaho Cobalt Belt has hardly been linear. Jervois spent eight years getting permits to mine and was scheduled to go into production in 2022, but the mine closed before any minerals were dug out of the earth. The markets were partially culpable—cobalt prices fell that year, partly thanks to Chinese stockpiling of the mineral. But another culprit was geology. The cobalt ore that Jervois was sitting on contained what a company report from 2020 described as "elevated" proportions of the poisonous metalloid arsenic, and it was therefore more expensive to process in an ecologically friendly way. "The arsenic is associated with the cobalt in the orebody," the report noted.

All of this meant that in June 2023, when I visited Idaho, the Cobalt Belt was not producing: The ore was lying there, waiting to be unearthed, and the mines were under care and maintenance.

At Bunker Hill, and in mines throughout the state and neighboring states, companies were having to deal with similar issues, as well as a central question: Did the U.S. want critical minerals enough to compete with China on the mining front? "We haven't had a real mining industry in the U.S. for four decades," said Ryan Melsert, the former Tesla employee who now heads the American Battery Technology Company, or ABTC. "We've outsourced most of that work. But there's nothing stopping us from reinitiating that."

In 2023, Melissa Sanderson, the former Freeport-McMoRan employee and official at the U.S. embassy in Kinshasa, took the helm of American Rare Earths, a firm that owns the giant Halleck Creek rare-earths mine in the neighboring state of Wyoming (she would later leave that position and join the company's board of directors). "Why is permitting of mines in America so complicated? It's because Americans hate mining," Sanderson mused. "I grew up in Cincinnati, Ohio, and I grew up hating mining because what I knew about mining was what we saw on television about coal companies strip-mining the heck out of West Virginia, destroying whole forests, et cetera." She continued: "I mean, powerful images remain, and they're part of the historic reality of mining in America."

The government, with its IRAs and security partnerships, was looking to change that. And when the Trump administration came into power, it, too, appeared to be keen on spurring U.S. mining. Its plans, however, were not particularly clear to anyone, including people in government. "We're a long way from getting domestic mining to where it needs to be," a senior U.S. official confided to me in Washington in March 2025.

COMPANIES LIKE ABTC LOOK AT MINING WHAT IS ALREADY INSIDE DEVICES that we use today in order to build the batteries of the future. In other words, they focus on recycling. Many companies grind batteries down into pulps called "black mass," and then use heat or water to refine them into an intermediary product. This product is then treated with chemicals and separated into its original materials, which are then reused in new batteries. It is not the only method. ABTC uses automated components to break the batteries to pieces. Another company, Redwood Materials, also helmed by a Tesla alum, JB Straubel, pulverizes batteries into mineral powders.

The former chief financial officer of Redwood, a bearded Tesla alum named Andy Stevenson, told me that the conditions were ripe for mass recycling of batteries in the U.S. A regulatory framework already existed for the handling, disposal, and treatment of hazardous materials, and batteries fit into that "pretty easily." But battery recycling was expensive, he said, and would probably need some external funding, possibly from the government, to really get off the ground. "Ultimately, somebody is going to have to pay if the value of the materials is not enough to cover the costs of processing and recovering," he noted. By 2024, Redwood was processing around sixty tons of lithium-ion batteries per day, the equivalent of 250,000 electric vehicles annually.

Nobody thinks that recycling is the entire solution. ABTC is recycling more than 115 percent of the number of batteries it originally planned to recycle. Redwood has among the most efficient recovery systems, one that reclaims 95 percent of nickel, cobalt, lithium, and copper. But, at present, only around 5 percent of batteries in the U.S. are recycled. And materials would need to be dug out of the ground to account for increased demand, even if every single battery was recycled and every single mineral recaptured. For us to fully rely on recycling, we would need a fully static population, fully static consumption trends, and a fully static economy. As Eric Frederickson, Call2Recycle's VP, put it: "That will not happen in my lifetime. That will probably not happen in my children's lifetime."

If the world wants batteries, it will still need mines.

———

AT BUNKER HILL, WILLIAMS AND ASH DIDN'T WANT TO END UP LIKE THE miners of the Idaho Cobalt Belt: huge amounts sunk into their mines but little to show for their money and efforts. "We negotiated a sensible purchase price," Williams told me. A sticking point, however, was that the EPA was still owed $19 million for the cleanup. They struck a deal with the EPA that would allow them to use cash flow generated by the mine to cover that cost. "We then spent an awful lot of time and money coming together," Williams recalled. They had "convinced the capital markets that it was worth funding and the EPA that we were trustworthy."

At first, when the Bunker Hill team went to community meetings, the project was met with apathy. At a town meeting, they were told, "You realize that you're about the tenth group to come in here and say you're going to restart the Bunker Hill Mine?" They understood that they had to "prove themselves" and slowly started to win over the local community by showing their progress.

In 2023, real momentum seemed to be building behind Uncle Bunker. The team had recently spoken to members of the Coeur d'Alene Tribe about their ongoing concerns around lead-laced water. "We have reached out to them; we have spent time with them and offered on every occasion to share data with them," Bradley Barnett, the vice president of sustainability at Bunker Hill, told me. "They know that we very much value clean water in the same way that they do." The firm was aiming for "extreme transparency," it said, and intended to share as much data as it could with stakeholders and the wider public in general. In early 2025, Teck, the Canadian mining firm, invested $40 million in the mine in anticipation of its production relaunch later that year.

Although some environmentalists were loath to accept new mining projects in the U.S., Bunker Hill was trying to show that environmental mining was possible and had a place in the U.S. "You see this all over the Western United States in all these communities where mines existed," Ash said. "It was a big part of middle America, and the middle class. And then it just—it just got obliterated." He blamed overzealous environmental regulation and people making rules far from the communities they affected for that obliteration. "They recognize that now," he continued. "The inevitable consequences of a lot of that legislation and enforcement was to push it into jurisdictions where they have no visibility and they have no control." Places, incidentally, like Congo.

Barnett knew the conundrum all too well. In another role, he ran a company called Critical Minerals International, and he had also been interested in investing in Congo. His efforts to build what he called a "model artisanal mine" near Kolwezi had been frustrated by government intransigence, but he still hoped that he could get some kind of project off the ground.

Barnett had also worked with Barrick Gold in Afghanistan, where he met Williams. At Bunker Hill, he told me, he was positive that what we were seeing was historic. The project was, he would write online a few months later, "the return of mining to the Western U.S., done right, by a team that is ready for a lot more."

SO THAT I COULD BETTER UNDERSTAND THE MINE, BARNETT SUGGESTED I go with him deep into Bunker Hill to see some of the tunnels. I donned a hard hat and a reflective vest, then clipped on a belt to which was attached a self-contained self-rescuer, or SCSR. Alan Longley, a health and safety manager, explained how the device worked, how once its cap had been removed, it would provide oxygen for at least an hour. It would be a lifesaver in the event of the release of poisonous carbon monoxide, as had happened at the Sunshine Mine in 1972, when ninety-one men suffocated after a fire. Longley remembered that time vividly. "Almost every family in the Silver Valley was, in one way or another, touched by that catastrophe," he said.

At the entrance to the mine, I boarded an all-terrain vehicle piloted by another Brit, Tom Francis, a former Royal Marine. We plunged into the mine, and the warmth was almost immediately sucked from the air. There were twenty-six levels to the mine, Francis explained, connected via the corkscrewing passage by which we were descending.

The walls were braced with rusted metal, putting me in mind of a howling mouth fitted with particularly unpleasant dental work. Mining equipment from thirty, fifty, eighty years before lay abandoned, crusted over with decades of mineral deposits in shades of brown, yellow, and green. The air was filled with the tang of metal.

All of a sudden, the atmosphere became more humid. Rusty-looking water was pooled on the floor. It was groundwater coming through billions of tiny fissures, and its color was a sure sign it needed to be "remediated"—plugged up by the team at Bunker Hill so that it didn't leak into the Silver Valley's water supply. Water flowed out of the old Bunker Hill Mine at four thousand gallons per minute.

When we were around half a mile into the hill, Francis slowed to a halt. The tunnels ahead hadn't been cleared. I felt like we were deep enough already. The weight of rock, and of history, bore down from above. I imagined Uncle Bunker filled with thousands of people, like it had been during its heyday, when it was churning out tons of silver for tables and lead for the battlefields of far-off wars. After a while, we had seen enough. Francis made a three-point turn and pointed the nose of our buggy back, then we lurched upward, heading for the mouth of the tunnel, where the air would be clear and we would see sky and trees and light.

Driving into the Bunker Hill Mine, 2023

POWER DREAMS

In many ways, history is repeating itself in Katanga—and in other places around the world where critical metals can be found. I had this revelation most powerfully in October 2021, at a schoolhouse beside the Berwinne River in Belgium, about a ten-minute drive from the Dutch border. It was early in the morning, and the *directrice* of the Dalhem Communal School, Séverine Botty, had welcomed me to the low-slung brick building. She showed me to the cafeteria. There, looking out from a commemorative placard affixed to a cream-colored cupboard, was Dalhem's most famous son, Albert Thys, "the creator of the railroad in Congo, businessman and one of the first Belgian financiers of the early twentieth Century."

In the dining hall, I examined the keepsakes of Thys's travels and exploits in the Congo Basin: hippo tusks and kudu horns, a carved wooden cup and a pipe fashioned into an effigy of the man himself. I understood that his was a story of subjugation, that in the objects I could read traces of killing on a genocidal scale and, by extension, the current dilemma over how to power the world.

On one wall, there was a photograph of Thys in the Congolese port city of Boma. In it, Thys is staring into the distance, belly protruding from beneath his tailcoat, hand proudly resting on a carved African staff, pith helmet in the crook of his arm.

Thys was born in 1849 into a Belgium that was not yet two decades old (the country had split from Holland in the 1830 revolution). His father was Dalhem's doctor, his mother the town's primary school teacher. Shortly after Thys's sixteenth birthday, King Leopold II, the country's second monarch, took power. Leopold was desperate for a colony, so he looked to Africa and to Congo.

Thys became embroiled in the Belgian colonial project in Congo before it even began. Thys joined the army in 1865. By 1876, when he was recommended to the king by one of his military school commanders, the monarch was exploring African territories that he could make his. "At the beginning, there was Leopold II, and then there was Thys," wrote one historian of Belgian mining in Congo, who explained how Thys's work was the "flowering progenitor" of the many companies through which Leopold ran the colony.

At an 1884 conference of colonial powers, Leopold used philanthropic

Nsala looks at the severed hand and foot of his daughter.

language to hoodwink the other European nations into allowing him to annex Congo as private property. His aim, of course, was not philanthropy but profit. To that end, in 1886, Thys created the very first of the corporate entities that would be used to cloak Leopold's barbarism in the guise of a business venture. Such companies would proliferate during Leopold's possession of Congo; they insulated the king and his fellow investors, mainly well-heeled Anglo-Saxons, from direct responsibility for the crimes that were being committed in the territory. They would be the first international firms to plunder Congo—a practice that, as we have seen, continued well into the twenty-first century.

By the last decade of the nineteenth century, the invention of the bicycle and the automobile had led to a boom in the demand for rubber (synthetic rubber was not invented until 1909). Rubber plantations are slow to grow, but Congo's forests were full of the vines. Soon, European overseers were press-ganging local men into harvesting rubber. If they refused to work, their wives and children were kidnapped as collateral. After Leopold colonized Congo, entire villages were enslaved in the quest for rubber, and mutilation and murder were used to ensure loyalty. One particularly haunting image shows a man named Nsala staring at a severed hand and foot on the ground. "He hadn't made his rubber quota for the day so the Belgian-appointed overseers had cut off his daughter's hand and foot. Her name was Boali. She was five years old. Then they killed her. But they weren't finished. Then they killed his wife too," Lady Alice Seeley Harris would remember years later. "Leopold had not given any thought to the idea that these African children, these men and women, were our fully human brothers, created equally by the same Hand that had created his own lineage of European Royalty."

Thys masterminded the use of corporate structures to carry out this plunder, harnessing not only Belgian greed but also the greed of shareholders in Europe and around the world. Like the lithium-ion battery, rubber tires were a globalized product that helped other people, in other places, to get around.

The shareholders' need for a constant stream of profit would push the men who had come to Congo, driving them to wipe out its elephants and jungles, enslave its population, and tear open its earth. Thys created companies for commerce, companies for industry, companies for railroads, companies for mining, and subcompanies for general stores and agricultural products. And Thys certainly didn't rest on his laurels: An old postcard pinned in one of the display cabinets at the school shows the splendid twin-turreted castle he built as his residence near Dalhem. Though slightly more elegant in form, it still reminded me of the oversize palaces of the corrupt and the villas of the magnates of today.

Some of the companies Thys created were the forebears of successful European businesses that still existed as I wrote this book. Umicore—a publicly traded Belgian-French materials technology and recycling company that had a 2024 market cap of nearly €5 billion and the ambition to become a "sustainability champion"—has its roots in a firm that Thys created: the Compagnie du Katanga.

This company, which was the progenitor of the Union Minière du Haut-Katanga, was key to the future financial success of the colony, and men like Thys dreamed of the riches it would bring them. By the 1890s, Leopold's agents and British colonialists were both trying to seize the southern province of Katanga, and the British colonialists under Cecil Rhodes almost took power there in the late nineteenth century. As a writer for the colonial *Journal de Congo* once put it, Katanga "would have been English if not for Thys"—Katanga, the richest part of the Congo Basin and the prototype for the overseas resource extraction that would continue long after colonialism had ended; Katanga, the province where Thys helped organize a series of progressively more murderous colonial expeditions, culminating in its seizure in 1892.

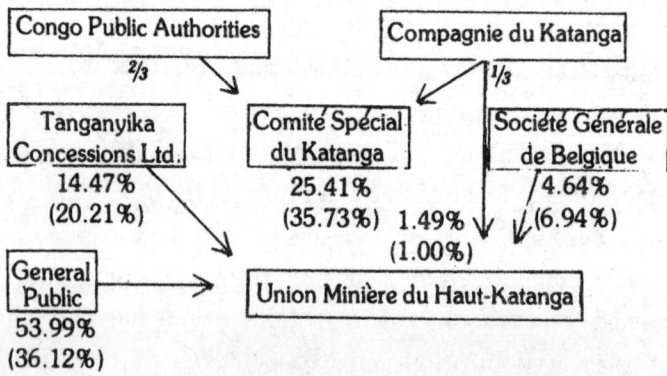

A snapshot of the Union Minière's many-tentacled ownership structure on the eve of independence, in 1960, shows just what an octopus Thys's corporations had evolved into since their inception in the late nineteenth century. The "General Public" were shareholders, Tanganyika Concessions was a British firm that shared an interlocking board with the Union Minière, and the Société Générale was Belgium's premier investment bank, so powerful at one point that it was called "the uncrowned queen of Belgium." The structure was designed to be complicated and opaque in order to protect the people who ultimately benefited from the extraction of Congo's minerals. And it uncannily resembled the web of companies that people like Gertler or Kabila would create in their quests for wealth.

IN MANY WAYS, OF COURSE, THINGS HAVE CHANGED IMMENSELY. THESE DAYS, Congo is, after all, an independent state, run by Congolese. But the current ownership structures are even more octopean, and they ensure that little wealth goes back to the Congolese state. The penumbral transactions, the web of transnational business dealings spun by foreign businessmen as they sell off Congo's minerals—all of this started with the Compagnie du Katanga. Then, shareholders were conveniently shielded from ultimate responsibility for what was going on in Congo. Now, in the twenty-first century, the convolutions of the supply chain comfortably remove by degrees the companies buying, selling, and making use of Congo's minerals from the harsh realities on the ground.

In 1887, when Thys made his first trip to Congo, the Belgian monarch put him in charge of building a railroad. As Adam Hochschild describes in *King Leopold's Ghost*, his seminal work on Congo's colonization, the toll on human life was horrendous—the railway was "a modest engineering success and a major human disaster." Men succumbed to heat, dehydration, malnutrition, malaria, shootings, floggings, and explosions of dynamite-laden cars. If they tried to escape, workers were chained together—conditions that recalled the worst excesses of slavery.

In 1892, Thys brought the first Chinese to Congo to work on the railroad. Leopold was apparently fixated on bringing people from China to Congo and creating five large Chinese settlements to stabilize his rule in the colony. This plan never came to pass, but Thys did end up employing 540 laborers from Macau for the rail project. Three hundred of them died, and after seeing the terrible conditions, several of them escaped, walking eastward in a desperate bid to make it home.

Undeterred, the Qing court in Beijing signed a treaty of commerce and friendship with Leopold's colonial state that gave Chinese the right to work and own property in Congo. Executed in 1898, it was the first treaty between China and any African polity (though Congo was, of course, under outside rule), and it portended the Sino-Congolese relations of the twenty-first century that would give Beijing unprecedented access to all kinds of materials it needed to power the future: coltan and lithium, cassiterite and tungsten, and, perhaps most significantly, copper and cobalt.

THE WORLD IS FACING THE BIGGEST SUPPLY-DEMAND DISLOCATION IN LIV-
ing memory with critical metals, and people have been desperately seeking other solutions. In the battery industry, everyone has their vision of the future. Some visions are limited—tweaks and changes to chemistry that will incrementally increase battery power and capacity. For example, battery makers in the twenty-first century have been testing materials like silicon as they strive to increase energy density in batteries using new anode materials. Other visions are more radical: The Japanese battery company official I met in Tokyo told me that the thorny problems around battery power wouldn't be solved until nuclear fusion had been perfected.

Solid-state batteries are another frontier. Such batteries use a solid electrolyte that allows ions to be transported more quickly between the electrodes, and thus these batteries can be more powerful. Firms outside of China, like LG, Samsung, and Toyota, have developed powerful solid-state batteries; inside China, BYD and CATL are competing against smaller firms like WeLion to begin producing solid-state on a mass scale.

By the mid-2020s, SAIC, the firm that Robert Aronson had clashed with back in the 1990s, was also making electric cars. They also owned MG, the former British carmaker, and the firm's general manager announced that they would launch an inexpensive model with a solid-state battery.

There are questions, however, about solid-state's sustainability. Eric Frederickson, Call2Recycle's vice president, told me that they are "virtually unrecyclable." Such batteries also often use materials like cobalt and lithium, which would continue creating the same demand for materials from complex parts of the world.

But what if you didn't have to use cobalt, or even lithium, for a battery? Marek Slavik, cofounder of Theion, a Berlin-based battery start-up that makes quasi-solid-state batteries, is asking that question. He told me in the summer of 2023 how his firm had made a powerful battery using sodium and sulfur. The materials needed to make sodium-sulfur batteries are cheap and common all around the world. "We have a very lightweight cathode material that is abundantly, democratically, and geographically available," Slavik said. "We need simply to also pull it out from the petrochemical industry. That is their waste product. We are transforming waste into cathode materials." Such batteries, however, are still barely out of the laboratory.

Yet more scientists homed in on sodium without sulfur. Sodium is cheap because it's abundant in the earth's crust and can even be extracted from seawater. As a result, sodium-ion batteries are much more cost-effective than their lithium-ion cousins. China, it must be said, is already ahead of the rest of the world when it comes to sodium-ion. At the Battery Show in Stuttgart in 2023, Shmuel De-Leon, an Israeli battery expert, said he was "shocked" by what he had seen at a Chinese battery fair two weeks earlier. Before going there, he had thought that only one company was making sodium-ion batteries. "I want to tell you that I found ten," he said. Later that year, Benchmark Mineral Intelligence counted thirty-six Chinese companies involved in sodium-ion cell production and research. In February 2023, a Chinese automaker called JAC had launched the first sodium-ion-powered car.

In Stuttgart, at a booth for AMTE Power, a British company that had created a sodium-ion battery, I held a cylindrical sodium-ion cell in my hands. Fergal Harrington-Beatty, the company's head of sales, told me he believed that the technology would eat into LFP's market share. It was marginally lower than LFP in terms of energy density, he said. Sodium-ion was far from a perfect solution: One of the most vaunted technologies, which used the pigment Prussian blue, produced a toxic waste that was hard to dispose of. The fact that yet another blue pigment was being used to produce power felt like a bit of a cosmic joke.

In the U.S. and Europe, however, the supply chain for sodium-ion was not established, and few people wanted to invest in the technology until it became cost-efficient. Scientists at research institutions like the University of Houston would continue to develop and refine powerful new types of sodium-ion batteries, even as businesses trying to put such technology into production faltered. AMTE Power was sold in 2024 after it failed to raise enough money. That was not the case in China, where state-run companies, and firms like BYD and CATL, were forging ahead in battery science and production, putting in the groundwork before the rest of the world even realized what had happened.

———

MAYBE THE ANSWER TO THE CRITICAL METALS CONUNDRUM IS IN FACT right in front of us, or, rather, in the ground beneath our feet. Over the years I researched this book, what became clear was that the Chinese were only the latest in a long stream of people who had exploited Katanga for its wealth. At the center of the Belgian dreams in Congo were vast riches, not the charity that Leopold had promised he would bestow upon the country. The land had long been in the grip of rulers who had prospered from its immense natural resources, and it would remain in their grip after the Belgians had left. In the town of Tervuren, near Brussels, Leopold built a pompous neoclassical museum to celebrate his conquest of Africa. Visiting in 2021, I saw copper crosses and huge, heavy I-shaped ingots that people had created out of Katanga's copper well before Belgium took control of Congo. I imagined ancient people lugging them across the savanna, moving them with little fuss along trade routes that crossed Africa. When Europeans later traversed these well-trod routes, they would be lauded by their own countrymen as intrepid explorers.

But the resources could also help Congo, not simply condemn the country to an endless "resource curse." Maybe we don't need new technology to help power our lives. Maybe Congo could even benefit from its own resources in the energy revolution. My experience at the Glencore mine had shown me how mining could be carried out in a modern manner, with an emphasis on safety. In plenty of places around the world—in Australia and Tanzania, for instance—sustainable mining has been shown to be possible. A trip to the Congolese town of Bunkeya in 2022 showed me how Congolese people—and, by extension, people living close to critical-metals mines around the world—could build better places to live from the resources beneath their feet.

Bunkeya is still governed by a descendant of Msiri, the Yeke king whom the Belgians, spurred on by Thys, killed to take Katanga. Remarkably, it is a community that has kept a strong sense of identity despite the years of violence and colonialism it has suffered. What's also remarkable is that the people of Bunkeya have managed to funnel their portion of the mining royalties from nearby Tenke Fungurume not into some corrupt pocket but back into agriculture, potable water, and the building of community infrastructure. To be sure, this money is only a small portion of the money made at TFM, but civil-society leaders in Lubumbashi took me through the numbers and showed me how the mwami, or king, had ensured that the community was given its share of the mineral wealth before it was stolen. According to them, the mwami had only managed to do so through vigorous advocacy in Kinshasa.

It was even more remarkable because elsewhere that money was being stolen. In February 2022, the government in Kinshasa had begun an investigation into Tenke Fungurume to try and ascertain the true size of the reserves and whether China Moly owed the state money, and suspended the firm's mining rights. They agreed to settle with China Moly in April 2023. The firm had to pay some $800 million over six years and a minimum of $1.2 billion in dividends over the mine's operational life, and said that it expected the money to "play a stronger role in promoting economic development and job creation." A senior U.S. official told me that even more money appeared to have been paid upfront by China Moly. He lamented that whatever was being paid was being poorly spent or "put into pockets" and stolen. He insisted that the Congolese government had to be transparent. "What are you doing with that?" the official questioned. "Show us that you're building something productive for your population—they couldn't."

I had come to Bunkeya in 2022 on the anniversary of the mwami's coronation, which takes place in July and is one of two yearly celebrations in the territory of the Yeke. I parked outside in the main square, crossed the main street, which is called Boulevard Msiri, and headed in through the red-and-white-painted gates of the royal enclosure of Mwenda Bantu Kaneranera Godefroid Munongo Jr., the current mwami. Unusually for Congo, no one came up to me, a foreigner, and pleaded, "*Makuta, makuta*" (Money, money). The sound of drumming and gunshots came from up ahead of me, as I passed a series of abodes, little thatched huts in miniature, homes for the spirits of each of the mwami's ancestors.

In a tented courtyard, the mwami sat enthroned in white robes as his ceremonial bodyguard, in red tunics, fired ancient muskets into the air. Someone told me that the guards' guns dated back to the colonial days. Traditional chiefs from across the country, the Mwaant Yav, and a Luba dignitary, as well as African aristocrats—a sultan from Chad, a king from Ghana, and a chief from the Republic of the Congo—lined up to salute the mwami and give him gifts.

Afterward, a foreign agriculture expert showed me how farming had been improved by the king, how he had constructed clinics and roads, even how he was thinking about building a model mine on a hill near town. Here was a community that stood in stark opposition to what was happening in many other towns throughout the South, where big companies and anarchic profit seekers were tearing the social and physical fabric of towns apart, a community that showed that money from cobalt could benefit the poorest in Congo.

In a way, Bunkeya has gotten lucky: There are few mines in the immediate vicinity of the town, and the people there have retained a strong sense of their own culture, something that has been stamped out in other parts of the South. This is not coincidental. By denying the history of Africans, European colonialists encouraged a view of the continent's populations as people in need of civilizing influences and deracinated them. As the development studies scholar Kevin C. Dunn has written of Congo, "External actors have frequently attempted to characterize the country as divided, chaotic, and lacking the ability of self-articulation, which in turn has allowed external actors to speak for it."

Such external articulations allowed the slave trade to flourish along the coast of West Africa, and later allowed explorers to claim huge swaths of land for European monarchs. The parallels with today are plain to see, in the condescension through which people speak about Africa and Africans; in the ignorance that is afforded to African affairs by major U.S. and European powers; and in the treatment of African workers by their Chinese bosses. Colonial myths are repackaged for the twenty-first century: that Africans are incapable of self-rule, that corruption is somehow a natural state on the continent, that they cannot manage their own resources. It is somehow fitting that the lithium-ion battery, which is in many ways the ultimate product of globalization, a thing that allows people to move power, and to take power from a desperately poor place and place it into the hands of the rich, is so sourced in Congo. And it is fitting, too, that localized measures might be the answers to the injustices in the battery supply chain.

Writing for *The New York Review of Books* in 2018, the journalist Howard W. French explained that one of the key rhetorical moves of colonialism has been "emphasizing the civilizational virtues that are being shared with less fortunate or explicitly inferior peoples, while minimizing one's own self-interested objectives and downplaying the violence and dispossession that are usually essential to the subjugation of others." The deletion of African history was essential in this project, French said, to establish mastery over people in the colonies. Any evidence of ancient civilization was ascribed to external influences. "To accord local agency in such matters would have undermined a long-standing narrative about the inherent inferiority or even subhuman nature of Africans, which was vital in giving Europeans license for their actions."

Bunkeya flies in the face of these assertions. In some ways, it always has. Msiri resisted colonial rule there in the 1880s, even though he was an invader himself. These days, under his distant descendent, it is trying once again to break free, resisting the control that has been foisted upon it since Msiri was murdered.

TO MEET CLIMATE GOALS THAT WILL MAKE A DENT IN GLOBAL WARMING, we need to massively invest in environmentally responsible mining. At the same time, much still needs to be done to focus the world's attention on the devastation wrought by many forms of mining—to our shared planet, to local lives and livelihoods—and how to mitigate it. Such goals might seem antithetical, but I believe there is hope.

Perhaps the power-storage space suffers most from a lack of imagination. Many mining companies seemed uninterested in the possibilities provided by clean energy. "The big miners haven't invested in the critical-metals projects because they're too small," Brian Menell, the TechMet CEO, told me. "One billion dollars barely moves the dial." But he also thought that few people had woken up to the opportunities offered by critical metals. "There is still a degree of naivete," he noted. The problem, Menell argued, was that businessfolk and politicians in Europe and the U.S. didn't look beyond the short term and invest in projects that were important for the future. "You need political will, which you don't have," he said. "We should have auto workers in Detroit demonstrating for more mining; we should have students demanding that their universities invest in clean mining. If investment doesn't happen, we won't make our climate change goals over the next twenty years."

One way to reconcile the goals of the new-energy revolution with the terrible toll it takes on communities is fairly simple: We should listen to the people from the places where we get our minerals. We must listen to them about pollution in their communities and consider their dreams for a healthier, more balanced world. Such people are commonly erased, thanks to big-stakes financial dealings and complex, often deadly geopolitical games in the cutthroat competition for resources, but citizens of wealthy countries cannot simply hope that innovation will save the planet, or ignore the horrific suffering that has come to be accepted as an unavoidable price for cleaner cities. To do so risks entrenching a system of cruelty and pollution that will eventually, in ways we are only just beginning to understand, prove as destructive as any hydrocarbon.

In the end, most places in southern Congo aren't like Bunkeya. Most places have been built on the buried dreams of miners, and upon a painful history of exploitation.

Sometimes, they have literally been built on human bodies.

On one of my early trips to Congo, I spoke to Charlotte "Maman Ocean" Cime Jinga, the parliamentarian who once served as Kolwezi's mayor. She complained about just how cavalier the government was being with miners' lives. The bodies of *creuseurs* who had died from a cave-in at KCC's mine had been unceremoniously dumped, she said. These were people like Françoise Ilunga's husband, Nitunga. "There is a mass grave, unmarked," she told me. Officially, forty-three people had suffocated or

The mass grave at the Cimetière Mwangeji, 2022

been crushed to death, but she contended that there had been many, many more. In an attempt to lower the death toll, officials had hastily ordered trenches be dug at the city cemetery and buried the bodies of those whose families hadn't come to claim them.

Maman Ocean's words haunted me throughout the COVID pandemic. I resolved to find the grave with Jeef Kazadi when I returned to Katanga in March 2022.

We made some inquiries and found out that such a mass grave does, indeed, exist. It lies in a far-flung part of the Cimetière Mwangeji.

We arrived at the cemetery in the early afternoon. Mwangeji sits on a giant plot in the middle of town next to a series of shops that sell wooden coffins. Marcel, a nineteen-year-old IT student in a blue suit, was headed for his lunch break. He worked as a gravedigger to make some extra cash, he said. He knew where the unknown *creuseurs* had been buried. There were about thirty of them, he remembered.

Marcel ushered us into the cemetery, which was quiet and huge and overgrown. Graves were everywhere in the soft soil, and only a few paths wended their way through them.

At the yard's southern wall, two men were digging. They shouted that I should get out. "You don't have the right authorization to be here," they told me. I needed to ask the new mayor for permission to enter, they said.

Marcel led me away, taking me on a roundabout route through wooden crosses staked into the ground. "Mwangeji is full," he said. "There is no *terre vièrge*." No virgin ground. The empty areas only appeared empty, because the tall grass and bushes that covered the cemetery were burned at the end of the dry season to free up space for more graves. The wooden crosses would burn in the flames.

He brought us into an area of thick bush. After a few feet, the foliage thinned out. It was here, he said. He pointed at a patch of scrub. Thirty men's bodies had been dumped here after the tunnel they were in collapsed.

There were no markings. Reed grass had grown from the soil, and it was red and long and feathery. It shuddered in the breeze.

Thirty bodies lay in the ground below my feet. They belonged to men who had been digging out the metals that keep our world powered, and they had been crushed, deep under the earth, in Congo.

ACKNOWLEDGMENTS

Munu umua kawena mua kukuma bionso. A finger, according to a Luba-Kasai proverb, cannot do everything by itself. And, though I do tend to type slowly using one finger on each hand, the same is true for a book like this, which has benefited from the input and thoughtfulness of scores of people. I am immensely grateful to everyone who spent the time helping me to learn about the issues surrounding critical minerals in the modern world, welcoming me into their homes and offices. Thanks especially to Odilon Kajumba, who opened up about the world of mining in Congo; to Sadip in Bahodopi, who spoke about how his land had been destroyed and served me delicious cooked bananas; and to Mohammad Aris in Ternate. I was lucky to meet Aminatou Haider in Las Palmas and I thank her for the time she gave me after a long trip.

I'd like to thank Jeeftour Kazadi Kamwanga, without whom this work would not have been possible, and who suffered the indignity of detention at the hands of the ANR. *Jeef, on reste ensemble.*

Thank you to my family for putting up with me during this, for listening to me drone on about mining and metals, and for all their support after I was arrested. To my mother and father, I'm sorry for all the frights, and to Lex and Inès, thanks for bearing with me.

Adam Eaglin at the Cheney Agency was the first person who encouraged me to look into critical minerals back in 2018 and helped shape my pitch and many drafts. At Penguin Press, thank you to Will Heyward, Natalie Coleman, Scott Moyers, Chelsea Cohen, and Lauren Morgan Whitticom. Kris French, Angely Mercado, Cadence Bambeneck, Aina de Lapparent Alvarez, and Amanda Braitman, thank you for your fine-toothed checking work. I'd like to thank Anthony Del Corso for all his help putting this book together, and Gurveen Kaur for being a great sounding board.

At *The New Yorker*, Daniel Zalewski first saw the potential for a long narrative about artisanal miners, and David Remnick took a chance on a piece from Congo (and then helped me get out of prison). David Rohde published my work on Western Sahara. Peter Canby helped me with key questions at various stages. Research and fact-checking by Katie Nodjimbadem, Lucie Kroening, and Han Zhang were essential. Thanks, too, to Fabio Bertoni, Willing Davidson, Hélène Werner, Virginia Cannon, Bruce Diones, Patrick Radden Keefe, Carolyn Kormann, Colin Stokes, Becky Cooper, Jon Blitzer, Steph Taladrid, Michael Agger, David Kortava, Camila Osorio, Anakwa Dwamena, Yasmine AlSayyad, and so many other checkers and editorial staff, past and present, whose encouragement kept me going.

My good friend Henry Steel first introduced me to the world of mining and provided key insight during the reporting: He also went to bat for me when I was imprisoned. Thanks to Jon Lee Anderson for passing the message on and Ben Taub for helping get in touch with my family.

Many people helped during my detention, and I'd like to thank them all. On Capitol Hill, Representative Chris Smith was essential in pushing for my release. The U.S. Embassy and State Department did incredible work. I'd like to thank Ambassador Michael Hammer and Molly Phee. Claire Sheldon, your appearance in the offices of the ANR with food will always be in my mind. Carol Cox, your words of encouragement on the way to the airport will always be lodged in my mind. D. D. and J. D., your help will always be remembered. Richard Golub and Milos Ivkovic. Victoria Dreesmann and Senator Chris Dodd. Thank you, too, to Piero Tozzi and all the staff of the Congressional-Executive Commission on China, who allowed me to brief Congress (at length, sorry!) on a topic close to my heart.

Don Guttenplan at *The Nation* published a piece on Indonesia that helped form the basis for my reporting on that part of the world in this book. Thanks also to Katrina vanden Heuvel, Christopher Shay, and Sarah Baum. At *The New York Review of Books*, thanks to Emily Greenhouse, Andrew Katzenstein, Ratik Asokan, and Max Nelson for letting me explore the world of battery-metals mining in an essay. At *Granta*, Tom Meaney, Tom Bolger, and Josie Mitchell helped bring Manono to the page.

Benoit Nyemba Basali, your help on my first trip to Congo was invaluable. Hans and Dany got me from Kamina to Manono on their bikes, Major helped me cross the Lualaba on his canoe, and Gaston Ntambo Nkulu got me out of there. I learned that *au Congo, la route est toujours bonne*. AVZ welcomed me to their facility, as did MMG and KCC: For this I thank them. Aigle at COMAKAT showed me around the mine site his firm had occupied at Kawama, as did SAEMAPE at Kasulo.

Thank you to Nima Elbagir, Erik Kennes, Clarissa Ward, Jonathan Rosen, Arnaud Froger, the Association Congolaise pour l'Accès à la Justice, the Committee to Protect Journalists, Reporters Without Borders, and the International Federation of Journalists for helping raise the alarm about Jeef's detention. At the ANR, I would like to thank Emmanuel, Papy, and Thierry, and everyone else who took pity on me during my hour of need. Patrick Masengo Kalasa, I hope your ordeal ends soon.

In Congo, I'd like to thank AFREWATCH, PODEFIP, Bon Pasteur, Schadrack Mukad Mway End Naw, Sister Catherine Mutindi, Steve Wembi, Vital Kamungu, Charles Tolchard, Mick Brown, and Jean Luc Kayoko, may he rest in peace. Still I Rise is an incredible organization: Thanks for allowing me in to speak to your students. And so many more people in Congo who inspired me every day with their acts of kindness—thank you.

In Indonesia, Richaldo Hariandja was a wonderful and thoughtful guide. Andreas Harsono, Sadam, Winwin, Upi, and Rabul provided vital assists. In Idaho, Brad Barnett, Richard Williams, Tom Carter, and of course Sam Ash, who welcomed me for the best tri-tip in the Silver Valley. In the Western Sahara, thanks to Mouloud Said, María Carrión, Mohamedsalem Werad, Hamahuallah Mohammed, Takween Mohamed Hajietou, and the late Ahmed Boukhari. In Japan, Peter Sayn Wittgenstein, Yurina Roche, Christopher Ax, Marc Luetten, Jeffrey Char, and Louis Chihara.

Several journalists and researchers provided me with essential insights—

Hugh Kinsella Cunningham served his regularly with lashings of wit. *N'oublie pas ton Go-Pass, mon ami, c'est très important.* Oz Woloshyn gave wise insights into early drafts and reporting plans. William Clowes, Emmet Livingstone, Michael Kavanagh, Giulia Paravicini, Rory Randall, Isaac Arnsdorf, Giles Clarke, Jacob Kushner, Valerie Hopkins, John Dell'Osso, Gabriel Bourdon-Fattal, and Adelle Hollmer all spent time listening to my questions and giving advice. Jean-Baptiste Gallopin, *merci.* Luis Aleman, Jennifer Fendrick, Joseph Mulala Nguramo, Milain Fayulu, J. S., M. B., P. T. M. H., D. W., Harry Mossop, Heyrick Bond-Gunning, Matthieu Bos, thanks all for your help along the way. Thanks to Charlie Watenphul for patiently listening to all my questions too. And to Omar Hilale, who was kind with his time in New York all those years ago.

Several people let me write at their homes along the way: Marilena and Alexandros Kedros, ευχαριστώ, especially for taking me in right after my detention. *Grazie*, Stefania Biondo and Giacomo Nurra, for your many kindnesses and letting me occupy your apartment and office. Thanks, too, to Christos, Seb, Olympia and J. C., Ike, Nikolai, Tassilo, Caspar and Sasha, Jacob, Jake, Paul, Jay, Zach, Lucas Z., Pippa B., Lucas W., Martin P., Juan S-C., Hugh M., Amber Bell, Anne Fadiman, Sanford Padwe, Judith Matloff, Walt Bogdanich, Moira Fradinger, John McKay, and of course my family writ large.

Thank you to the conferences that welcomed me for my reporting: the Battery Show, DRC Mining Week, and the Cobalt Conference. Thanks especially to Marina Demidova for all your help. And thanks to the New Energy Industry Development Conference, and I'm sorry circumstances did not permit me to attend. My thanks to the Mudd Library at Princeton, the École communal de Dalhem, and the State Archives of Belgium.

To the people who read early drafts of this book and gave me valuable insights, endless gratitude. Thank you, Shirley Meng, Andrew Gulley, Patrick Alley, and Paola Subacchi.

My penultimate thanks go out to the two people who passed my letters to the U.S. Embassy. Without you, I still might be languishing in captivity. I owe a great deal to the kindness of you strangers.

And finally, to my wife, Malù, who dealt with my mad hours and strange trips. Thank you and *t.a.b.s.t.f.*

NOTE ON SOURCES

This book relies on hundreds of interviews and conversations conducted over six years over the phone and in person in the United States, Belgium, the United Kingdom, France, Switzerland, Germany, the Democratic Republic of the Congo, Zambia, Kenya, Indonesia, Algeria, the Western Sahara, Mauritania, Morocco, China, and Japan. Some of the people I spoke to requested to speak anonymously: If they had a compelling reason to do so, I agreed to not reveal their names.

I have also relied on written sources to corroborate the information that was given to me. I waded through archives and thousands of pages of court documents, some of which, like the Gertner-Gertler arbitration, were not public. I lost some of the documents I had on my computer when it was confiscated by the Congolese police, and it took a while to reassemble my reporting, but luckily I had backed up many of my interviews and notes. I still have the diary I took in match-ash during my days in the Kinshasa detention site.

A book like this rests on the shoulders of giants, and I have indicated where I have learned facts and quoted from texts in the endnotes. Seth Fletcher's *Bottled Lightning* and Charles Murray's *Long Hard Road* are indispensable histories of the lithium-ion battery. Henry Sanderson's *Volt Rush* was an excellent meditation on the necessities and exigencies of critical-metals mining. Ernest Scheyder's *The War Below* helped me through questions around metals mining in the U.S. Siddharth Kara's *Cobalt Red* raised awareness at a vital time. Howard French's *China's Second Continent* is a wonderful book and a great guide to China in Africa. I only read Vince Beiser's *Power Metal* as I was finishing final proofs of this book, but it helped refocus my work. Lukasz Bednarski's *Lithium* also was a guide to the white metal. Adam Hochschild's *King Leopold's Ghost* remains a masterwork. *The World for Sale* by Jack Farchy and Javier Blas taught me about the world of commodities trading. Tom Burgis's *The Looting Machine* remains a master class on corruption. The work of Michela Wrong has also been a guide as I have worked in Central Africa. David Van Reybrouck's *Congo* and *Revolusi* are two of my favorite books I have ever read and signposted me on the history of both Congo and Indonesia.

The work of several journalists was invaluable as I researched this story. Michael J. Kavanagh and Will Clowes at Bloomberg; Daniel Balint-Kurti at several places; Aaron Ross at Reuters; Eric Lipton and Dionne Searcey at *The New York Times*; and Andrew Maykuth at *The Philadelphia Inquirer* all come to mind.

Several parts of this book are based on reporting in various different magazines. The kernel of this book was "Buried Dreams," a long *New Yorker* piece published in 2021. Other scenes are based on "Dirty Nickel, Clean Power: Making the Ocean Bleed Red," a 2023 piece for *The Nation*, and "Power Metals," a piece for *Granta*.

While I thank everyone involved in helping me with this book, any and all errors in it are my own.

NOTES

Introduction: The New Power

1 **For the iPhone 17 Air:** Johanna Romero, "This iPhone 17 Air Leak Suggests You May Not Have to Worry About It Having Horrific Battery Life," PhoneArena, March 6, 2025, phonearena .com/news/this-iphone-17-air-leak-suggests-you-may-not-have-to-worry-about-it-having -horrific-battery-life_id168270/.

2 **not for nothing:** Brian Menell, interview with the author, November 2023.

2 **70 to 90 percent:** Varun Sivaram, Noah Gordon, and Daniel Helmeci, *Winning the Battery Race: How the United States Can Leapfrog China to Dominate Next-Generation Battery Technologies*, Carnegie Endowment for International Peace, October 21, 2024, carnegieendowment.org/research /2024/10/winning-the-battery-race-how-the-united-states-can-leapfrog-china-to -dominate-next-generation-battery-technologies?lang=en/.

3 **an outrageous $288 million:** Trisha Thadani, Clara Ence Morse, and Maeve Reston, "Elon Musk Donated $288 Million in 2024 Election, Final Tally Shows," *Washington Post,* January 31, 2025, washingtonpost.com/politics/2025/01/31/elon-musk-trump-donor-2024-election.

3 **a 2023 shareholder meeting:** Elon Musk, "Tesla 2023 Annual Shareholder Meeting," speech, Austin, TX, May 16, 2023.

3 **Tesla provided only:** Alan Ohnsman, "Elon Musk's Laughable New Solution to Tesla's Child Labor Worries," *Forbes,* July 2, 2024, forbes.com/sites/alanohnsman/2024/07/02/elon-musks -laughable-new-solution-to-teslas-child-labor-worries/.

3 **Then, on a Wednesday:** *Special Report,* aired March 19, 2025, on Fox News.

4 **Tshisekedi had a problem:** The rebels, who belonged to a group called the March 23 movement, or M23, and the Alliance Fleuve Congo were both proxies for Rwandan interests in the region and harbored grievances based on years of discrimination and venal national politics. They were profiting from metals too: The M23 had seized mines and were funding their rebellion with them. In one area alone, they were collecting at least $800,000 a month from mining and taxing minerals in transit by the end of 2024. See: United Nations, "Letter Dated 27 December 2024 from the Group of Experts on the Democratic Republic of the Congo Addressed to the President of the Security Council," UN Document S/2024/969, p. 14, https:// documents.un.org/doc/undoc/gen/n24/373/37/pdf/n2437337.pdf.

4 **Karl Von Batten:** Karl Von Batten, interview with the author, March 2025.

5 **"Are we being played?":** Senior congressional official, interview with the author, March 2025.

6 **Peter Sahlas, one of:** Peter Sahlas, interview with the author, January 2025.

7 **Elon Musk had raised $2.7 billion:** "Tesla's Performance Gives Elon Musk Much to Think About," *Economist*, June 15, 2019.

8 **tried to go back:** In mid-2023, I registered for a Shanghai conference on lithium-ion batteries and hoped to meet battery makers there. I applied for my visa in New York, in person, but the lady behind the desk said that there was a problem with my application. I was later told that I needed to sit for an interview if I wanted to travel to China. I agreed to an interview and dialed in to a Zoom meeting. During the meeting, a lady from the consulate who identified herself as "Madam Bai" asked me pointed questions about what I thought of China, and what I thought of trade embargoes. She told me that protectionist U.S. policy was an obstacle to progress, especially with regard to green technology that could be used to slow climate change. I told her—truthfully—that it was remarkable what China had done to create a thriving battery industry. Seeming satisfied, Madam Bai said she would send my application to Beijing to be approved. Weeks passed with no response, and the visa never came.

9 **the Pentagon focus:** *Securing Defense-Critical Supply Chains: An Action Plan Developed in Response to President Biden's Executive Order 14017* (Department of Defense, February 2022), 10, apps.dtic .mil/sti/pdfs/AD1163223.pdf.

9 **one former Defense Department:** Matthew D. Zolnowski, interview with the author, May 2024.

10 **A tourist in London:** "42% of the City Buses Registered in Europe in 2023 Are Zero Emission," Sustainable Bus, February 14, 2024, sustainable-bus.com/news/electric-bus-market-2023 -registrations-man-solaris-yutong-wrightbus/.

11 **"It's always going to be":** Melissa Sanderson, author interview, August 2023.

12 **forty million electric:** Coral Davenport and James Ewing, "Can Trump Really Slam the Brakes on Electric Vehicles?," *New York Times*, May 27, 2024.

12 **they were set to grow:** "Electric Vehicle Sales Headed for Record Year but Growth Slowdown Puts Climate Targets at Risk, According to BloombergNEF Report," *BloombergNEF*, June 12, 2024, about.bnef.com/blog/electric-vehicle-sales-headed-for-record-year-but-growth-slowdown -puts-climate-targets-at-risk-according-to-bloombergnef-report/.

12 **longtime EV skeptic:** David Shepardson, "Trump Says He Would Consider Ending $7,500 Electric Vehicle Credit," Reuters, August 19, 2024; and Todd Lassa, "Trump Plans to Nix Biden's EV Tax Credits—and Why It's Good for Musk," *Autoweek*, November 21, 2024.

12 **Apple's CEO, Tim Cook:** Yang Jie, "Building Apple Products Has Become a Side Hustle for China's Biggest EV Maker," *Wall Street Journal*, December 2, 2024, wsj.com/tech/building -apple-products-has-become-a-side-hustle-for-chinas-biggest-ev-maker-f26e251c.

12 **"Last night Teslas":** "Wieder brennen Teslas auf Berlins Straßen," *Kontrapolis*, June 19, 2024, kontrapolis.info/13354/.

13 **the "real villains":** Ashleigh Fields, "Musk: Trump Will 'Go After' People 'Pushing the Lies' About Tesla," *The Hill*, March 28, 2025, https://thehill.com/homenews/administration/5219374 -elon-musk-donald-trump-response-tesla-vandalism/.

13 **If you count:** Shannon Osaka, "The 'Greenest' Car in America Might Surprise You," *Washington Post*, February 29, 2024; André Thomas, "Automobile: L'hybride rechargeable serait moins polluante que l'électrique, assure une étude," *Ouest France*, October 22, 2022, ouest-france.fr /economie/automobile/automobile-l-hybride-rechargeable-serait-moins-polluante-que-l -electrique-assure-une-etude-c33e1b94-4f97-11ed-9919-8fbf073b2344.

14 **That's to say:** Paul Lienert, "US Consumers Keep Vehicles for a Record 12.5 Years on Average— S&P," Reuters, May 15, 2023.

PART 1: FUNDAMENTS

15 **"When Katanga is hurt":** Conor Cruise O'Brien, *To Katanga and Back: A UN Case History* (Simon & Schuster, 1962), 261.

Chapter 1: A Bend in the Lufilian Arc

18 **"The worst souvenir":** Odilon Kajumba Kilanga, interview with the author, conducted by Jeef Kazadi Kamwanga in Kolwezi, 2024. Because I had been banned from Congo, I sent questions to Kazadi, who conducted the interview with Kajumba in Kolwezi.

18 **"I would never accept":** Kajumba, interview.

19 **children might be:** AFREWATCH, "361 000 enfants dans les mines de cuivre et cobalt de Haut Katanga et Lualaba en 2024: La société civile émet une réserve au chiffre de l'UNICEF et demande les informations complémentaires," press release, June 1, 2024, afrewatch.org /communique-de-presse-n01-06-2024-361-000-enfants-dans-les-mines-de-cuivre-et-cobalt-de -haut-katanga-et-lualaba-en-2024-la-societe-emet-une-reserve-au-chiffre-de-lunicef-et/.

20 **exposure to dust:** Dr. Billy Mukong, interview with the author, March 2022. He also blamed a traditional practice in which women eat soil in order to quell menstrual bleeding. In Kolwezi, he said, the practice is very unsafe, not only because so much of the soil has been contaminated from years of mining but also because of the sheer volume of metals that are naturally present in the soil.

20 **The respiratory effects of such:** E. J. Bigwood, *Du problème alimentaire des travailleurs de l'U.M.H.K. et de leurs familles*, 1965, 20–21, box 2, folder 718, Union Minière Collection: Archives sur l'Exploitation des Mines de Cuivre Katangaises, Premier Série, Quatrième Partie, Archives de l'État en Belgique, Dépôt Joseph Cuvelier, Brussels, Belgium.

20 **"had metal concentrations that":** Daan Van Brusselen et al., "Metal Mining and Birth Defects: A Case-Control Study in Lubumbashi, Democratic Republic of the Congo," *Lancet: Planetary Health* 4, no. 4 (2020): 158–67.

21 **"the vilest scramble":** Joseph Conrad, "Geography and Some Explorers," in *Last Essays* (Doubleday, Page & Company, 1926), 17.

21 **By the 1930s:** Editors of Encyclopedia Britannica, "Copperbelt," *Encyclopedia Britannica*, January 6, 2014, britannica.com/place/Copperbelt-region-Africa.

21 **Kajumba was one:** "The World Bank in DRC," World Bank, April 8, 2024, worldbank.org/en /country/drc.

21 **Forty-three percent:** "Democratic Republic of Congo: Overview," World Bank, 2023, worldbank .org/en/country/drc/overview.

21 **worked as a tax:** Patrick Masengo Kalasa, interview with the author, April 2024.

22 **a series of leaks:** Louis Colart, "Congo Hold-Up: Le clan Kabila a détourné au moins 138 millions de dollars de fonds publics," *Le Soir*, November 19, 2021, lesoir.be/407338/article/2021 -11-19/congo-hold-le-clan-kabila-detourne-au-moins-138-millions-de-dollars-de-fonds.

22 **The car's price point:** International Monetary Fund, *Democratic Republic of the Congo: 2019 Article IV Consultation—Press Release; Staff Report; and Statement by the Executive Director for the Democratic Republic of the Congo*, IMF Country Report No. 19/285 (IMF, 2019), 4. In 2019, the IMF said that per capita income for the average Congolese was around USD $470.

23 **A 2021 analysis:** Jon Lynch, "Copper's Role in Growing Electric Vehicle Production," CME Group, May 5, 2021, cmegroup.com/openmarkets/commodities/2021/copper-role-in-electric -vehicle-production.html.

23 **"Copper is the metal":** Pippa Stevens, "A Coming Copper Shortage Could Derail the Energy Transition, Report Finds," CNBC, July 14, 2022, cnbc.com/2022/07/14/copper-is-key-to -electric-vehicles-wind-and-solar-power-were-short-supply.html.

24 **According to the ASM Database:** "Democratic Republic of the Congo," ASM Database, last modified October 27, 2022, artisanalmining.org, http://artisanalmining.org/InventoryData /doku.php/country:congo_drc.

24 **told me that around 170,000:** Richard Muyej Mangez Mans (governor of Lualaba), interview with the author, May 2019.

24 **6 million metric tons:** "Global Cobalt Reserves 2024, by Country," Statista, March 10, 2025, statista.com/statistics/264930/global-cobalt-reserves/.

Chapter 2: The Great Rock That Spreads All Over the Lands

28 **"The plains pullulate":** Gaston-Denys Périer, *Moukanda*, 2nd ed. (Office de Publicité, J. Lebègue, 1924), 43.

28 **In the *miombo*:** Pierre Meerts and Michel Hasson, *Arbres et arbustes du Haut-Katanga* (Jardin Botanique Meise, 2016), 288–89.

28 **"You who have preceded":** Eugenia W. Herbert, *Red Gold of Africa: Copper in Precolonial History and Culture* (University of Wisconsin Press, 1984), 34.

29 **was a hunter who:** Jan Vansina, *Kingdoms of the Savannah* (University of Wisconsin Press, 1966), 71.

29 **short mnemonic poems:** Thomas Q. Reefe, *The Rainbow and the Kings: A History of the Luba Empire to 1891* (University of California Press, 1981), 30. These realms functioned very differently

than modern states, with villages in their domain essentially paying tribute to a king but largely left to run their own affairs.

30 **Other groups of people:** Vansina, *Kingdoms of the Savannah*, 14. *Empire* and *kingdom* are helpful terms, as they put us in mind of the rise and fall of empires, precisely what happened many times over in Katanga, where kingdoms rose and fell with leaders and trade. As the historian Jan Vansina writes in *Kingdoms of the Savanna*, a 1966 book that is still the most authoritative work on the region's precolonial history, the word *tribe* suggests a stable set of people that is more or less perennial. "This notion of the perennial tribe is meaningless," Vansina writes. "It can easily be shown that tribes are born and die, sometimes without displacement of populations or even without changes in the objective cultures of the communities involved."

30 **his work on mining:** Claude Iguma Wakenge, *Eating the Congo: Unveiling State Governance of Copper and Cobalt Mining in Former Katanga* (Lambert Academic Publishing, 2019), 136.

30 **As one local administrator:** Iguma, *Eating the Congo*, 134.

30 **"green stones (malachite)":** B. A. Beadle, trans., "Journey of the Pombeiros: P. J. Baptista and Amaro José, Across Africa from Angola to Tette on the Zambeze," in *The Lands of Cazembe: Lacerda's Journey to Cazembe in 1798*, trans. and annot. Captain R. F. Burton (John Murray, 1873), 188.

30 **the first written instance:** Herbert, *Red Gold of Africa*, 23.

30 **When the German explorers:** Thomas Pakenham, *The Scramble for Africa, 1876–1912* (Random House, 1991), 400.

30 **Msiri's empire was known:** David Van Reybrouck, *Congo: The Epic History of a People*, trans. Sam Garrett (HarperCollins, 2014), 118.

30 **Other missions ensued:** *Centenary of the First Studies on Shaba Geology (Zaïre): Strata-Bound Copper Deposits and Associated Mineralizations, Proceedings of the International Cornet Symposium*, ed. Jean-Marie Charlet (Royal Academy of Overseas Sciences, 1997), https://www.kaowarsom.be /documents/PUBLICATIONS/COLLOQUE%20INTERNATIONAL%20CORNET.pdf.

31 **The British also had designs:** J. B. Thomson, *Joseph Thomson, African Explorer* (London: Sampson Low, Marston and Company, 1896), 239. In the spring of 1890, Cecil Rhodes, Britain's arch-colonialist in Africa, had written a telegram urging the explorer Joseph Thomson to "go and get Katanga." The telegram from Rhodes is reproduced in full in Thomson's biography, written after his death by his brother. Thomson received the telegram in Blantyre—now the second city of Malawi—while recovering from an expedition to Lake Bangweulu. ("Portuguese slave traders had transformed the country into a pathless and silent wilderness, into which no man would venture as their guide.") Thomson, recovering from a smallpox outbreak that had killed some of his best men ("such, however, are the ups and downs of African travel," he wrote, reflecting on the death of "a fine young fellow" called Wilson), was in no state to conduct the expedition but went about making preparations to do so. Rhodes's letter, he wrote to a friend, "rang in my ears like a bugle call to an old cavalry horse." The British government, however, finally weighed in and prohibited him from proceeding. For "scramble for Katanga," see, e.g., S. E. Katzenellenbogen, *Railways and the Copper Mines of Katanga* (Clarendon Press, 1973), 1.

31 **From his homeland:** *The Cambridge History of Africa*, ed. John E. Flint, vol. 5, *From c. 1790 to c. 1870*, ed. J. D. Fage and Roland Oliver (Cambridge University Press, 1976), 245. Msiri's roots were among the Nyamwezi people, who lived across Lake Tanganyika.

31 **intrusion into their ancestral lands:** Giacomo Macola and James Hogan, "Guerrilla Warfare in Katanga: The Sanga Rebellion of the 1890s and Its Suppression," *Small Wars and Insurgencies* 30, no. 4–5 (2019): 873. The scholar Giacomo Macola has done excellent work on this period. He called this "a grassroots rebellion against an oppressive foreign elite."

31 **Arab slavers were cutting:** Abdul Sheriff, *Slaves, Spices and Ivory in Zanzibar: Integration of an East African Commercial Empire into the World Economy, 1770–1873* (Ohio University Press, 1987), 109. As the Tanzanian historian Abdul Sheriff notes, clove plantations had "begun to encroach seriously on areas better suited to other crops and to undermine the islands' self-sufficiency in foodstuffs." The work was labor-intensive, and the demand for slaves shot up. Slaves from all over Africa were also traded more generally across the Indian Ocean during Msiri's rule.

32 **Msiri welcomed European:** Frederick Arnot, letter, August 11, 1886, as cited in Robert I. Rotberg, "Plymouth Brethren and the Occupation of Katanga, 1886–1907," *Journal of African History* 5, no. 2 (1964): 288.

32 **king or no king:** Matthew Craven, "Between Law and History: The Berlin Conference of 1884–1885 and the Logic of Free Trade," *London Review of International Law* 3, no. 1 (2015): 31–59.

32 **They used imported labor:** Sandrine Vinckel, "Violence and Everyday Interactions Between Katangese and Kasaians: Memory and Elections in Two Katanga Cities," *Africa: Journal of the International African Institute* 85, no. 1 (2015): 81.

32 **Brasseur was known:** Macola and Hogan, "Guerrilla Warfare in Katanga," 881.

33 **The Belgian masters:** "And should this White man have two rooms, enough to give both to his employee, the boy, he puts a lock on it [to make it] his chicken coop or rabbit hutch." See André Yav, *Vocabulaire de ville de Elisabethville: A History of Elisabethville from Its Beginnings to 1965*, trans. and ed. Johannes Fabian with Kalundi Mango, in *Archives of Popular Swahili* 4, no. 2 (2001), lpca.socsci.uva.nl/aps/vol4/facsimile/toc.html.

33 **frequently chartered private planes:** South African mine engineer, interview with the author, November 2019.

Chapter 3: The Beginnings of a Battery

35 **The ink was barely:** Katie Ellis, "Campus and Community Fete Nobel Laureate," *BingU-News*, October 19, 2019, binghamton.edu/news/story/2098/campus-and-community-fete-nobel-laureate.

35 **"We were looking":** M. Stanley Whittingham, interview with the author, February 2020.

36 **"Part of that was tied":** M. Stanley Whittingham, interview with the author, February 2020.

36 **work on fast-ion transport:** Yung-Fang Yu Yao and J. T. Kummer, "Ion Exchange Properties of and Rates of Ionic Diffusion in Beta-Alumina," *Journal of Inorganic and Nuclear Chemistry* 29, no. 9 (1967): 2453–75.

36 **Ford's battery, however:** M. Stanley Whittingham and Robert A. Huggins, "Beta Alumina—Prelude to a Revolution in Solid State Electrochemistry," in *Solid State Chemistry: Proceedings of the 5th Materials Research Symposium Sponsored by the Institute for Materials Research, National Bureau of Standards, October 18–21, 1971, Held at Gaithersburg, Maryland*, ed. Robert S. Roth and Samuel J. Schneider Jr. (U.S. Department of Commerce and National Bureau of Standards, 1972), 139–54.

37 **"The original Belgirate Meeting":** Bruno Scrosati et al., *Fast Ion Transport in Solids* (Springer Netherlands, 1993), xiii.

Chapter 4: The Land of the Three Kings

38 **Kufi exemplified success:** Elaine Sullivan, University of Johannesburg, conversation with the author, September 2022. There is a debate about whether Hemba as a category was created by colonial art historians to define a specific type of Luba art, but there is evidence that the Hemba were quite distinct from their Luba neighbors.

38 **"My family are Hemba":** Gaylord Kilanga, interview with the author, February 2024.

39 **scholars also trace their roots:** Tshilemalema Mukenge, *Culture and Customs of the Congo* (Bloomsbury Academic, 2002), 16.

39 **The catalog entry:** Hemba artist, *Commemorative Portrait of a Chief (Singiti)*, 19th–early 20th century, wood, 78 × 19.5 × 20.3 cm, Metropolitan Museum of Modern Art, New York, object no. 2015.119, metmuseum.org/art/collection/search/320673.

39 **The education of Black Africans:** Joe Trapido, "Africa's Leaky Giant," *New Left Review*, no. 92 (March/April 2015): 15.

40 **swiftly electrifying world:** André Vène, "Les perspectives du marché du cuivre et les grandes valeurs du Centre-Afrique," *Le Monde*, August 29, 1955. In 1954, for example, the company produced 223,791 tons of copper—more than 8 percent of the global supply—and more than 8,500 tons of cobalt.

40 **"The colonial economy":** Emizet François Kisangani, *The Belgian Congo as a Developmental State: Revisiting Colonialism* (Routledge, 2023), 15.

40 **"Unemployment has appeared":** Raymond Bertieaux, "The Economy of the Belgian Congo in 1958–59," *Civilisations* 9, no. 3 (1959): 385.

41 **As the sociologist Claude:** Claude Iguma Wakenge, *Eating the Congo: Unveiling State Governance of Copper and Cobalt Mining in Former Katanga* (Lambert Academic Publishing, 2019), 81.

41 **The seeds of an ethnic:** Katharine Frederick and Elise van Nederveen Meerkerk, "From Temporary Urbanites to Permanent City Dwellers? Rural-Urban Labor Migration in Colonial Southern Rhodesia and the Belgian Congo," in *Migration in Africa: Shifting Patterns of Mobility from the 19th to the 21st Century*, ed. Michiel de Haas and Ewout Frankema (Routledge, 2022),

256–81. The Belgians also brought in labor from their neighboring territories of Ruanda-Urundi and other places in Africa.

42 **scores of African:** Sandrine Vinckel, "Violence and Everyday Interactions Between Katangese and Kasaians: Memory and Elections in Two Katanga Cities," *Africa: Journal of the International African Institute* 85, no. 1 (2015): 82.

42 **Foremost among them:** Author email correspondence with Thomas Q. Reefe, April 14, 2023.

42 **a Belgian scholar described:** Raoul Van Caenegem, "Les langues indigènes dans l'enseignement," *Zaïre*, July 1950, 717, as cited in Thomas Bakajika Banjikila, *Épuration ethnique en Afrique: "Les Kasaïens" (Katanga 1961–Shaba 1992)* (L'Harmattan, Études Africaines, 1997).

42 **Union Minière employed a staff:** Erik Kennes and Miles Larmer, *The Katangese Gendarmes and War in Central Africa: Fighting Their Way Home* (Indiana University Press, 2016), 33.

42 **called themselves "autochthonous":** Kennes and Larmer, *Katangese Gendarmes and War in Central Africa*, 33.

42 **Luba-Kasai were "foreigners":** A. Rubbens, "Political Awakening in the Belgian Congo," *Civilisations* 10, no. 1 (1960): 68, https://www.jstor.org/stable/41230425.

42 **Tshombe founded the Confederation:** Rubbens, "Political Awakening in the Belgian Congo," 63–76. For the other ethnic groups in Katanga, the Kasaian issue was underlined in 1957, when the Belgians held what they called "consultations." These were, in effect, municipal elections that allowed Africans, for the first time, to run for the seat of *bourgmestre*, or mayor. Luba-Kasai candidates won three out of the four seats in Katanga, spurring soul-searching among other ethnic groups. "The racial tensions are due in some measure to the industrialisation of the High Katanga," Antoine Rubbens, a Belgian magistrate in Elisabethville who had also served as a colonial administrator, wrote in 1960. This, he noted, required Luba-Kasai laborers and, later, Luba-Kasai clerks. "As a result, about one half of the population in the industrial towns of the High Katanga (Elisabethville, Jadotville and Kolwezi) comes from the Kasai, while an even greater proportion of the highest posts open to natives in public and private administrations are occupied by Kasai tribemen."

42 **The party's leaders cherry-picked:** Kennes and Larmer, *Katangese Gendarmes and War in Central Africa*, 35.

42 **"When the first white explorers":** Godefroid Munongo, "Comment est né le nationalisme katangais," Elisabethville, June 16, 1962, as cited in Kennes and Larmer, *Katangese Gendarmes and War in Central Africa*, 36–37. Munongo was also a Belgian territorial secretary and the first president of Conakat until the colonial authorities urged him to step down.

43 **invited Lumumba and African activists:** "'Hands Off Africa!!' The 1958 All African People's Conference: Its Impact Then and Now—All Parts," Institute of Commonwealth Studies, School of Advanced Study: University of London, archived March 23, 2023, at web.archive.org/web/20230323075328/https://commonwealth.sas.ac.uk/podcasts/hands-africa-1958-all-african-peoples-conference-its-impact-then-and-now-all-parts.

43 **He inveighed against "colonialism":** "Discours prononcé par Patrice Lumumba, président du Mouvement National Congolais à la Conférence d'Accra, 11 Décembre 1958," in *La pensée politique de Patrice Lumumba*, ed. Jan Van Lierde (Présence Africaine, 2003), 17.

44 **Kufi looked up to:** Gaylord Kilanga, interview with the author, February 2024.

Chapter 5: The Prime Minister's Tooth

45 **"The basic idea was to":** Albert Kalonji Ditunga Mulopwe, *Congo 1960: La sécession du Sud-Kasaï* (L'Harmattan, 2005), 80.

45 **South Kasai would be:** Bill Berkeley, "Zaire: An African Horror Story," *Atlantic*, August 1993.

46 **Eisenhower became the first:** Ben Quinn, "MI6 'Arranged Cold War Killing' of Congo Prime Minister," *Guardian*, April 1, 2013. The CIA sent an agent carrying vials of poison to use on Lumumba. Britain's intelligence service reportedly also planned to assassinate him. Neither plot was carried out.

46 **A Belgian adviser suggested:** Stuart A. Reid, *The Lumumba Plot: The Secret History of the CIA and a Cold War Assassination* (Alfred A. Knopf, 2023), 383.

46 **Devlin did not protest:** Reid, *Lumumba Plot*, 385.

47 **It had been yanked:** Reid, 399. It was from Gerard Soete's collection of grisly memorabilia that the tooth was recovered in 2016. Another tooth and a finger had also been removed by the brothers, but they were said to have been thrown into the North Sea long before.

47 **the tooth was retrieved:** At the ceremony, a group of women carrying LUNDA EMPIRE banners were present to celebrate the man that Tshombe, a Lunda royal, had seen fit to murder. See

"RDC: Le cercueil de Patrice Lumumba à Shilatembo, le lieu du crime," posted June 27, 2022, by AfricaNews, YouTube, youtube.com/watch?v=82Wtah8laFY.

Chapter 6: A Patriot with a Cause

49 **funneled mining revenue:** *Économie katangaise et économie congolaise à la veille de l'indépendance*, SD No. 61/5, 1 July 1961, box 4, folder 854, Union Minière Collection: Archives sur l'Exploitation des Mines de Cuivre Katangaises, Premier Série, Quatrième Partie, Archives de l'État en Belgique, Dépôt Joseph Cuvelier, Brussels, Belgium. The mines were "the departure point of the stimulation of the economic development of Katanga," trumpeted an internal note about the breakaway province that circulated at Union Minière's headquarters in July 1961.

49 **He also purged:** Thomas Bakajika Banjikila, *Épuration ethnique en Afrique: "Les Kasaïens" (Katanga 1961–Shaba 1992)* (L'Harmattan, Études Africaines, 1997).

49 **A UN envoy decried:** Rajeshwar Dayal, *Report of Recent Developments in Northern Katanga from the Special Representative of the Secretary-General*, No. S/4691/Add.2 (United Nations Security Council, February 1961), digitallibrary.un.org/record/630681?v=pdf.

49 **studying in Elisabethville:** Kufi was lucky not to be arrested and imprisoned for his activism. His ethnicity helped elevate him above suspicion: Lots of other Hemba had joined Tshombe's secessionists. The Hemba lived close to the Balubakat, and Gaylord Kilanga, Kufi's son, told me that many of them had "petty rivalries" with the Luba, causing them to join Tshombe and the secessionists. Kufi, the student from a minority ethnic group down in Lubumbashi, had managed to place himself once more in a minority: Only a sliver of a percentage of Hemba supported a larger unitary project in Congo. "He was an exception among the Bahemba."

49 **Throughout the Katangese secession:** Note sur les "Grandes entreprises, seul soutien actuel de l'économie congolaise," 4 June 1962, box A4, folder 869, Union Minière Collection: Archives sur l'Exploitation des Mines de Cuivre Katangaises, Premier Série, Quatrième Partie, Archives de l'État en Belgique, Dépôt Joseph Cuvelier, Brussels, Belgium. A Union Minière report from 1962 showed that the production of copper had actually increased between 1959 and 1961. Independent Katanga and Kasai were doing quite well. See "Les causes de la situation économique et financière désastreuse au Congo," 1962, box A4, folder 870, Union Minière Collection: Archives sur l'Exploitation des Mines de Cuivre Katangaises, Premier Série, Quatrième Partie, Archives de l'État en Belgique, Dépôt Joseph Cuvelier, Brussels, Belgium.

49 **the company made commercial:** "NOTE concernant les contrats de raffinage Union Minière / Hoboken," UMHK, 23 February 1960, box C4, folder 1002, Union Minière Collection: Archives sur l'Exploitation des Mines de Cuivre Katangaises, Premier Série, Quatrième Partie, Archives de l'État en Belgique, Dépôt Joseph Cuvelier, Brussels, Belgium.

50 **a copy of the note:** P. L. Mathieu, "Résumé de l'étude 'African Development,'" 1960, box A4, folder 843, Union Minière Collection: Archives sur l'Exploitation des Mines de Cuivre Katangaises, Premier Série, Quatrième Partie, Archives de l'État en Belgique, Dépôt Joseph Cuvelier, Brussels, Belgium.

50 **crushed the fledgling Katangese:** David Van Reybrouck, *Congo: The Epic History of a People*, trans. Sam Garrett (HarperCollins, 2014), 315.

50 **After thirty-six hours:** Lloyd Garrison, "Tshombe Offers to End Secession Under Amnesties," *New York Times*, January 17, 1963. The negotiations were coordinated with Brussels through the Union Minière radio, allowing the company to have an up-to-the-minute understanding of what was happening.

50 **Tshombe's gendarmes melted:** Erik Kennes and Miles Larmer, *The Katangese Gendarmes and War in Central Africa: Fighting Their Way Home* (Indiana University Press, 2016), 66. The historians Erik Kennes and Miles Larmer have detailed the world of the gendarmes in this excellent book.

50 **The central Congolese government:** Kennes and Larmer, *Katangese Gendarmes and War in Central Africa*, 88.

50 **a minister who now led:** David H. Shinn and Joshua Eisenman, *China and Africa: A Century of Engagement* (University of Pennsylvania Press, 2012), 471. The Chinese government quickly pledged some $2.8 million in aid (around $30 million in 2024 dollars).

51 **"In their concerted action":** China Secretariat of the General Political Department, "The Congo Situation and Its Development," in *The Politics of the Chinese Red Army*, ed. and trans. J. Chester Cheng (Hoover Institution, 1966), 180–81. The report argued that China was "giving great support to the Congolese people."

51 **another Chinese intelligence:** China Secretariat of the General Political Department, "The Present Situation in the Congo and the New Schemes of American Imperialism," in Cheng, *Politics of the Chinese Red Army*, 398–400.

51 **Mulele fled Congo:** Peer Schouten, *Roadblock Politics: The Origins of Violence in Central Africa* (Cambridge University Press, 2022), 220. Such myths would persist well into the next century: According to one analysis, in 2022, roughly half of the 120 rebel groups in Congo were known as Mai-Mai, a concept that had evolved from, and in some cases could be traced quite clearly back to, Mulele's strategy of mixing spiritual practice with guerrilla warfare. His guerrillas would run into battle screaming "Mai Mulele!" (Mulele's water!), terrifying many Congolese army troops who believed in such incantations.

51 **He elevated the struggle:** Schouten, *Roadblock Politics*, 217.

51 **"excellent revolutionary situation":** Ernest Lefever, *Crisis in the Congo* (Brookings Institution, 1965), 134.

51 **"Anything black was killed":** Michel Honorin, "Horreurs et duperies congolaises," *Historia* 406 bis (1980): 41–53, as cited in Olivier Lanotte, "Chronology of the Democratic Republic of Congo/Zaire (1960–1997)," SciencesPo, April 6, 2010, sciencespo.fr/mass-violence-war-massacre-resistance/en/document/chronology-democratic-republic-congozaire-1960-1997.html. These mercenaries would go on to be lionized as the "Wild Geese" in a novel by Wilbur Smith and a popular action movie.

51 **The government troops rounded:** J. Anthony Lukas, "500 Are Executed as Congo Rebels," *New York Times*, January 10, 1965.

52 **One young man in sunglasses:** "European Mercenaries Interviewed About Killing Congolese," interview by the BBC, posted July 25, 2018, by Emma Goldman, YouTube, youtube.com/watch?v=iOBfBrvWiAc&list=PLCPp4qoEy-dnBaPmaSLhGsYkSAHPC6cuS.

53 **China rapidly began:** Mohamed A. el-Khawas, "China's Changing Policies in Africa," *Issue: A Journal of Opinion* 3, no. 1 (1973): 25. In 1973, Mohamed A. el-Khawas argued that Beijing was "eager to set up working arrangements with moderate, conservative, and radical governments alike."

53 **A CIA report from 1972:** Directorate of Intelligence, *China's Role in Africa*, Special Report Weekly Review (Central Intelligence Agency, February 1972), cia.gov/readingroom/docs/CIA-RDP08S02113R000100080001-0.pdf. The report was declassified on August 24, 2012.

53 **"The Congo was utterly lacking":** U.S. intelligence officer, correspondence with author, 2023.

53 **New laws were:** Wolf Radmann, "The Nationalization of Zaire's Copper: From Union Minière to Gecamines," *Africa Today* 25, no. 4 (1978): 38.

54 **The dictator still needed:** P. De Vos, "Kinshasa s'oppose au rapatriement de quarante-trois femmes et enfants d'agents de l'Union Minière," *Le Monde*, February 1, 1967.

54 **There he met Mao:** Shinn and Eisenman, *China and Africa*, 471.

54 **After Mobutu's visit:** Osita G. Afoaku, "The U.S. and Mobutu Sese Seko: Waiting on Disaster," *Journal of Third World Studies* 14, no. 1 (1997): 65–90.

55 **traveled to Kinshasa to sign:** Radmann, *Nationalization of Zaire's Copper*, 40–41.

55 **Mobutu maintained power over:** Machiavelli talks about how the difficulty of attacking a country like Turkey arises from the fact that there "are no barons to invite you in, and you can't expect anyone to make your life easier by rebelling against the king." See Niccolò Machiavelli, *The Prince*, trans. Tim Parks (Penguin Classics, 2014), 17. For Mobutu's fascination with Machiavelli, see Michela Wrong's deeply reported account of his rule and the downfall of Zaire, *In the Footsteps of Mr. Kurtz: Living on the Brink of Disaster in Mobutu's Congo* (Fourth Estate, 2000), 100.

55 **such a rotation:** V. S. Naipaul, "A New King for the Congo," *New York Review of Books*, June 26, 1975.

56 **"Authentic" music was promoted:** Kevin C. Dunn, *Imagining the Congo: The International Relations of Identity* (Palgrave Macmillan, 2003), 119. In *Imagining the Congo*, Dunn argues that Mobutu "re-employed past discourses used to justify Belgian colonialism in order to justify his own repressive rule."

56 **Even Mobutu changed:** Or, as the journalist Michela Wrong has pointed out, referring to Mobutu's sexual antics with his ministers' wives, "the cock who covers every chicken." See Wrong, *In the Footsteps of Mr. Kurtz*, 100.

56 **"The evidence is legion":** Edouard Nyindu, "Devoir de mémoire / Bandundu ville: Le patriarche Kufi Kilanga honoré à sa juste valeur," *Nzadi News*, November 19, 2019, nzadinews.net/devoir-de-memoire-bandundu-ville-le-patriarche-kufi-kilanga-honore-a-sa-juste-valeur/.

56 **According to a declassified:** Directorate of Intelligence, *Zaire: Mobutu and the Military* (Central In-

telligence Agency, August 1982), cia.gov/readingroom/docs/CIA-RDP83S00855R000100080001 -6.pdf. This intelligence assessment was declassified on July 30, 2008.

57 **Those who critiqued Mobutu:** Glenn Frankel, "Political Repression Charged in Zaire," *Washington Post*, May 23, 1985.

57 **"Generals had become businessmen":** George Arthur Forrest, *Un siècle de rêves: Ensemble, bâtissons l'avenir* (Le Cherche Midi, 2022), 59.

57 **Mobutu neatly acknowledged:** Keith B. Richburg, "Mobutu: A Rich Man in Poor Standing," *Washington Post*, October 2, 1991.

Chapter 7: Liberty in a Wasteland

58 **Polling data showed:** Richard Reeves, *President Nixon: Alone in the White House* (Simon & Schuster, 2001), 163. During the run-up to the 1968 presidential election, which saw Richard M. Nixon return to power, the environment was barely mentioned by either the Republican or Democratic campaign. Nixon understood that it was a divisive issue and tried to co-opt the movement.

58 **"Strip mining in our steep":** Chad Montrie, "'To Have, Hold, Develop, and Defend': Natural Rights and the Movement to Abolish Strip Mining in Eastern Kentucky," *Journal of Appalachian Studies* 11, no. ½ (2005): 67.

59 **"We have seen industrial":** United States Congress, Senate, *Strip Mining and Its Impact: Hearings Before the Senate Committee on Interior and Insular Affairs.* 90th Cong., 87-95, 5855–856 (1968), (testimony of Harry M. Caudill).

59 **"There is still only one":** John C. Whitaker, *Striking a Balance: Environment and Natural Resources Policy in the Nixon-Ford Years* (AEI-Hoover Policy Studies, 1976), 27.

59 **the Democratic senator Edmund:** John C. Nagle, "The Earth Day Pioneer Nobody Remembers," *Scientific American*, April 22, 2016.

59 **California governor Ronald:** Ronald Reagan, "Our Environment Crisis," *Nation's Business*, February 1970, 27. Reagan's sentiment would not last: By the time he became president, a decade later, he would do more than anyone to prop up fossil-fuel businesses.

59 **he told his general counsel:** Reeves, *President Nixon*, 163.

60 **"I realize that the argument":** Richard Nixon, "1970 Annual Message to the Congress on the State of the Union," January 22, 1970, American Presidency Project, transcript and video, presidency.ucsb.edu/node/241063.

60 **superpower was also being questioned:** Much ado was made, for example, when it was reported that the Soviet Union had outstripped the U.S. in "National Material Capability," a composite rating developed by J. David Singer, a political scientist at the University of Michigan. See J. David Singer, as cited in Mark Harrison, "The Soviet Economy, 1917–1991: Its Life and Afterlife," *Independent Review* 22, no. 2 (2017): 202–3.

60 **After a decade of crisis:** The price of oil rose by 300 percent around the world and even higher in the U.S. In Washington, there was a fear that oil would be used "as an economic weapon" by rogue states. See Eric Pace, "Arabs Halt Oil to Portugal, Rhodesia and South Africa," *New York Times*, November 29, 1973.

60 **Companies started to look:** In the early 1970s, this was one thing that the hawks and the hippies seemed to be aligned on. The confluence of environmental issues and increasing geopolitical uncertainty came to be known as the "energy-environmental balance," a concept that had at its root the kind of Cold War gamesmanship that had offed Lumumba and led the U.S., the U.S.S.R., and China to support autocratic regimes around the world. According to this theory, competition for resources would intensify as resources became scarcer, and energy would thus need to be conserved and expended with care. As the environmental historian Martin V. Melosi would note a decade later, "Policy makers began to realize that the interplay between energy policy and environmental protection could be the key to the future of both." See Martin V. Melosi, "Energy and Environment in the United States: The Era of Fossil Fuels," *Environmental Review* 11, no. 3 (Autumn 1987): 174.

60 **In 1881, another French scientist:** Kevin Desmond, *Gustave Trouvé: French Electrical Genius (1839–1902)* (McFarland & Co., 2015), 50.

61 **first car to drive faster:** Petrol cars lagged at 22 percent. Steam-powered cars accounted for the other 40 percent. See Ken W. Purdy et al., "Automobile," *Encyclopedia Britannica*, November 2, 2021, britannica.com/technology/automobile. Incidentally, Thomas Edison was obsessed with creating an electric car, and in 1908, he finally launched a successful version whose battery

used lithium hydroxide, which prevented the battery's strength from being diminished by unintended chemical reactions. (Edison "had no clue why it worked, and he probably didn't care.") See Seth Fletcher, *Bottled Lightning: Superbatteries, Electric Cars, and the New Lithium Economy* (Hill and Wang, 2011), 17.

61 **Even Henry Ford's wife:** "1914 Detroit Electric Model 47 Brougham, Personal Car of Clara Ford," Henry Ford Museum, thehenryford.org/collections-and-research/digital-collections /artifact/209957/.

61 **U.S. had fallen in love:** There were a few notable exceptions to this rule, including electric milk floats in Britain and, in Nazi-occupied France, the Peugeot VLV, a city runabout with a top speed of twenty-one miles an hour.

61 **prompted by oil shortages:** The most popular electric vehicle of the 1970s was the CitiCar, a wedge-shaped car with sliding windows and a top speed of around thirty-eight miles per hour. Until 2012, when the Tesla Model S came along, it was the bestselling electric car of the postwar era. Twenty-three hundred of them were sold between 1974 and 1977. See Máté Petrány, "Florida's Hopeful EV from the 1970s Is a Fantastic First Car," *Road & Track*, November 27, 2017, roadandtrack.com/car-culture/car-design/a13931959/floridas-hopeful-ev-from -the-1970s-is-a-fantastic-first-car/.

61 **Even Greece produced:** Thanos Pappas, "The Story of Enfield Neorio 8000," *Neos Kosmos*, March 6, 2018, https://neoskosmos.com/en/2018/03/06/life/technology/the-story-of-enfield -neorio-8000/.

62 **Aronson became a salesman:** Paal Kvamme, "Pioneren utviklet elbil hjemme i garasjen og verdens første stasjonskjede for hurtiglading," *Teknisk Ukeblad*, May 30, 2021.

62 **"The energy crisis has turned":** Robert W. Irvin, "The Revival of Electric Vehicles: Passing Fancy or Car of the Future?," *New York Times*, April 7, 1974.

62 **As Aronson's firm:** "New Battery for Electric Cars: U.S. Patent 7,037,620 B2, May 2, 2006, Multi-Cellular Battery with Lead Foam," Apollo Energy Systems, archived at electricauto .com/_pdfs/new_batt_Ecar_Whitepaper.pdf.

62 **six fifty-kilowatt:** Barry Iseard, author interview, Apollo Energy Systems, May 2022.

62 **Americans just didn't want to:** *U.S.-China Trade Relations and Renewal of China's Most-Favored-Nation Status*, 104th Cong. 132 (1995) (testimony of Robert R. Aronson). In 1974, when he was speaking to the *Times*, Aronson had only sold around sixty of his cars; he would sell forty more in the next two decades.

63 **The high-water mark:** Electric and Hybrid Vehicle Research, Development, and Demonstration Act, H.R. 8800, 94th Cong. (1975–1976) (enacted); and Electric and Hybrid Vehicle Research, Development, and Demonstration Act, Pub. L. No. 94-413, 90 Stat. 1260 (1976).

63 **everyone seemed to be:** A year later, Carter established an award to "encourage citizen participation in the national drive toward greater energy efficiency." See White House Press Office, "President's Award for Energy Efficiency Announcement of 25 Award Recipients in the Field of Transportation," press release, July 22, 1980, presidency.ucsb.edu/node/250996.

63 **"capable of providing":** "Small, Battery-Powered Automobile Recently Completed by Toyota Motors," *New York Times*, April 12, 1975, 35.

Chapter 8: Intercalation Station

64 **Exxon had begun:** Kevin Desmond, *Innovators in Battery Technology: Profiles of 95 Influential Electrochemists* (McFarland, 2016), 238.

64 **In addition to high-powered:** Michael Goodwin, "Exxon's Innovative Little Offshoots," *New York Times*, March 14, 1976.

64 **at his Exxon lab:** Karl Kordesch and Waltraud Taucher-Mautner, "History: Primary Batteries," in *Encyclopedia of Electrochemical Power Sources*, ed. Bruno Scrosati et al. (Elsevier, 2009), 561.

65 **rechargeable carbon fluoride battery:** Desmond, *Innovators in Battery Technology*, 239.

65 **Whittingham's early work:** Seth Fletcher, *Bottled Lightning: Superbatteries, Electric Cars, and the New Lithium Economy* (Hill and Wang, 2011), 28.

65 **But tantalum was too heavy:** Desmond, *Innovators in Battery Technology*, 238.

65 **work was fast-paced:** M. Stanley Whittingham, interview with the author, February 2020.

65 **"it describes the reversible":** M. Stanley Whittingham and Allan J. Jacobson, eds., *Intercalation Chemistry* (Academic Press, 1982), 1.

66 **devised a chemical process:** Martin B. Dines, Method for Lithiating Metal Chalcogenides

and Intercalated Products Thereof, U.S. Patent 3,933,688A, filed September 30, 1974, and issued January 20, 1976, patents.google.com/patent/US3933688A/en.

66 **created a workable battery:** M. Stanley Whittingham, "Lithium Titanium Disulfide Cathodes," *Nature Energy* 6, no. 2 (2021): 214, doi.org/10.1038/s41560-020-00765-7.

66 **Whittingham filed a patent:** M. Stanley Whittingham, Chalcogenide Battery, U.S. Patent 4,009,052A, filed April 5, 1976, and issued February 22, 1977. The patent was twice updated before finally being granted in 1977.

66 **As the philosopher:** Ivan Illich, *Energy and Equity* (Harper & Row, 1974), 3. Illich argued, as in the quote in the epigraph, that society should slow down, lest it become a destructive technocracy. "High speed," he wrote, "is the critical factor which makes transportation socially destructive. A true choice among political systems and of desirable social relations is possible only where speed is restrained. Participatory democracy demands low energy technology, and free people must travel the road to productive social relations at the speed of a bicycle."

Chapter 9: A Spy in Priest's Clothing

67 **John Stockwell was dressed:** John Stockwell, *In Search of Enemies: A CIA Story* (Norton, 1978), 139.

67 **a rebel named Jonas:** This was the British tycoon and corporate raider Roland W. "Tiny" Rowland, who was for many years a go-between for Savimbi and Western governments. When Rowland called on Downing Street to advocate for Savimbi, a civil servant wrote that he was "an old supporter of Savimbi and has invested a good deal of assistance in him." See "Angola: Howe PS Letter to No. 10 ('Angola: Call by Mr. Tiny Rowland: 1 August') ['Savimbi's Prospects and Intentions'] [declassified January 2014]," 29 July 1983, PREM 19, Records of the Prime Minister's Office, U.K. National Archives, Kew, accessed at https://www.margaretthatcher.org/document/220839.

68 **a Scottish entrepreneur had built:** Lobito Corridor Investment Promotion Authority, *What It Is and Why It Matters*, January 2024, 7, https://www.lobitocorridor.org/_files/ugd/9fa7ad_700894b8a8b9427faec094b5fbd0f5fc.pdf. The Belgians would later take over ownership, and the Société Générale de Belgique would nominally own the railroad until 2001.

68 **Until the late '60s:** Jean Dusausoy, *Kolwezi 1977: Un technicien belge dans les mines du Katanga* (Éditions Luc Pire, 2018), 26.

68 **"I projected my mind":** Stockwell, *In Search of Enemies*, 150.

68 **"They wanted Mobutu":** Jean Dusausoy, interview with the author, November 2021.

68 **Roving work gangs:** Central Intelligence Agency National Foreign Assessment Center, "Angola: UNITA vs. the Benguela," November 30, 1978, declassified December 2, 2004, cia.gov/readingroom/docs/CIA-RDP80T00634A000500010013-7.pdf.

69 **"poor guys, lost":** Jean Dusausoy, fact-checking email, March 2025.

69 **"When you knew it":** CIA, "Angola: UNITA vs. the Benguela."

70 **mineral was a small fry:** Former Marc Rich & Co. trader, conversation with author, February 2022.

70 **depths of the freezing London winter:** For traders, January in London was "gout season," as one trader would tell the journalist A. Craig Copetas a few years later, "the time of year when industrial doges discussed the effects of ballooning cobalt prices with struggling African economic ministers over glass after glass of French Grand Cru poured by the waiters at Langan's Brasserie in Mayfair." It is a masterful description. See A. C. Copetas, *Metal Men: Marc Rich and the 10-Billion-Dollar Scam* (Harper & Row, 1985), 17.

70 **building up its stockpiles:** Bernard D. Nossiter, "Soviets Reportedly Bought Up Cobalt Before Zaire Invasion," *Washington Post*, May 24, 1978.

70 **Cobalt may have been niche:** Nossiter, "Soviets Reportedly Bought Up Cobalt Before Zaire Invasion."

71 **amounts that the Soviets:** Nossiter, "Soviets Reportedly Bought Up Cobalt Before Zaire Invasion." Moscow's traders had used such stealth buying strategies before. During the "Great Grain Robbery" of 1973, after a drought in the U.S.S.R., Nikolai Belousov, a Soviet bureaucrat, flew to New York and negotiated private purchases of grain that eventually caused global food prices to shoot up by almost 30 percent.

71 **The overall market was:** Former cobalt trader, interview with the author, May 2023.

71 **buyers also bargained hard:** Former cobalt trader, interview.

71 **Soviet Union engineered the attack:** Erik Kennes and Miles Larmer, *The Katangese Gendarmes and War in Central Africa: Fighting Their Way Home* (Indiana University Press, 2016), 128.

71 **several hundred civilians were:** George Arthur Forrest, *Un siècle de rêves: Ensemble, bâtissons l'avenir* (Le Cherche Midi, 2022), 57.

71 **Those who survived were left:** International opinion at the time placed the blame on the Katangese gendarmes for "massacres" of civilians, but the poorly trained Zairean army soldiers were probably responsible, according to Kennes and Larmer, *Katangese Gendarmes and War in Central Africa*, 138. Jean Dusausoy, who had left Zaire by that point, told me that his community of ex-Gécamines friends blames Zaire's armed forces for the killings.

72 **so ineffective at defending:** Andrew L. Gulley, "One Hundred Years of Cobalt Production in the Democratic Republic of the Congo," *Resources Policy* 79 (2022): art. 103007, doi.org/10.1016/j.resourpol.2022.10300.

72 **was even profitable to fly:** Alan Cowell, "Zaire's Bloody Past Makes Cobalt's Future Uncertain," *New York Times*, August 30, 1981.

72 **The stockpile reached:** Former cobalt trader, interview; and Mark Burton et al., "US Moves to Restore Stockpiling 'Panic Button' in EV Metals Fight with China," *Bloomberg*, February 19, 2024.

72 **Some businessmen suggested:** Cowell, "Zaire's Bloody Past Makes Cobalt's Future Uncertain."

PART 2: TRADE AND WAR

Chapter 10: Putting Out Fires

75 **New York from New Delhi:** "In Memory of Fazil Khan," Ever Loved, everloved.com/life-of/fazil-khan/obituary/.

75 **"It is an enormous":** Tom Outerbridge, interview with the author, September 2024.

76 **"Keeping batteries out":** Eric Frederickson, interview with the author, September 2024.

76 **bedeviled lithium-ion batteries:** M. Stanley Whittingham, "Nobel Prize in Chemistry Recognizes Lithium Battery Discoveries," interview by Jenni Doering, *Living on Earth*, October 25, 2019, loe.org/shows/segments.html?programID=19-P13-00043&segmentID=5.

76 **One early version:** Seth Fletcher, *Bottled Lightning: Superbatteries, Electric Cars, and the New Lithium Economy* (Hill and Wang, 2011), 31.

76 **Some experimental vehicles used:** Joseph T. Kummer and Neill Weber, "A Sodium-Sulfur Secondary Battery," *SAE Transactions* 76 (1968): 1003–7, 1023–28, jstor.org/stable/44564986.

77 **thirty-five years later:** Nathalie Pereira et al., "Lithium–Titanium Disulfide Rechargeable Cell Performance After 35 Years of Storage," *Journal of Power Sources* 280 (April 2015): 18–22.

77 **A small solar-powered:** Catherine Meyers, "How Some Nobel Prize–Winning Battery Research Weathered the Test of Time," *Inside Science*, October 11, 2019, archived May 30, 2024, web.archive.org/web/20240530064822/insidescience.org/news/how-some-nobel-prize-winning-battery-research-weathered-test-time.

77 **"They said to us":** Neela Banerjee, "For Exxon, Hybrid Car Technology Was Another Road Not Taken," Inside Climate News, October 5, 2016, insideclimatenews.org/news/05102016/exxon-climate-change-hybrid-cars-technology-another-road-not-taken-electric-vehicle-toyota-prius/.

77 **The Iranian Revolution:** For example, in the North Sea, in Venezuela, in the U.S.S.R., and in Nigeria.

77 **Gulf oil producers ramped up:** Dermot Gately, "Lessons from the 1986 Oil Price Collapse," *Brookings Papers on Economic Activity* 1986, no. 2 (1986): 237.

77 **energy dried up:** Others, like Hamlen, recalled that Exxon was searching for even bigger markets—around $1 billion a year. See Fletcher, *Bottled Lightning*, 36.

77 **Exxon sold off:** *1984 Annual Report* (Exxon Corporation, 1985), 2, as cited in J. A. Pratt, "Exxon and the Control of Oil," *Journal of American History* 99, no. 1 (2012): 145–54.

78 **Enterprises had made investments:** Michael Goodwin, "Exxon's Innovative Little Offshoots," *New York Times*, March 14, 1976.

78 **The firm's oil sales:** "Exxon to Sell Unit," UPI, June 9, 1981.

78 **Exxon later decided to:** Fletcher, *Bottled Lightning*, 36. Oxford University, whose laboratories had produced this research, were not even interested in applying for a patent.

Chapter 11: A Cobalt Cathode and a Carbon Anode

79 **took much interest:** John B. Goodenough, *Witness to Grace* (PublishAmerica, 2008), 72.

80 **In his memoir:** Goodenough, *Witness to Grace*, 51.

80 **"We found that over":** Goodenough, 72.

81 **The government's Ministry:** Chalmers Johnson, *MITI and the Japanese Miracle: The Growth of Industrial Policy, 1925–1975* (Stanford University Press, 1982), 16; and Masanao Itoh et al., "Bank of Japan's Monetary Policy in the 1980s: A View Perceived from Archived and Other Materials," *Monetary and Economic Studies* 33 (2015): 106. The first oil crisis, in 1973, affected Japan the most. According to a Bank of Japan monograph: "Firms and households responded in anticipation of higher inflation in the future. They created speculative demand based on speculative accumulation of stocks unrelated to economic activity, which brought about tight demand and supply conditions. Also, substantial pay rises put pressure on corporate profits, which resulted in a subsequent reduction in employment and a decline in capital investments." Price increases were coupled with job cuts and cost-savings measures by companies. During the second crisis, in 1978, the country was not growing as quickly as it had been in 1973, the central bank moved fast to tighten fiscal policy, and there was less speculation.

81 **boatloads of licensing agreements:** Leonard Lynn, "Japanese Technology: Successes and Strategies," *Current History* 82, no. 487 (1983): 366.

81 **The concerns voiced:** Lynn, "Japanese Technology," 366.

82 **levying tariffs on Chinese:** Jim Tankersley and Mark Landler, "Trump's Love for Tariffs Began in Japan's '80s Boom," *New York Times*, May 15, 2019; and Jonathan Soble and Keith Bradsher, "Donald Trump Laces into Japan with a Trade Tirade from the '80s," *New York Times*, March 7, 2016. When Trump first ran for president, in 2016, he began making the same complaints, which confused people.

82 **summer of 1982:** Lynn, "Japanese Technology," 366.

82 **headlines read HOW:** Agathe Demarais, "How the U.S.-Chinese Technology War Is Changing the World," *Foreign Policy*, November 19, 2022.

82 **laboratory at Asahi Kasei:** "They Developed the World's Most Powerful Battery," Royal Swedish Academy of Sciences, nobelprize.org/uploads/2019/10/popular-chemistryprize2019 .pdf.

82 **Sony sold ten:** Meaghan Haire, "The Walkman," *Time*, July 1, 2009.

82 **"The battery problem is especially":** Andrew Pollack, "Battery Pollution Worries Japanese," *New York Times*, June 25, 1984.

82 **"It was completely disorganized":** Akira Yoshino, interview with the author, January 2022.

82 **a silvery-gray substance:** Tomoki Sawai, "The Invention of Rechargeable Batteries: An Interview with Dr. Akira Yoshino, 2019 Nobel Laureate," *WIPO Magazine*, September 2020; and Akira Yoshino, "Nobel Lecture: Akira Yoshino, Nobel Prize in Chemistry 2019," December 8, 2019, Royal Swedish Academy of Sciences, video and transcript, nobelprize.org/prizes/chemistry /2019/yoshino/lecture/. As Yoshino put it, "It could store a lot of electricity, and its performance is very, very stable."

82 **"plastic that conducts electricity":** Royal Swedish Academy of Sciences, "Plastic That Conducts Electricity," press release, October 10, 2000, nobelprize.org/prizes/chemistry/2000/press -release/.

83 **"I needed a positive":** Yoshino, interview.

83 **He had understood that lithium:** "Building on Work of Others Was Key to Lithium-Ion Batteries," *Asahi Shimbun*, October 10, 2019, asahi.com/ajw/articles/13059462.

83 **When he finally applied:** Akira Yoshino et al., Secondary Battery, Japanese Patent 1,989,293, filed May 10, 1985, and issued November 8, 1995; and Akira Yoshino et al., Secondary Battery, U.S. Patent 4,668,595, filed May 9, 1986, and issued May 26, 1987.

83 **"I just followed the way":** Yoshino, interview.

83 **end of December 1983:** Yoshino, "Nobel Lecture."

83 **vapor-grown carbon fiber:** Gary G. Tibbetts, "Vapor-Grown Carbon Fibers," in *Carbon Fibers Filaments and Composites*, ed. J. L. Figueiredo et al., NATO ASI Series, vol. 177 (Springer, 1990), 73; and Munehiro Ishioka et al., "Electrical Resistivity, Magnetoresistance, and Morphology of Vapor-Grown Carbon Fibers Prepared in a Mixture of Benzene and Linz– Donawitz Converter Gas by Floating Catalyst Method," *Journal of Materials Research* 8, no. 8 (1993): 1866.

83 **"grown" in a laboratory:** Yoshino et al., Secondary Battery, U.S. Patent 4,668,595.

84 **China placed export:** Tony Alderson (Benchmark Mineral Intelligence), interview with the author, December 2024. See also Nicolas Niarchos, "Beijing Calls Washington's Bluff on Strategic Metals," *Nation*, January 2, 2025, thenation.com/article/world/china-export-minerals-trump -tariffs/.

84 **When the slug came:** "Building on Work of Others Was Key to Lithium-Ion Batteries," *Asahi Shimbun*.

Chapter 12: The Milking Cow Falls III

85 **once won a scholarship:** Augustin Katumba Mwanke, *Ma vérité* (EPI, 2013), 32.

85 **"He was very quiet":** Mwamba Wanzala, interview with the author, July 27, 2023.

85 **"Life at that time":** Katumba, *Ma vérité*, 39.

86 **even Gécamines orchestras:** Agence France-Presse, "RDC: La Gécamines, de la petite musique du déclin à l'espoir bleu cobalt," posted March 26, 2021, YouTube, https://www.youtube .com/watch?v=a6MyRT22aYk.

86 **"He was God":** Katumba, *Ma vérité*, 39.

86 **Gécamines was still:** "Zaire: Inciting Hatred: Violence Against Kasaiens in Shaba," *News from Africa Watch* 5, no. 10 (June 1992): 19, hrw.org/sites/default/files/reports/ZAIRE936.PDF.

86 **dictator's personal piggy bank:** Jimmy Burns et al., "How Mobutu Built Up His $4 Billion Fortune," *Financial Times*, May 12, 1997.

86 **according to an analysis:** Steve Askin and Carole Collins, "External Collusion with Kleptocracy: Can Zaïre Recapture Its Stolen Wealth?," *Review of African Political Economy*, no. 57 (1993): 72–85.

87 **as *bouffer l'argent*:** Eric Lipton and Dionne Searcey, "Congo Ousts Mining Leader in a Cloud of Corruption Claims," *New York Times*, December 3, 2021.

87 **"Layer by layer":** George Arthur Forrest, *Un siècle de rêves: Ensemble, bâtissons l'avenir* (Le Cherche Midi, 2022), 102.

87 **"Gécamines was facing real":** Forrest, *Un siècle de rêves*, 103. Forrest's phrase—"Petits gisements, petits rendements" in the original French—recalls King Leopold II's famous put-down of Belgium: "Petit pays, petit esprit" (Small country, small spirit).

87 **Forrest was in charge:** Andrew L. Gulley, "One Hundred Years of Cobalt Production in the Democratic Republic of the Congo," *Resources Policy* 79 (2022): fig. 1a, doi.org/10.1016/j.resourpol .2022.10300.

87 **known as "cobaltists":** By the early 1980s, Étienne Tshisekedi had broken with Mobutu Sese Seko, who imprisoned him and forced his family into exile, first in Kasai, then in Belgium. In the post–Cold War era, Tshisekedi tried to push for democracy, briefly becoming prime minister three times in 1991 and 1992.

   In 1960 and 1961, Tshisekedi had been closely involved in the project of Kasaian secession— in fact, he was a minister in that region's secessionist government. Later, he joined the national government in Kinshasa and supported Mobutu as the dictator consolidated power and executed rivals. Historians are divided on whether this was opportunism or something more genuine. See Benjamin Rubbers, "La dislocation du secteur minier au Katanga (RDC)," *Politique Africaine* 1, no. 93 (2004): 28.

88 **He was a recognizable:** Philipp Sandner, "Obituary: Étienne Tshisekedi, 84," Deutsche Welle (DW), February 2, 2017, dw.com/en/obituary-etienne-tshisekedi-84/a-37380033.

88 **"The Kasaians are foreigners":** "Zaire: Inciting Hatred," *News from Africa Watch*.

89 **"These events were disastrous":** Forrest, *Un siècle de rêves*, 59.

89 **At least 661:** Olivier Lanotte, "Chronology of the Democratic Republic of Congo/Zaire (1960– 1997)," *Mass Violence & Resistance*, April 6, 2010, sciencespo.fr/mass-violence-war-massacre -resistance/en/document/chronology-democratic-republic-congozaire-1960-1997.html. For more on the massacre of the Kasaians, see Thomas Bakajika Banjikila, *Épuration ethnique en Afrique: "Les Kasaïens" (Katanga 1961–Shaba 1992)* (L'Harmattan, Études Africaines, 1997).

Chapter 13: A Battery and a Bubble

90 **"Sony's yearning for":** "Chapter 13: Recognized as an International Standard," Sony Corporation, accessed March 6, 2025, https://www.sony.com/en/SonyInfo/CorporateInfo/History /SonyHistory/2-13.html.

90 **"By shooting three":** "Chapter 13: Recognized as an International Standard," Sony Corporation.

90 **Sony had almost:** Charles Murray, *Long Hard Road: The Lithium-Ion Battery and the Electric Car* (Purdue University Press, 2022), 127.

91 **"The question of how":** Murray, *Long Hard Road*, 129.

91 **battery would be called:** Murray, 135.

91 **would be a long time:** And, indeed, a long time before airport security measures that made it so lithium-ion batteries were forbidden from being transported in checked luggage. Such rules resulted in the seizure of a rechargeable lithium-ion battery pack at Zurich's airport during a reporting trip for this book.

92 **no one seemed particularly:** K. B. Shedd, "Cobalt," in *Metal Prices in the United States Through 2010: Scientific Investigations Report 2012–5188*, ed. U.S. Geological Survey National Minerals Information Center Staff (U.S. Department of the Interior, U.S. Geological Survey, 2013), 37–40; and Andrew L. Gulley, "One Hundred Years of Cobalt Production in the Democratic Republic of the Congo," *Resources Policy* 79 (2022): art. 103007, doi.org/10.1016/j.resourpol.2022.10300.

92 **the Soviet Union collapsed:** During the Cold War, lithium was stockpiled to be used in the production of tritium, an isotope of hydrogen used in nuclear fission. Lithium will also be vitally important to the creation of fusion power in the future. A fusion reactor, which creates energy in much the same way as the sun or stars, would rely on lithium-produced tritium. The website for the ITER's nuclear fusion project is informative on this front: "Lithium from proven, easily extractable land-based resources would provide a stock sufficient to operate fusion power plants for more than 1,000 years. What's more, lithium can be extracted from ocean water, where reserves are practically unlimited (enough to fulfill the world's energy needs for ~6 million years)." Is there currently enough lithium around to fuel fusion reactors? In a 2022 article for *Science*, the science journalist Daniel Clery said that he thought so, but online, some skeptics have raised the specter of a shortage. See ITER, "Fuelling," iter.org/sci/FusionFuels; Daniel Clery, "Out of Gas," *Science* 376, no. 6600 (June 2022): 1372–76; and Steven B. Krivit, "#97 Lithium, Lithium, Everywhere, and None to Use for Fusion Reactors," *New Energy Times*, updated January 27, 2022, news.newenergytimes.net/2022/01/08/lithium-lithium-everywhere-and-none-to-use-for-fusion-reactors/.

92 **the material was mainly:** Joyce A. Ober, "Lithium," in *Minerals Yearbook: Minerals and Metals*, ed. U.S. Geological Survey and U.S. Department of the Interior (U.S. Government Printing Office, 1998), 477.

92 **"lithium prices were generally":** Alessio Miatto et al., "The Rise and Fall of American Lithium," *Resources, Conservation and Recycling* 162 (November 2020): art. 105034, doi.org/10.1016/j.resconrec.2020.105034.

93 **"The idea of Japanese economic":** David Pilling, *Bending Adversity* (Penguin Books, 2014), 97–98.

93 **its third-generation battery:** Murray, *Long Hard Road*, 147.

93 **the spring of 1996:** Murray, 148.

Chapter 14: Getting Rich Is No Sin

94 **Iseard met Aronson when:** Barry Iseard (Apollo Energy Systems), interview with the author, May 2022.

94 **China had started pursuing:** "The goal of the contract system," Frank Dikötter writes, "besides introducing greater flexibility, was to achieve greater growth, not to improve development." See Frank Dikötter, *China After Mao: The Rise of a Superpower* (Bloomsbury, 2022), 60–61, 69.

94 **As Deng told the CBS:** Evelyn Iritani, "Great Idea but Don't Quote Him," *Los Angeles Times*, September 9, 2004.

95 **Encouraged by their reports:** *House Hearing: U.S.-China Trade Relations and Renewal of China's Most-Favored-Nation Status*, 104th Cong., 152–60 (1995) (testimony of Robert R. Aronson).

95 **"It's very slow but":** Douglas B. Feaver, "McDonnell Douglas, China Sign Pact," *Washington Post*, April 13, 1985.

95 **the plant was up:** *House Hearing: U.S.-China Trade Relations and Renewal of China's Most-Favored-Nation Status*, 152–60.

95 **like Elon Musk and many:** *House Hearing: U.S.-China Trade Relations and Renewal of China's Most-Favored-Nation Status*, 152–60.

95 **Investors may have been:** Iritani, "Great Idea but Don't Quote Him."
96 **transformation of state-owned:** Hiroki Takeuchi, "Political Economy of Trade Protection: China in the 1990s," *International Relations of the Asia-Pacific* 13, no. 1 (2013): 1–32, jstor.org/stable /26155971.
96 **The battery factory was seized:** Dexter Roberts, "Cheated in China?," *Bloomberg,* October 6, 1997.
96 **"We began learning that China":** *House Hearing: The Future of U.S.-China Relations and the Possible Accession of China into the World Trade Organization,* 105th Cong., 106 (1995) (testimony of Robert R. Aronson).
96 **The company was acquired:** Yochi J. Dreazen, "McDonnell Douglas Will Pay Fine for Problems with Sale to China," *Wall Street Journal,* November 15, 2001.

Chapter 15: Twilight of the Big Vegetables

98 **which was finished in 1993:** Gérard Prunier, *Africa's World War: Congo, the Rwandan Genocide, and the Making of a Continental Catastrophe* (Oxford University Press, 2009), 139.
98 **"The country's formal economy":** U.S. Embassy Kinshasa, *FY1997 Country Commercial Guide: Zaire* (Bureau of Economic and Business Affairs, August 1996), 1997-2001, state.gov/about_state /business/com_guides/1997/africa/zaire97.html.
98 **huge Kamoto mine:** Andrew L. Gulley, "One Hundred Years of Cobalt Production in the Democratic Republic of the Congo," *Resources Policy* 79 (2022): art. 103007, doi.org/10.1016/j .resourpol.2022.10300.
98 **and fifteen million tons:** *Kamoto Copper Company Technical Report* (Katanga Mining Limited, November 2019), 64.
99 **Copper production was just:** Jason Stearns, *Dancing in the Glory of Monsters: The Collapse of Congo and the Great War of Africa* (PublicAffairs, 2011), 164.
99 **Cobalt production had almost:** Prunier, *Africa's World War,* 139.
99 **"You have guns":** "Congo's War Was Bloody. It May Be About to Start Again," *Economist,* February 15, 2018.
99 **"The informal economy":** U.S. Embassy Kinshasa, *FY1997 Country Commercial Guide: Zaire.*
99 **he had legalized small-scale:** Raphael Deberdt, *Baseline Study of Artisanal and Small-Scale Cobalt Mining in the Democratic Republic of the Congo* (Responsible Sourcing Network and UBC Anthropology, July 2021), securityhumanrightshub.org/media/pdf/resources/Baseline+Study +Cobalt+ASM+Mining+-+7.28.2021.pdf.
101 **the dictator's China policy:** Francis Boulle, interview with the author, July 2023. "Mobutu was getting too close to China," Boulle told me. "They wanted someone else to run things." Boulle, a mining entrepreneur, is the son of Max Boulle, a mine investor who, along with his brother Jean-Raymond, would aid the rebellion in search of diamonds and other raw materials.
101 **mine entrepreneur of the era:** Congo mine entrepreneur, interview with the author, August 2023.
101 **his fellow rebel leaders:** *How Kabila Lost His Way: The Performance of Laurent-Désiré Kabila's Government,* ICG Democratic Republic of Congo Report No. 3 (IGC, May 1999), crisisgroup.org /africa/central-africa/democratic-republic-congo/how-kabila-lost-his-way.
101 **"like an ebony Buddha":** Vincent Hugeux et al., "L'obscur M. Kabila," *L'Express,* June 25, 1998, lexpress.fr/monde/afrique/l-obscur-m-kabila_493335.html.
101 **he had joined:** For a discussion of the pair's travails in the Congolese bush, see Jon Lee Anderson's excellent *Che Guevara: A Revolutionary Life* (Grove Press, 1997).
101 **"it is essential to have":** Che Guevara, *Congo Diary: Episodes of the Revolutionary War in the Congo* (Seven Stories Press, 2011), 312.

Chapter 16: Building Dreams

103 **"Dynamic and intelligent":** "BYD Seal," BYD, https://www.byd.com/fr/car/seal.
104 **BYD's salesman didn't:** Author's web conversation with "Charles" on BYD.com.
104 **BYD would surpass:** Daniel Ren, "BYD Overtakes Tesla as World's Largest Maker of Pure Electric Cars in Fourth Quarter," *South China Morning Post,* January 4, 2025, scmp.com/business /china-evs/article/3293269/byd-overtakes-tesla-worlds-largest-maker-pure-electric-cars-fourth -quarter.
104 *Wuwei* **means something:** I am indebted to the Medium writer Guangxizhang and Kevin Xu on Substack for their overviews on Wang Chuanfu's life; both gave me a good basis from which

to conduct further research. See Guangxizhang, "The Legendary Life of Wang Chuanfu, the Founder of China's Trillion-Dollar Car Company, Was Highly Respected by Munger Buffett," Medium, July 20, 2013, medium.com/@guangxizhang1207/the-legendary-life-of-wang-chuanfu-the-founder-of-chinas-trillion-dollar-car-company-was-highly-76433dbe4be5; and Kevin Xu, "Wang Chuanfu: A Name Everyone in the West Should Know," *Interconnected*, February 8, 2024, interconnect.substack.com/p/wang-chuanfu-a-name-everyone-in-the.

104 **The tune of his upbringing:** Around 790 million people lived in rural China in 1978, and this number would not decline until the late 1980s, when the country urbanized. See Hua Zhang et al., "Evolution and Influencing Factors of China's Rural Population Distribution Patterns Since 1990," *PLOS One* 15, no. 5 (May 2020), doi.org/10.1371/journal.pone.0233637.

105 **"Importing batteries from Japan":** Marc Gunther, "Warren Buffett Takes Charge," *Fortune*, April 13, 2009, money.cnn.com/2009/04/13/technology/gunther_electric.fortune/.

106 **"He renounced a comfortable":** Li Daqian, *Wang Chuanfu: Créateur Innovant* (Infini Découverte, 2013), 12.

106 **Wang didn't even understand:** "察｜比亚迪：从电池作坊到新能源汽车巨头 多次跨界 新域" [BYD: From a Battery Workshop to a New Energy Vehicle Giant: Multiple Crossovers into New Fields], *Sohu News*, November 13, 2018, sohu.com/a/275199569_275361.

106 **Their stakes would later:** "察｜比亚迪：从电池作坊到新能源汽车巨头 多次跨界 新 域," *Sohu News*.

106 **In the absence of mechanical:** River Davis and Selina Cheng, "How China's BYD Became Tesla's Biggest Threat," *Wall Street Journal*, October 4, 2023.

107 **"invented a production pattern":** Li Daqian, *The Creative Wisdom of Wang Chuanfu*, trans. Denis Mair (China Intercontinental Press, 2013), 19.

107 **the firm was able to:** Li, *Wang Chuanfu*, 22.

107 **"So we started to invest":** Robert S. Huckman and Alan D. MacCormack, "BYD Company, Ltd.," Harvard Business School Case No. 606-139 (HBS Case Collection, April 2006, revised September 15, 2009), 2, hbs.edu/faculty/Pages/item.aspx?num=33206.

107 **the end of the 1980s:** Christopher Wood, *The Bubble Economy: Japan's Extraordinary Speculative Boom of the '80s And the Dramatic Bust of the '90s* (Solstice, 2006), 2.

108 **Thanks to innovations like:** "Sony to Establish Lithium-Ion Polymer Rechargeable Battery Plant in China," Sony, September 7, 2007, sony.com/en/SonyInfo/News/Press/200009/00-039E/.

108 **the company was not immune:** David Pilling, *Bending Adversity* (Penguin Books, 2014), 170. According to Pilling, "Sony's biggest failing was its inability to navigate the industry's transformation from analogue to digital." Akio Morita, the peppy cofounder of Sony, had pushed for digitalization, but his engineers had rebelled.

108 **Chinese companies were making:** "Sony to Establish Lithium-Ion Polymer Rechargeable Battery Plant in China."

108 **Samsung was producing:** Emi Emoto and Tim Kelly, "Exclusive: Banks Offer to Help Sony Offload Battery Unit—Sources," Reuters, November 27, 2012.

109 **The economics in Korea:** Robin Harding, "Beware the Great Battery Industry Fallacy," *Financial Times*, February 2, 2023.

109 **"the mentality is different":** Japanese battery executive, interview with the author, November 2022.

109 **BYD reverse engineered:** "Innovations and IPRs Boost BYD's Profits," *Global Times*, October 8, 2010, www.globaltimes.cn/content/579843.shtml.

110 **Zeng's factory was located:** Henry Sanderson, *Volt Rush: The Winners and Losers in the Race to Go Green* (Oneworld, 2022), 39.

110 **an estimated 95 percent:** Seth Fletcher, *Bottled Lightning: Superbatteries, Electric Cars, and the New Lithium Economy* (Hill and Wang, 2011), 59.

110 **In Wang's conception of BYD:** Li, *Creative Wisdom of Wang Chuanfu*, 38.

110 **The company had grown yearly:** *2002 Annual Report* (BYD Company Ltd., 2003), 5–6.

111 **But BYD had also:** "China: New Energy Vehicle (NEV) Policy," DieselNet, dieselnet.com/standards/cn/nev.php.

111 **Chinese state prioritized:** "The 10th Five-Year Plan for Economic and Social Development of the People's Republic of China (2001–2005)," International Energy Agency, iea.org/policies/1736-the-10th-five-year-plan-for-economic-and-social-development-of-the-peoples-republic-of-china-2001-2005.

111 **BYD profited from government:** Monica Miller, "BYD: The Top Electric Car Maker That Is Not Tesla," *BBC News*, October 19, 2023; "China: New Energy Vehicle (NEV) Policy"; and

"High Tech Research and Development (863) Programme," Consulate General of the People's Republic of China in New York, October 21, 2003, newyork.china-consulate.gov.cn/eng/xbwz/kjsw/zgkj/200310/t20031021_5431224.htm. In 2001, electric vehicles were included in the Chinese government's 863 Program, a research-and-development scheme whose goal was, according to the Chinese government, to "select several high technologies in which China has superiority for the breakthrough of industrialization, support the establishment and development of high-tech industries, and make the 863 Programme a starting Point of high-tech industries." It allocated 880 million renminbi (almost $195 million in 2025 U.S. dollars) to research into strategic new technologies.

111 **he was elected deputy:** *Green Tech for Tomorrow, Annual Report 2008* (BYD Inc., 2009), 17.

111 **Wang was joking:** Huckman and MacCormack, "BYD Company, Ltd.," 2.

111 **Zeng's ATL had also:** Edward White et al., "China's 'Battery King' Faces Scrutiny over EV Market Dominance," *Financial Times*, April 4, 2023.

111 **licensed from Bell Labs:** Christopher Chico, "CATL's Success Story: A Strategic Journey of Innovation and Expansion," The Battery Chronicle, February 2, 2025, https://christopherchico.substack.com/p/catls-success-story-a-strategic-journey.

111 **grow to become the world's:** Amy Hawkins, "CATL, the Little-Known Chinese Battery Maker That Has the US Worried," *Guardian*, March 18, 2024.

111 **the dichotomy was underscored:** Yan Zhang and Kevin Krolicki, "China Battery Giant CATL Would Build US Plant If Trump Allows It," Reuters, November 13, 2024, reuters.com/business/autos-transportation/china-battery-giant-catl-would-build-us-plant-if-trump-allows-it-2024-11-13/.

111 **bought 77 percent:** Joann Muller, "Thanks, Now Move Over," *Forbes*, July 26, 2004.

112 **"Due to the limited oil":** *Interim Report* (BYD Inc., June 2003), 4.

112 **"a whole-industry chain":** Li, *Creative Wisdom of Wang Chuanfu*, 22–23.

112 **It was perhaps inevitable:** Li, 27.

112 **Part of Wang's strategy:** Li, 23.

112 **BYD was also hard:** Li, 20.

112 **"We can learn a lot":** Muller, "Thanks, Now Move Over."

113 **was "no longer content":** David Barboza, "China's Ambition Soars to High-Tech Industry," *New York Times*, August 1, 2008, nytimes.com/2008/08/01/business/worldbusiness/01factory.html.

113 **at the Beijing:** "China: New Energy Vehicle (NEV) Policy."

113 **The car would combat:** Richard S. Chang, "A Plug-In Hybrid Goes on Sale, in China," Wheels, *New York Times*, December 18, 2008, archive.nytimes.com/wheels.blogs.nytimes.com/2008/12/18/a-plug-in-hybrid-goes-on-sale-in-china/.

Chapter 17: Fire Sale at the Karavia

114 **Katumba was a young accountant:** Bruce Jewels, interview with the author, July 26, 2023.

114 **Katumba's old classmate:** Wanzala would later work on supply-chain logistics for UPS, Coca-Cola, and Alstom, among other firms, and settle in Georgia. Other classmates of his and Katumba's from the Collège Imara Saint François de Sales went on to work for the EU and the U.S. Food and Drug Administration. Some became pilots. One even worked for NASA. "Thirty or forty of us succeeded so much in life," he told me. "We are spread across the world." One need only consider classes like Katumba's to realize the entrepreneurial and innovative potential that the children in Congo—and, indeed, *anywhere*—hold. It becomes all the more devastating to think of those whose lives will be spent and cut short working in artisanal mines.

115 **Even as their war:** A theory popular in some corners of the internet has it that U.S. mining interests unseated Mobutu for their own gain, but this doesn't stand up to much scrutiny. See Gérard Prunier, *Africa's World War: Congo, the Rwandan Genocide, and the Making of a Continental Catastrophe* (Oxford University Press, 2009), 139.

116 **"We want to know how":** Jean-Philippe Ceppi, "'Business as Usual': Attirés par la richesse du sous-sol zaïrois, les hommes d'affaires étrangers courtisent Kabila," *Libération*, April 19, 1997.

116 **The convention gave Anvil:** John Cumming, "Anvil Expands Dikulushi Copper-Silver Mine," *Northern Miner*, October 18, 2004.

116 **Kabila himself arrived:** John Pitman, "Kabila Arrives in Lubumbashi," *Voice of America*, April 14, 1997.

117 **"This is the first time"**: "Washington Quandary: The U.S. Government Juggles Carrot and Stick While Business Scoops a Mineral Map," *Africa Confidential* 38, no. 19 (September 26, 1997), africa-confidential.com/home/issue/id/434.

117 **the U.S. engineering and construction behemoth**: Robert Block, "U.S. Firms Seek Deals in Central Africa—Bechtel Woos Congo as Region's Conflicts Continue to Fester," *Wall Street Journal*, October 14, 1997.

117 **"I think those who"**: *A State Affair: Privatizing Congo's Copper Sector* (Carter Center, November 2017), 18, cartercenter.org/resources/pdfs/news/peace_publications/democracy/congo-report -carter-center-nov-2017.pdf.

117 **buffet felt like a battlefield**: Or at least it did to Andrew Maykuth, a *Philadelphia Inquirer* journalist who had flown in with twenty-five businessmen claiming to be the first outside investors in rebel-held territory. See Andrew Maykuth, "Outside Mining Firms Find Zaire an Untapped Vein," *Philadelphia Inquirer*, May 11, 1997.

117 **"You need strong nerves"**: Maykuth, "Outside Mining Firms Find Zaire an Untapped Vein."

118 **Jewels had come**: The concession on which Tilwezembe found itself would be acquired in 2004 by Nikanor. The main stakeholders in the firm were Dan Gertler and Beny Steinmetz, two Israeli investors. Gertler was later sanctioned by the U.S. Treasury Department. Steinmetz was convicted in Geneva, Switzerland, on corruption charges related to businesses in Guinea. (Steinmetz appealed, but the conviction was upheld in 2023.) Starting in 2007, Tilwezembe became part of Glencore's Katanga mining operation after the company's takeover of Nikanor. The concession, however, was overrun by artisanal miners and has been criticized for particularly lax labor practices. The artisanal mining collectives, however, were connected to the Katangese power structure (former Governor Richard Muyej's son Yves managed the offtake through a company called Empire Mining SARL), and Glencore distanced itself from the site while retaining the mineral rights. Glencore has always denied it bought hand-mined cobalt from Tilwezembe.

118 **"Think of your country"**: Augustin Katumba Mwanke, *Ma vérité* (EPI, 2013), 62–63.

118 **The banker did feel**: Katumba, *Ma vérité*, 62–63.

118 **To the banker's reckoning**: Katumba, 61.

119 **"We've made a conscious decision"**: At Kisangani, the rebels had asked for international firms to reopen their diamond-trading businesses there. De Beers, Anglo American's sister company, still had mines in the western province of Kasai, in Mobutu-controlled territory, so it demurred. The Boulles jumped at the opportunity, throwing a $1 million payment—"advance taxes," it was called—to the rebels. When the rebel forces took Kasai, they had to bid against the Boulles for diamonds that they considered theirs, and then they were informed that they would be kicked off the mine. "They just took those diamonds away and didn't put anything back in the community," one of Kabila's rebel ministers said of De Beers. "It was exploitation. That's why we prefer bids." The rebels were cleaning up shop, and even if the deals they were doing were opaque, the sheer scale of Mobutu's theft provided a justification for almost anything in those early days. See Maykuth, "Outside Mining Firms Find Zaire an Untapped Vein."

119 **looking a little stupid**: Maykuth, "Outside Mining Firms Find Zaire an Untapped Vein."

119 **the Boulles had just signed**: "Kolwezi Tailings," *Economist*, January 15, 1998.

119 **The Belgian social scientist**: Erik Kennes, "Le secteur minier au Congo: 'Déconnexion' et descente aux enfers," in *L'Afrique des Grands Lacs: Annuaire 1999–2000*, ed. Filip Reyntjens and Stefaan Marysse (L'Harmattan, 2000), 305–48, 312.

120 **the Boulles brought**: "Kolwezi Tailings."

120 **"a must-have operator"**: Benjamin Rubbers, "La dislocation du secteur minier au Katanga (RDC)," *Politique Africaine* 1, no. 93 (2004): 24.

120 **Forrest's family had been**: George Arthur Forrest in *Katanga Business*, directed by Thierry Michel (Les Films de la Passerelle, 2009).

120 **"He didn't compensate us"**: Hubert Leclercq, "RDCongo: George Forrest, une histoire congolaise de Kasa Vubu à Tshisekedi," *Le Libre Afrique*, April 16, 2023, afrique.lalibre.be/76887 /rdcongo-george-forrest-une-histoire-congolaise-de-kasa-vubu-a-tshisekedi/. Forrest has inveighed against foreign NGOs writing about Congolese businesses and is generally unresponsive to media queries. He did not respond to my repeated requests for an interview, and when I visited the Lubumbashi headquarters of the Entreprise Générale Malta Forrest in 2022, I was met with blank stares. The slip of paper I filled out for the attention of Forrest never received a response.

121 **"For reasons that have"**: H. W. French, "U.S. Apathy Paved the Way for China in Africa," *Foreign Policy*, May 22, 2023, foreignpolicy.com/2023/05/22/congo-mining-batteries-china-biden-climate-change-us-africa-policy/.

Chapter 18: Plans on the Back of a Comet

122 **Gertler was twenty-four**: Mathieu Olivier and Romain Gras, "Exclusif—Dan Gertler: 'Tous étaient effrayés par le Congo, mais pas moi,'" *Jeune Afrique*, August 1, 2022.

122 **The precious stones were**: Nicole Gaouette, "Inside Israel's Diamond Trade: A Family Affair," *Christian Science Monitor*, February 21, 2002.

123 **his tenure, exports**: "Diamond News: MOSHE SCHNITZER 1921–2007," Israel Diamond Exchange, archived June 14, 2012, at web.archive.org/web/20120614132205/http://www.israelidiamond.co.il/english/news.aspx?boneid=918&objid=2492.

123 **"There was no real separation"**: Gaouette, "Inside Israel's Diamond Trade: A Family Affair."

123 **Gertler remembered getting up**: Franz Wild et al., "Dan Gertler Earns Billions as Mine Deals Fail to Enrich Congo," *Washington Post*, December 29, 2012.

123 **Moshe's Irgun comrades**: Leah Granof, "Are Diamonds Forever?," *Jerusalem Post*, February 1, 1997.

123 **The charm wasn't universal**: Mining executive, interview with the author on background, February 2022.

123 **"he looks at a wooden chair"**: Olivier and Gras, "Exclusif—Dan Gertler."

124 **streets were patrolled by child**: Michela Wrong, *In the Footsteps of Mr. Kurtz: Living on the Brink of Disaster in Mobutu's Congo* (Fourth Estate, 2000), 30. As Wrong reported, these soldiers, who had swelled Kabila's ranks, would boast that they had walked to Kinshasa from Kampala, "artlessly spilling the beans on Uganda's involvement in the rebel uprising."

125 **"he was dreaming up plans"**: Augustin Katumba Mwanke, *Ma vérité* (EPI, 2013), 64.

125 **the rabbi arranged a meeting**: Wild et al., "Dan Gertler Earns Billions as Mine Deals Fail to Enrich Congo."

126 **"That's what kept us going"**: James C. McKinley Jr., "A Fallen City, Seeking Peace, Greets Rebels," *New York Times*, March 17, 1997.

126 **The president would soon call**: "He was seconded and kept his benefits until Kabila called me to request that we approve his transfer to Governor of Katanga at which time Katumba had to terminate his relationship with the bank," Jewels explained. Bruce Jewels, fact-checking email, March 2025.

Chapter 19: The Nightmare

127 **"We talked about life"**: Mathieu Olivier and Romain Gras, "Exclusif—Dan Gertler: 'Tous étaient effrayés par le Congo, mais pas moi,'" *Jeune Afrique*, August 1, 2022.

127 **was meeting with ministers**: Christian Dietrich, "Have African-Based Diamond Monopolies Been Effective?," *Central Africa Minerals and Arms Research Bulletin*, International Peace Information Service (IPIS), June 18, 2001, 6.

128 **The ensuing conflict would kill**: This figure is hotly debated; a 2008 report by the International Rescue Committee suggested that 5.4 million people had been killed, but a review published by the Human Security Report Project in 2010 challenged the IRC's methodology. All agree, however, that the death toll from conflict in Congo since 1998 has been staggering. For more on this debate, see "How Many Have Died Due to Congo's Fighting? Scientists Battle Over How to Estimate War-Related Deaths," *Science*, January 21, 2010, science.org/content/article/how-many-have-died-due-congos-fighting-scientists-battle-over-how-estimate-war-related.

128 **"It was a way to thank"**: George Arthur Forrest, *Un siècle de rêves: Ensemble, bâtissons l'avenir* (Le Cherche Midi, 2022), 104.

128 **De Beers had left**: "Addendum to the Report of the Panel of Experts on the Illegal Exploitation of Natural Resources and Other Forms of Wealth of the Democratic Republic of the Congo," letter from the Secretary-General to the President of the Security Council, 10 November 2001, United Nations Document No. S/2001/1072, 15, securitycouncilreport.org/un-documents/document/drc-s-2001-1072.php.

128 **The diamond trade was pushed**: Dietrich, "Have African-Based Diamond Monopolies Been Effective?," 5.

128 **"He didn't have much knowledge"**: Forrest, *Un siècle de rêves*, 113.

129 **To avoid conflicts of interest**: Forrest, 105.

129 **asked for $20 million**: Franz Wild et al., "Dan Gertler Earns Billions as Mine Deals Fail to Enrich Congo," *Washington Post*, December 29, 2012.

129 **A UN report would later**: "Addendum to the Report of the Panel of Experts on the Illegal Exploitation of Natural Resources and Other Forms of Wealth of the Democratic Republic of the Congo," 15.

130 **The money went to**: Franz Wild et al., "Dan Gertler Earns Billions as Mine Deals Fail to Enrich Congo," *Washington Post*, December 29, 2012.

130 **"This is the optimum way"**: Sharon Berger, "Congo Signs $700M Agreement with IDI Diamonds," *Jerusalem Post*, August 2, 2000.

130 **Gertler maintained that he paid**: Dietrich, "Have African-Based Diamond Monopolies Been Effective?," 4.

130 **"deal turned out"**: *Report of the Panel of Experts on the Illegal Exploitation of Natural Resources and Other Forms of Wealth of the Democratic Republic of the Congo*, United Nations Document No. S/2001 /357 (United Nations Security Council, April 2001), 33, securitycouncilreport.org/atf/cf /%7B65BFCF9B-6D27-4E9C-8CD3-CF6E4FF96FF9%7D/DRC%20S%202001%20357.pdf.

130 **North Korean military trainers**: Anna Kuchment, "Kabila and the North Koreans, Nuclear Dread in South Africa," *Newsweek*, October 17, 1999.

130 **Kabila Sr. had also arranged**: Mamadou Faye, "Laurent-Désiré Kabila: retour sur une mort mystérieuse et tragique," *BBC News Afrique*, January 16, 2021, https://www.bbc.com/afrique /monde-55675241.

131 **"IDI agreed to arrange"**: "Addendum to the Report of the Panel of Experts on the Illegal Exploitation of Natural Resources and Other Forms of Wealth of the Democratic Republic of the Congo," 15. Reports also surfaced that Gertler was involved in deals in Liberia and Sierra Leone, countries that were both in the middle of bloody civil wars. These deals had reportedly included an arms dealer named Yair Klein, who was wanted in the U.S. for training Medellín drug-cartel paramilitaries, and had involved the trading of Israeli weapons and military training in exchange for diamonds. See Dietrich, "Have African-Based Diamond Monopolies Been Effective?," 7. In the Congo, Gertler had "done that deal, diamonds for weapons, or weapons for diamonds," a mining company official who worked with Gertler once told me. Interview with mining company official, March 2025.

131 **The report alleged that diamonds**: "Addendum to the Report of the Panel of Experts on the Illegal Exploitation of Natural Resources and Other Forms of Wealth of the Democratic Republic of the Congo," 15.

131 **"the Russian Military Brotherhood"**: Nicole Gaouette, "Inside Israel's Diamond Trade: A Family Affair," *Christian Science Monitor*, February 21, 2002.

132 **"philosophy" was never to hide**: Gaouette, "Inside Israel's Diamond Trade."

132 **Cohen and Gertler had met**: Gur Megiddo, "Mossad Chief Cohen Kicked Out of DRC, on a Mission That Could Jeopardize Israel," *Haaretz*, May 19, 2022; and E. Bronner, "Israel Pushed Trump Officials to Lift Sanctions on Dan Gertler," *Bloomberg*, March 19, 2021.

132 **"I will keep you informed"**: Augustin Katumba Mwanke, *Ma vérité* (EPI, 2013), 160.

133 **his father had agreed to**: Olivier and Gras, "Exclusif—Dan Gertler."

133 **"Remember that you are guests"**: Melissa Sanderson, interview with the author, August 2023.

133 **a compensation deal**: Olivier and Gras, "Exclusif—Dan Gertler."

133 **"Our termination of his contract"**: Katumba, *Ma vérité*, 200–201.

134 **He set up meetings**: Katumba, 195.

134 **"Chinese official had a bitter"**: Katumba, 195.

134 **"rather than succumb"**: Joseph Kabila, letter to George W. Bush, April 6, 2002, int.nyt.com /data/documenttools/2002-04-gertler-bush-drc-kabila/1139881eff0eab6a/full.pdf.

134 **"a disruptive force"**: Dan Gertler, "Summary of Meetings and Correspondence on Behalf of the DRC in the White House, Washington, DC, April–May 2002," int.nyt.com/data/documenttools /2002-04-gertler-bush-drc-kabila/1139881eff0eab6a/full.pdf.

135 **Gertler turned to Katumba**: Katumba, *Ma vérité*, 201.

135 **Katumba wrote Gertler**: Mathieu Olivier and Romain Gras, "RDC: Dan Gertler, l'irrésistible ascension du businessman de Kabila," *Jeune Afrique*, August 2, 2002.

135 **"between two men"**: Katumba, *Ma vérité*, 201.

135 **The Rwandan president assured**: Katumba, 189.

135 **the "final act":** United Nations, *Inter-Congolese Political Negotiations: The Final Act* (Sun City, South Africa, April 2, 2003), peacemaker.un.org/sites/default/files/document/files/2024/05/cd030402suncityagreement.pdf.

136 **"a war in which DRC":** "Biography," Dan Gertler: Philanthropist & Businessman (personal website), dan-gertler.com/biography.html.

136 **clearly of the opinion:** Wild et al., "Dan Gertler Earns Billions as Mine Deals Fail to Enrich Congo."

136 **"increased seven-fold from 1996":** Andrew L. Gulley, "One Hundred Years of Cobalt Production in the Democratic Republic of the Congo," *Resources Policy* 79 (2022): 1, doi.org/10.1016/j.resourpol.2022.10300.

137 **money borrowed from banks:** Forrest, *Un siècle de rêves*, 116. The recipe for success? "Will, economic vision, business sense, but also a sense of patriotism," Forrest later wrote.

137 **invested in companies that controlled:** Wild et al., "Dan Gertler Earns Billions as Mine Deals Fail to Enrich Congo."

137 **"This was approved by":** Katumba, *Ma vérité*, 203.

137 **"debt monster":** Gertner and Gertner v. Gertler, פסק בוררות בפני כבוד הבורר, השופט (בדימוס) איתן אורנשטיין [Arbitration Award Before the Honorable Arbitrator, Judge (Ret.) Eitan Orenstein], Arb. Eitan Orenstein, April 22, 2024, 97.

137 **"beyond the natural order":** *Gertler*, 112.

137 **"If there is no Katumba":** *Gertler*, 724.

137 **The Congolese politician:** *Gertler*, 722, 725.

138 **"Katumba was the one who dealt":** *Gertler*, 721.

138 **"Despite the massive sacrifices":** Forrest, *Un siècle de rêves*, 116.

PART 3: BATTERY BOOM

Chapter 20: Accelerating the Transition

142 **"a sexy product":** Tad Friend, "Plugged In," *New Yorker*, August 17, 2009.

143 **"robbed of their livelihoods":** "Block Tesla—Disrupt Elon," Disrupt, archived May 10, 2024, at web.archive.org/web/20240510121822/https://disrupt-now.org/en/disrupt-tesla/.

143 **social-media app:** Elon Musk (@elonmusk), "Why do the police let the left-wing protestors off so easily?," Twitter (now X), May 10, 2024, 4:29 p.m., https://twitter.com/elonmusk/status/1788939318720950772.

144 **proposed battery factory:** Jochen Knoblach, "Tesla: Batteriefabrik in Grünheide bringt weitere 10.000 Jobs," *Berliner Zeitung*, November 28, 2020.

144 **a German court allowed:** "Tesla Wins Court Approval to Build Gigafactory in Germany," Deutsche Welle (DW), February 20, 2020, dw.com/en/tesla-wins-court-approval-to-build-gigafactory-by-clearing-forest-in-germany/a-52454649.

144 **"We need to keep some":** "Nach Tesla-Klage gibt es eine Rebellion innerhalb der Grünen Liga," *Berliner Zeitung*, February 19, 2020, berliner-zeitung.de/wirtschaft-verantwortung/nach-tesla-klage-gibt-es-eine-rebellion-innerhab-der-gruenen-liga-li.76367.

144 **Twelve thousand people worked:** Ilona Wissenbach and Christoph Steitz, "Tesla Targeting About 400 Voluntary Job Cuts in Germany," Reuters, April 23, 2024.

144 **The factory is meant:** Maria Merano, "Tesla Giga Berlin's First Goal Is to Produce One Model Y Body Every 45 Seconds," Teslarati, October 9, 2021, teslarati.com/tesla-model-y-production-45-seconds-giga-berlin/.

145 **an illegal diesel station:** Christian Esser et al., "Bizarrer Vorfall in Teslas Gigafactory: Wie eine illegale Tankstelle unter einem Partyzelt verschwand," *Stern*, October 5, 2023, stern.de/wirtschaft/tesla-recherche--wie-eine-illegale-tankstelle-unter-einem-partyzelt-verschwand--33868106.html.

145 **leaked feces into the water:** Daniel Wüstenberg, "Illegale Tankstelle und ausgelaufene Fäkalien: Diese Vorfälle deckte der *Stern* bei Tesla auf," *Stern*, October 5, 2023, stern.de/wirtschaft/tesla—diese-unglaublichen-vorfaelle-deckte-der-stern-auf-33861518.html.

145 **Tesla denied the claims:** Eva Fox, "Tesla Responds to Workplace Safety Allegations in Giga Berlin," Tesmanian, September 20, 2023, www.tesmanian.com/blogs/tesmanian-blog/tesla-responds-to-workplace-safety-allegations-in-giga-berlin.

145 **a "toxic" culture:** Bryce Covert, "The Toxic Culture at Tesla," *Nation*, April 9, 2024.

145 **Musk had publicly mused:** For BYD batteries, see "Exklusiv: Tesla Startet Produktion von Model Y Mit Akku von BYD in Deutscher Gigafactory," *Teslamag*, May 8, 2023, teslamag.de /news/exklusiv-tesla-model-y-rwd-deutsche-fabrik-struktur-akku-byd-58282; and for CATL batteries, see Cristian Agatie, "CATL Will Be the Battery Cell Supplier for Tesla's Next-Generation Compact EV," Autoevolution, March 28, 2024, autoevolution.com/news/catl -will-be-the-battery-cell-supplier-for-tesla-s-next-generation-compact-ev-231514.html.

145 **where molten aluminum:** Maria Merano, "Tesla Begins Installing IDRA Giga Press Casting Machine at Gigafactory Berlin," Teslarati, March 21, 2021, teslarati.com/tesla-gigafactory-berlin -idra-giga-press-video/.

145 **Tesla can trim:** Evannex, "Learn About Tesla's Giga Press from IDRA: Five-Part Documentary," Inside EVs, January 25, 2022, chap. 1, 2:20, insideevs.com/news/563005/tesla-giga-press -idra-documentary/.

Chapter 21: Breaking the ICE

147 **especially the flagship Prius:** Peter Bohan, "Detroit Gets a Buzz from Electric Cars," Reuters, January 21, 2007.

147 **GM had occupied:** Seth Fletcher, *Bottled Lightning: Superbatteries, Electric Cars, and the New Lithium Economy* (Hill and Wang, 2011), 110.

147 **"Toyota was the darling":** "Bob Lutz at #NAIAS 2019," posted January 26, 2019, by Detroit PBS, YouTube, youtube.com/watch?v=bfmnD1j7nFw&t=40s.

148 **a Los Angeles engineer:** *Who Killed the Electric Car?*, directed by Chris Paine (Sony Pictures Classics, 2006).

148 **"very viable solution":** Alan Cocconi, "Alan Cocconi (BS '80), Electrical Engineer," interview by David Zierler, Caltech Heritage Project, December 8 and 15, 2021, posted July 5, 2022, heritageproject.caltech.edu/interviews-updates/alan-cocconi.

148 **a newly minted:** "Electric Car tZero 0-60 3.6 sec Faster than Tesla Roadster," posted August 22, 2010, by Evans Electric, YouTube, youtube.com/watch?v=gb9E222QsM0.

148 **$2.79 billion from Tesla:** Kah Seng Tay (Quora), "How Much Equity Did Elon Musk Get from Investing in Tesla's Series A?," *Forbes*, December 29, 2014, forbes.com/sites/quora/2014 /12/29/how-much-equity-did-elon-musk-get-from-investing-in-teslas-series-a/.

148 **the company's shareholders approved:** Aimee Picchi, "Tesla Shareholders Approve $46 Billion Pay Package for CEO Elon Musk," CBS News, August 22, 2023, cbsnews.com/news/elon -musk-pay-package-vote-cbs-news-explains/.

149 **plan the Gigafactories:** Ben Tarnoff, "Ultra Hardcore," *New York Review of Books*, January 18, 2024, nybooks.com/articles/2024/01/18/ultra-hardcore-elon-musk-walter-isaacson/.

149 **"their heads down":** Ryan Melsert (American Battery Technology Company), interview with the author, August 2023.

149 **"Way beyond technology":** Mujeeb Ijaz, interview with the author, June 2023.

149 **"Whether Tesla is ever":** "Electric Cars," *Charlie Rose*, November 9, 2011, charlierose.com /videos/15404.

149 **"only way to stop them":** Bob Lutz, interview with the author, December 2022.

150 **"If we are going to":** Lutz, interview.

150 **cure the "range anxiety":** "2011 Volt," Chevrolet (website), archived February 23, 2011, at web.archive.org/web/20110223173940/http://www.chevrolet.com/volt/features-specs/.

150 **"total crock of shit":** "GM Exec Stands by Calling Global Warming a 'Crock,'" Reuters, February 22, 2008, reuters.com/article/economy/gm-exec-stands-by-calling-global-warming-a -crock-idUSN22372976/.

150 **"This is a real program":** Peter Bohan, "Detroit Gets a Buzz from Electric Cars," Reuters, August 9, 2007, reuters.com/article/sports/motor-sports/detroit-gets-a-buzz-from-electric -cars-idUSNOA839010/.

Chapter 22: Le Petit

151 **"batteries are the foot-soldiers":** "Hooked on Lithium," *Economist*, June 22, 2002, economist .com/technology-quarterly/2002/06/22/hooked-on-lithium.

152 **"nobody knows where":** Todd Pitman, "Uranium Flows from Congo Mines—in Burlap Bags," Associated Press, June 1, 2004, spokesman.com/stories/2004/jun/01/uranium-flows-from-congo -mines-in-burlap-bags/.

153 **wood and spice:** Richard Wachman, "How a Dubai-Based Businessman Has Emerged as a Key Player in the World's Cobalt Supplies," *Arab News*, April 19, 2018.

153 **now ran a company:** "Local European operators tend to establish themselves as 'brokers' with international companies," the Belgian academic Benjamin Rubbers observed in 2004, "because they are introduced into the political world that holds the keys to the Katangese underground." See Benjamin Rubbers, "La dislocation du secteur minier au Katanga (RDC)," *Politique Africaine* 1, no. 93 (2004): 37.

153 **a "charismatic child prodigy":** Gertner and Gertner v. Gertler, איתן אורנשטיין (בדימוס) פסק בוררות בפני כבוד הבורר, השופט (בדימוס) איתן אורנשטיין [Arbitration Award Before the Honorable Arbitrator, Judge (Ret.) Eitan Orenstein], Arb. Eitan Orenstein, April 22, 2024, 86.

153 **"Dan and Katumba have":** *Gertler*, 718.

154 **described as a "logistics genius":** Glencore trader, interview with the author, June 2019; and Glencore official, interview with the author, March 2022.

154 **"People were pouring in":** Norbert Nawiji, interview with the author, October 2019.

154 **antelope and elephant:** Aigle Mujinga (COMAKAT), interview with the author, March 2019.

154 **a Congolese investigative reporter:** Jonas Kiriko, "'New Energy' Rush Is Stripping DRC's Natural Assets," Oxpeckers, February 5, 2024, oxpeckers.org/2024/02/new-energy-rush-drc/.

155 **Gécamines technically possessed:** "Contrat de création de société entre la Générale de Carrières et de Mines et SAMREF-Congo pour l'exploitation du gisement de Mutanda ya Mukonkota," May 2001, No. 474/10300/SG/GC/2001, Kinshasa, Democratic Republic of the Congo.

155 **such junior companies:** Hence the mining engineer John Hays Hammond's 1911 admonishment that a mine is "a hole in the ground sold by a lying promoter to a stupid investor." See J. H. Hammond, "Good Advice to Mine Investors," *Gilpin Observer*, March 30, 1911.

155 **Brown agreed to sell:** Brown met with Hamze on May 6, 2004, in the South African city of Johannesburg to sign over his shares. After buying more shares the next year, Hamze gained control of the mine.

155 **coercion to wrest away:** Clowes and Wilson, "Glencore Faces New Legal Challenge Against Congo Cobalt Mine."

155 **died shortly after filing suit:** "Jugement RAC 122," 212–14.

156 **a mine in Zambia:** Javier Blas and Jack Farchy, *The World for Sale: Money, Power, and the Traders Who Barter the Earth's Resources* (Oxford University Press, 2021), 226.

156 **acquired in 1998:** "Government Gives Mopani Copper Mines a Friday Deadline," *Lusaka Times*, March 12, 2009, lusakatimes.com/2009/03/12/government-gives-mopani-copper-mines-a-friday-deadline/.

156 **"concentrates and copper":** John Ross and David de Vries, "Mufulira Smelter Upgrade Project" (Mopani Copper Mines, PLC, Glencore Technologies, 2005), glencoretechnology.com/.rest/api/v1/documents/3f912fe463bebc7f0de40789ad95b319/XTpaper_Mopani_MSUPyromet05.pdf.

156 **"a very smart guy":** Former Glencore trader, interview with the author, July 2019.

156 **known for heterogenite:** Rubbers, "La dislocation du secteur minier au Katanga," 27.

156 **"A comparison of official":** International Crisis Group, "Katanga: The Congo's Forgotten Crisis," *Africa Report No. 103*, January 9, 2006, 9, crisisgroup.org/sites/default/files/katanga-the-congo-s-forgotten-crisis.pdf.

156 **vowed to stop the minerals:** Moïse Katumbi Chapwe, interview with the author, May 2019. And see Claude Iguma Wakenge, *Eating the Congo: Unveiling State Governance of Copper and Cobalt Mining in Former Katanga* (Lambert Academic Publishing, 2019), 45.

157 **exported with special permits:** "Loi n° 007/2002 du 11 juillet 2002 portant code minier," *Journal Officiel*, no. spécial du 15 juillet 2002 (Kinshasa, 2002), 38, leganet.cd/Legislation/Droit%20economique/Code%20Minier/cd-codeminier.pdf.

157 **building processing plants in Congo:** Rubbers, "La dislocation du secteur minier au Katanga," 45n37.

157 **Accusations were rife:** International Crisis Group, "Katanga: The Congo's Forgotten Crisis," 6.

157 **Bazano began to signal:** It did continue to buy and process artisanally mined ore, however. See Andrew Gulley, "China, the Democratic Republic of the Congo, and Artisanal Cobalt Mining from 2000 Through 2020," *Proceedings of the National Academy of Sciences* 120, no. 26 (June 2023): art. e2212037120, doi/10.1073/pnas.2212037120.

157 **Hamze's companies finally:** Gécamines and Samref CONGO Sprl., "Contrat no. 474/10300/SG/GC/2001 du 16 mai 2001 conclu entre GECAMINES et SAMREF CONGO Sprl portant création de la société MUTANDA YA MUKONKOTA MINING (MUMI)," Annexe I au procès-verbal synthétique de la réunion extraordinaire du Conseil d'Administration du 3 décembre 2008, https://congomines.org/system/attachments/assets/000/000/322/original/PV-Dec-2008-KCC-DCP.pdf?1430928476.

157 **Tilwezembe was known:** Chantal Peyer and François Mercier, *Glencore in the Democratic Republic of Congo: Profit Before Human Rights and the Environment* (Swiss Catholic Lenten Fund, 2012), 12.

158 **"creating environmental problems":** Peyer and Mercier, *Glencore in the Democratic Republic of Congo*, 16.

158 **former associate of Hamze's:** John Sweeney, "Mining Giant Glencore Accused in Child Labour and Acid Dumping Row," *Observer*, April 14, 2012.

158 **"an extremely tough man":** Peyer and Mercier, *Glencore in the Democratic Republic of Congo*, 12.

158 **Ismaël's firm processed:** Peyer and Mercier, *Glencore in the Democratic Republic of Congo*, 12.

158 **firm had signed a contract:** Peyer and Mercier, *Glencore in the Democratic Republic of Congo*, 14.

158 **Bazano was long gone:** Tilwezembe *creuseur*, interview with the author, 2022.

Chapter 23: Into the Pits

159 **someone they loved:** According to the U.S. Centers for Disease Control and Prevention, malaria is the reason for 60 percent of all hospital visits in Congo. See "CDC in the Democratic Republic of the Congo," Centers for Disease Control and Prevention, 2022, archived December 16, 2022, at web.archive.org/web/20221219200136/https://www.cdc.gov/globalhealth/countries/drc/pdf/DRC_2022.pdf.

161 **"This man was a brother":** Moïse Katumbi Chapwe, interview with the author, May 2019.

161 **mother was a princess:** Cnaan Liphshiz, "Son of Greek Jewish Holocaust Refugee Now One of Most Powerful Leaders in Congo," *Times of Israel*, February 18, 2021.

161 **supplied UNITA rebels:** *Storm Clouds over Sun City: The Urgent Need to Recast the Congolese Peace Process*, ICG Africa Report No. 44 (International Crisis Group, May 2002), 12, crisisgroup.org/africa/central-africa/democratic-republic-congo/storm-clouds-over-sun-city-urgent-need-recast-congolese-peace-process.

161 **backed anti-Kabila rebels:** Colette Braeckman, "Moïse Katumbi, Katanga Big Boss," *Le Soir*, May 6, 2009.

161 **jumped on the Kabila bandwagon:** Katumbi, interview. "I lent the AFDL twenty million dollars, from me, in 1997, because I was a very prosperous businessman in Zambia," Katumbi told me when we spoke. He was still bitter about not being repaid. "They didn't even reimburse to me even ten percent of my money up to now."

161 **monopolistic food supplier:** David Pilling, "Is Trucking Tycoon Moïse Katumbi the Man to Rescue Congo?," *Financial Times*, July 7, 2017.

161 **In 2006, Katumbi:** Braeckman, "Moïse Katumbi, Katanga Big Boss."

161 **number had climbed:** "Katumbi: The Moses of Katanga," *African Business*, April 4, 2013.

161 **MCK was signed over:** Braeckman, "Moïse Katumbi, Katanga Big Boss."

162 **"keeping the money":** Joe Bavier, "Chinese Firms Face New Reality of Congo Mining," Reuters, August 10, 2007.

162 **All of a sudden:** Moïse Katumbi Chapwe, interview with the author, May 2019. When we spoke, Katumbi remembered how the export ban had been key to driving higher revenues for the province. At first there was resistance, he told me. "Everybody said, 'No, no. Maybe this governor wants the money, and not small money, big money,'" he recalled. A former minister in the regional government came and offered him a bribe. "I didn't accept that; I closed the company of these guys. I arrested him, and I said, 'Nobody's going to move out one kilogram of unprocessed material.' That made us more successful."

162 **profited from the minerals:** Bavier, "Chinese Firms Face New Reality of Congo Mining."

162 **headquarters in the Swiss:** Former Glencore employees, interviews with the author, 2019, 2020, 2022, 2023.

162 **whose eponymous founder:** Eric N. Berg, "Marc Rich Indicted in Vast Tax Evasion Case," *New York Times*, September 20, 1983.

162 **later controversially pardoned:** Lior Dattel and Ronit Domke, "Marc Rich, the Man Who Sold Iranian Oil to Israel," *Haaretz*, June 27, 2013.

162 **booted him out and formed:** Javier Blas and Jack Farchy, *The World for Sale: Money, Power, and the Traders Who Barter the Earth's Resources* (Oxford University Press, 2021), 127.

162 **They meant that the market:** They also meant that cobalt was a sideshow to copper, a metal that traders have called "Doctor Copper"; by following up- and downswings in demand for the red metal. Some say it is possible to make calls on the direction of the global economy. See "Why It Is Time to Retire Dr. Copper," Buttonwood, *Economist*, October 19, 2023.

163 **feared running afoul:** Glencore official, interview with the author, February 2022.

163 **a young Glencore employee:** Former Glencore trader, interview with the author, March 2022.

163 **"spend half the day there":** Former Glencore employee, interview with the author, November 2023.

163 **Gertners' lawyers would:** Michael J. Kavanagh, "Gertler, a 'King' in Congo, Describes Mine Payments in Arbitration Testimony," *Bloomberg*, July 13, 2025, https://www.bloomberg.com /news/articles/2025-07-14/billionaire-gertler-describes-congo-mine-payments-in-arbitration -testimony.

164 **"Almost everybody used agents":** Former Glencore trader, interview with the author, February 2022.

164 **two giant concessions:** Augustin Katumba Mwanke, *Ma vérité* (EPI, 2013), 203.

164 **The record of the arbitration:** While the arbitrator writes he has not discovered evidence of bribery (it was out of the scope of the proceedings to determine whether bribes were in fact paid), he states that "it was necessary to conceal the identity of Mr. Katumba and those acting on his behalf" (Gertner and Gertner v. Gertler, פסק בוררות בפני כבוד הבורר, השופט (בדימוס) איתן אורנשטיין [Arbitration Award Before the Honorable Arbitrator, Judge (Ret.) Eitan Orenstein], Arb. Eitan Orenstein, April 22, 2024, 691). On the question of bribes, the arbitrator is hardly conclusive. "On the face of it, payments to a private individual, even if he previously worked in government, are not bribery," he writes. "He is not the government official whose authority the bribe is intended to influence. The plaintiffs have not proven with convincing evidence that illegal payments were actually made to government officials in Congo, as they claim. Neither to President Kabila nor to any other person holding an official position." But Katumba was far from a private individual: He held the title of Itinerant Ambassador for much of the time period they were discussing and was elected as a parliamentary deputy in 2006.

165 **"What does he":** *Gertler*, 718.

165 **"local people":** *Gertler*, 718.

165 **Shell companies and code words:** "We acted according to the instructions he gave us," Gertler said at one point during the testimony (*Gertler*, 724). The businessmen used various code words: Jaynet Kabila, for example, was "our lady friend" (*Gertler*, 667), and Katumba was "local sources" (*Gertler*, 20). See also *Gertler*, 1,145: "We treat our friends with discretion," a Gertler associate explained at one point. "Katumba is a discrete man."

165 **"these two monkeys":** Former Glencore official, interview with the author, July 2022.

165 **Gertler kept asking for money:** Former Glencore official, interview with the author, February 2022.

165 **Forrest was better liked:** Former Glencore official, interview.

165 **Gertler advised Katumba:** Katumba, *Ma vérité*, 204.

166 **Katumba fell into a coma:** The passage reads oddly, especially because Katumba said he received treatment at the Amselia Medical Center in Tel Aviv. There is no record of an Amselia medical center in Israel. Perhaps it was merely a slip of the convalescent mind.

166 **"my life," Katumba writes:** Katumba, *Ma vérité*, 203.

166 **"a new family":** Katumba, 206.

166 **"despite everything that appears":** Katumba, 208.

166 **"just about money anymore":** Melissa Sanderson, interview with the author, August 2023.

166 **echoing the use:** See, for example, *A State Affair: Privatizing Congo's Copper Sector* (Carter Center, November 2017), cartercenter.org/resources/pdfs/news/peace_publications/democracy /congo-report-carter-center-nov-2017.pdf.

166 **"Wolves don't eat":** Claude Iguma Wakenge, *Eating the Congo: Unveiling State Governance of Copper and Cobalt Mining in Former Katanga* (Lambert Academic Publishing, 2019), 111.

166 **"a far cry from"**: *A State Affair,* 21.

166 **"He became the shadow"**: Benoit Nyemba Basali, conversation with the author, March 2019.

Chapter 24: Oozing Evil

167 **potentially radioactive ores:** Claude Iguma Wakenge, *Eating the Congo: Unveiling State Governance of Copper and Cobalt Mining in Former Katanga* (Lambert Academic Publishing, 2019), 87.

167 **elites who controlled:** "Congo-K: Les mines aux cœur des réseaux ethniques," *Africa Intelligence,* July 26, 2013, ingeta.com/wp-content/uploads/2014/02/Afr-mining-intelligence-novembre-2013.pdf.

168 **leased its office:** *Anvil Mining Limited and the Kilwa Incident: Unanswered Questions* (Rights and Accountability in Development, October 2005), 10, raid-uk.org/wp-content/uploads/2023/04/qq-anvil.pdf.

168 **leased machinery to the company:** *Katanga: The Congo's Forgotten Crisis,* Africa Report No. 103 (International Crisis Group, January 2006), 10, crisisgroup.org/africa/central-africa/democratic-republic-congo/katanga-congo-s-forgotten-crisis.

168 **a smattering of weapons:** *Anvil Mining Limited and the Kilwa Incident,* 10.

168 **nearby town of Kilwa:** *Katanga: The Congo's Forgotten Crisis,* 11.

168 **abetting the uprising:** *Sixteenth Report of the Secretary-General on the United Nations Mission in the Democratic Republic of the Congo,* S/2004/1034 (UN Security Council, December 2004), 4.

168 **At least seventy-three:** "Military Court Delivers a Not Guilty Verdict in Kilwa Trial," Rights and Accountability in Development (RAID), 2007, raid-uk.org/wp-content/uploads/2023/04/pr-verdict.pdf.

168 **struggle to control resources:** United Nations Organization Mission in the Democratic Republic of Congo (MONUC), *Report on the Conclusions of the Special Investigation into Allegations of Summary Executions and Other Violations of Human Rights Committed by the FARDC in Kilwa (Province of Katanga) on 15 October 2004* (MONUC, 2005), 11.

168 **Anvil denied wrongdoing:** Canadian Press, "Congolese Raise Mining Lawsuit in Supreme Court," CBC, March 26, 2012.

168 **Nkambo Gédéon Kyungu:** Junior Kanyiki, "Mitwaba: La présence de Gédéon Kyungu Mutanga et de son épouse confirmée pour 'déclarer l'indépendance du Katanga' (société civile)," *Enquête,* January 29, 2022, enquete.cd/2022/01/29/mitwaba-la-presence-de-gedeon-kyungu-mutanga-et-de-son-epouse-confirmee-pour-exiger-lindependance-du-katanga-societe-civile/.

169 **Gédéon's former spokesman:** Thierry Mukelekele, interview with the author, March 2019.

169 **pointed a finger at:** John Numbi, interview with the author, November 2023.

169 **two thousand fighters strong:** Christophe Rigaud, "RDC: Gédéon, le Mauvais Génie du Katanga," Mediapart, October 19, 2020, blogs.mediapart.fr/afrikarabia/blog/191020/rdc-gedeon-le-mauvais-genie-du-katanga.

169 **accused of cannibalism:** Eva Gilliam, "DRC: Mai Mai Leader Gedeon of Manono Territory—Known 'Good Guy,' Accused Cannibal," MONUC, April 14, 2004.

169 **"They sometimes had human fingers":** Eva Gilliam, interview with the author, September 2023.

169 **scare people away:** Gilliam, "DRC: Mai Mai Leader Gedeon of Manono Territory."

170 **"He's the most promising":** "Sifting Through a Dark Business," *Newsweek,* updated March 13, 2010, newsweek.com/sifting-through-dark-business-131907.

170 **"I should get a Nobel":** Franz Wild et al., "Gertler Earns Billions in Mine Deals as Congo Remains Poorest," *Bloomberg,* December 5, 2012, bloomberg.com/news/articles/2012-12-05/gertler-earns-billions-as-mine-deals-leave-congo-poorest.

170 **funded by the European Union:** Bryan Mealer, *All Things Must Fight to Live: Stories of War and Deliverance in Congo* (Bloomsbury, 2008), 100.

170 **"Key political and economic":** Cable "05KINSHASA731_a," April 29, 2005, 16:16 (Friday), WikiLeaks, wikileaks.org/plusd/cables/05KINSHASA731_a.html.

171 **a hefty profit:** Wild et al., "Gertler Earns Billions as Mine Deals Leave Congo Poorest."

171 **"grabbing and flipping":** James Wood et al., "Tokyo Sexwale and the DRC's Mr. Grab," *Mail & Guardian,* August 17, 2012.

171 **president would spend it:** John Prendergast and Sasha Lezhnev, "Five Reasons Why Biden's Move Against Corrupt Billionaire Dan Gertler Matters," Sentry, March 9, 2021, thesentry

.org/2021/03/09/5591/medium-op-ed-five-reasons-bidens-move-corrupt-billionaire-dan-gertler
-matters/.

171 **"The Congolese way":** Gertner and Gertner v. Gertler, פסק בוררות בפני כבוד הבורר, השופט (בדימוס) איתן אורנשטיין [Arbitration Award Before the Honorable Arbitrator, Judge (Ret.) Eitan Orenstein], Arb. Eitan Orenstein, April 22, 2024, 305.

171 **"I am a king":** Kavanagh, "Gertler, a 'King' in Congo, Describes Mine Payments in Arbitration Testimony."

171 **Belgian documentarian Thierry:** George Arthur Forrest in *Katanga Business*, directed by Thierry Michel (Les Films de la Passerelle, 2009).

172 **"I think we always":** Financier, interview with the author, May 2024.

172 **A fund manager:** Fund manager, interview with the author, April 2024.

172 **unnamed Congolese official:** Plea Agreement, United States v. OZ Africa Management GP, LLC, 16-CR-515 (NGG) (E.D.N.Y. 2016), Exhibit 3 (Statement of Facts), 5, justice.gov/criminal /criminal-fraud/file/900276/dl?inline.

172 **media outlets and NGOs:** *A State Affair: Privatizing Congo's Copper Sector* (Carter Center, November 2017), 11, cartercenter.org/resources/pdfs/news/peace_publications/democracy/congo -report-carter-center-nov-2017.pdf. For the subsequent civil lawsuit, see: Opinion and Order, Menaldi v. Och-Ziff Capital Management Group LLC, no. 1:2014cv03251 - Document 136 (S.D.N.Y. 2017), Background at 1, 6.

172 **$150 million worth:** Plea Agreement, Exhibit 3 at 9, *OZ Africa Management*, 16-CR-515 (NGG) (E.D.N.Y. 2016).

172 **$254 million in loans:** Securities and Exchange Commission v. Michael L. Cohen and Vanja Barros, Complaint, no. 17-CV-00430 (E.D.N.Y. Jan. 26, 2017), Summary at 6, 5-6, https://www .sec.gov/files/litigation/complaints/2017/comp-pr2017-34.pdf.

173 **"would be used to bribe":** Complaint, Summary at 6, Securities and Exchange Commission v. Michael L. Cohen and Vanja Barros, 17-CV-00430 (E.D.N.Y. Jan. 26, 2017).

173 **"a bigger picture":** Plea Agreement, Exhibit 3 at 10, *OZ Africa Management*, 16-CR-515 (NGG) (E.D.N.Y. 2016).

173 **"case of bribery":** Former Och-Ziff fund manager, conversation with the author, March 2021.

173 **disagreed and frequently squabbled:** "Our engineers had a shit fight with his engineers," the former Glencore official recalled. Glencore official, interview with the author, February 2022.

173 **"Alex bought a lot":** South African mine engineer, interview with the author, September 2019.

173 **A Gertler company soon acquired:** "After Criminal Complaint by Public Eye: Glencore Convicted Following Corrupt Mine Deals in the DRC," Public Eye, accessed August 22, 2024. www.publiceye.ch/en/topics/commodities-trading/after-criminal-complaint-by-public-eye -glencore-convicted-following-corrupt-mine-deals-in-the-drc.

174 **another Hamze company:** Clara Ferreira-Marques and Jonny Hogg, "Glencore Takes Control of Mutanda with $480 Million Deal," Reuters, May 22, 2012.

174 **the merger was:** "Fleurette and Glencore Complete Merger of Mutanda and Kansuki Mining Operations," Fleurette Group, PR Newswire, July 25, 2013, prnewswire.co.uk/news-releases /fleurette-and-glencore-complete-merger-of-mutanda-and-kansuki-mining-operations -216882041.html.

174 **having lost family money:** Former Glencore Official, author interview, February 2022. "We diluted the shit out of him," the former Glencore official told me. "In the end of the day, Dan didn't make that much money in Katanga." The official noted that Gertler was complaining of losing $200 million of his family's money on the deal. Neither Gertler nor his family members would speak to me for this book.

174 **He built a hospital and a park:** Bazano, "Response to the Article Published by PREMICONGO on the Subject Mining Exploitation, Environmental Management, and Social Responsibility of Mining Companies in Katanga," January 2013, 3–5, congomines.org/system/attachments /assets/000/000/477/original/MINING-in-DRC-Response-of-GROUPE-BAZANO-to-the -report-of-PREMICONGO-January-2013.FV_.English-1.pdf?1430928905.

176 **care who he was:** Melissa Sanderson, interview with the author, August 2023.

176 **Maltese planning applications:** The Government of Malta, *The Malta Government Gazette* no. 20,866, May 25, 2022, 8692, gov.mt/en/Government/DOI/Government%20Gazette/Documents /2022/05/Government%20Gazette%20-%2025th%20May%20PA.pdf.

176 **"I saw him out one night":** U.S. official, conversation with the author, April 2024.

176 **spies had fought hard:** See Susan Williams, *Spies in the Congo: America's Atomic Mission in World War II* (PublicAffairs, 2016).

Chapter 25: A New Cathode

177 **"Lithium-ion is the reason"**: Larry Edsall, *Chevrolet Volt: Charging into the Future* (Motorbooks, 2010), 49.

178 **Osaka City University**: Charles Murray, *Long Hard Road: The Lithium-Ion Battery and the Electric Car* (Purdue University Press, 2022), 192.

178 **researchers at the Argonne:** That story, and the subsequent development of an even more powerful type of NMC battery, is told in detail in *The Powerhouse*, an excellent 2015 book by Steve Levine.

178 **tetrahedron-shaped ion arrangements:** Steve Levine, *The Powerhouse: America, China, and the Great Battery War* (Penguin Books, 2015), 31–32.

178 **When Thackeray first suggested:** "Battery Heroes: Michael Thackeray," *Batteries International*, April 28, 2015, batteriesinternational.com/2015/04/28/michael-thackeray.

178 **"may have commercial significance":** Michael M. Thackeray, *Running with Lithium—Empowering the Earth: A Personal Journey* (Archway Publishing, 2019), 60.

178 **"a blizzard to pale":** Thackeray, *Running with Lithium*, 86.

178 **team threw together:** Michael M. Thackeray et al., Lithium Metal Oxide Electrodes for Lithium Cells and Batteries, U.S. Patent 6,677,082 B2, filed June 21, 2001, and issued January 13, 2004, patents.google.com/patent/US6677082B2/en; and "IMLB X—Lithium 2000," Tenth International Meeting on Lithium Batteries, May 28–June 2, 2000.

179 **a longer cycle life:** Seth Fletcher, "GM's New Battery Chemistry? It's Already in the Chevy Volt," *Popular Science*, January 7, 2011, popsci.com/cars/article/2011-01/gms-new-battery-chemistry-its-already-chevy-volt/.

179 **"superior" cathode material:** Levine, *Powerhouse*, 45.

179 **position of chief scientist:** Shirley Meng, interview with the author, May 2020.

179 **administration was "gambling away":** Deepa Seetharaman and Ayesha Rascoe, "Battery Maker A123 Systems Files for Bankruptcy," Reuters, October 12, 2012.

180 **CPI was positioning:** Bill Cooke, "Profile: Li-ion Provider Compact Power, Inc. Focusing on the Automotive and Vehicle Markets," *Green Car Congress*, February 17, 2019, greencarcongress.com/2009/02/profile-li-ion.html.

180 **Compact Power had also:** "Governor Granholm Joins President Obama in Celebrating New Advanced Battery Manufacturing Plant in West Michigan," Michigan.gov, July 15, 2010, michigan.gov/formergovernors/recent/granholm/press-releases/2010/07/15/joins-president-obama-in-celebrating-new-advanced-battery-manufacturing-plant-in-west-michigan.

180 **"stability and resistance":** "Governor Granholm Joins President Obama in Celebrating New Advanced Battery Manufacturing Plant in West Michigan."

180 **Jon Lauckner noted:** Fletcher, "GM's New Battery Chemistry?"

180 **NMC battery output:** Jörn Jürgens, "This Is Why NCM Is the Preferable Cathode for Li-ion Batteries," LG Energy Solution, August 29, 2019, lghomebatteryblog.eu/en/this-is-why-ncm-is-the-preferable-cathode-material-for-li-ion-batteries/.

180 **Goodenough's original LCO:** "Batteries and Electric Vehicles," Cobalt Institute, cobaltinstitute.org/essential-cobalt-2/powering-the-green-economy/batteries-electric-vehicles/. For the percentages of cobalt in batteries, there are various forms of NMC available today. NMC111 (or NMC333) contains 33 percent nickel, 33 percent manganese, and 33 percent cobalt. NMC532 has 50 percent nickel, 30 percent manganese, and 20 percent. NMC811, the most advanced form of the material available at the time of this writing, has 80 percent nickel, 10 percent manganese, and 10 percent cobalt. See, for example, J. J. Hocken, "Lithium Nickel Manganese Cobalt Oxide (NMC)," *Runaway Review*, April 17, 2023, mitsubishicritical.com/resources/blog/the-runaway-review/lithium-nickel-manganese-cobalt/; and "Lithium Cobalt Oxide (LiCoO2) Powder," Ossila, ossila.com/products/lithium-cobalt-oxide-powder.

181 **costliest material used in cathodes:** Vehicle Technologies Office, "FOTW #1228: Cobalt Is the Most Expensive Material Used in Lithium-ion Battery Cathodes," U.S. Department of Energy, March 7, 2022, energy.gov/eere/vehicles/articles/fotw-1228-march-7-2022-cobalt-most-expensive-material-used-lithium-ion.

181 **also pointed out cobalt's:** M. Stanley Whittingham, interview with the author, February 2020.

181 **took to calling NMC:** Sourav Mallick et al., "Low-Cobalt Active Cathode Materials for High-Performance Lithium-Ion Batteries: Synthesis and Performance Enhancement Methods," *Journal of Materials Chemistry A* 11 (2023): 3789.

181 **"it was scientists at Argonne":** Argonne scientist, interview with the author, September 2022.

181 **still had many advantages:** Nadim Maluf, "LCO, LFP, NMC . . . Cryptic Lives of the Cathode," Qnovo, December 6, 2014, qnovo.com/blogs/lco-lfp-nmc-cryptic-lives-of-the-cathode.

Chapter 26: Nickel from the Forest

182 **Chinese magazine *Caijing*:** "Tsingshan's Indonesia Morowali Industrial Park: Build, and They Will Come," *Caijing*, no. 30 (2019), archived March 12, 2023, at web.archive.org/web /20230325135815/business.hsbc.com.cn/en-gb/campaigns/belt-and-road/story-5.

183 **town of Falconbridge:** Maya Bilbao, "The Hidden History of Thomas Edison in Canada," *Northern Ontario Travel*, January 13, 2022, northernontario.travel/sudbury/thomas-edison -sudbury-ontario. Edison's quest for nickel also would have connections to Congo: The company that did manage to discover nickel in the area, Falconbridge, was later acquired by Xstrata, which later merged with Glencore, the owner of two of the biggest Congolese copper-and-cobalt mines. Another firm that tried to acquire Falconbridge was the International Nickel Corporation, INCO, which provided Robert Friedland with his first "big score," a nickel bonanza at Voisey's Bay. One of Friedland's most successful projects in recent years is Kamoa-Kakula, a giant copper mine near Kolwezi in Congo. See Jacquie McNish, *The Big Score: Robert Friedland, INCO, and the Voisey's Bay Hustle* (Doubleday Canada, 1998).

183 **"Indonesia's nickel strategy":** Isabelle Huber, *Indonesia's Nickel Industrial Strategy* (Center for Strategic and International Studies, December 2021), csis.org/analysis/indonesias-nickel-industrial -strategy.

183 **"We will control the world":** Wilda Asmarini and Fergus Jensen, "Construction on $4B Indonesia EV Battery Project Begins Jan 2019—Minister," Reuters, November 30, 2018.

183 **nickel laterite production:** Divya Karyza, "Luhut Launches Integrated EV Battery Plant in Morowali," *Jakarta Post*, September 27, 2022.

183 **Indonesia had become the largest:** Harry Dempsey and Mercedes Ruehl, "Indonesia Considers OPEC-Style Cartel for Battery Metals," *Financial Times*, October 31, 2022.

184 **dramatic spine of mountains:** Yuri Choi et al., "Geochemical and Mineralogical Characteristics of Garnierite from the Morowali Ni-Laterite Deposit in Sulawesi, Indonesia," *Frontiers in Earth Science* 9 (November 2021): 48. When Mount Soputan erupted in 2018, ash was thrown four thousand meters into the sky. See "Indonesia Rocked by Volcanic Eruption After Devastating Sulawesi Quake," *Bloomberg*, October 3, 2018.

184 **Connecticut was grafted:** Choi et al., "Geochemical and Mineralogical Characteristics."

185 **The limonite can be worked:** Gustaw Konopka et al., "Ni-Co Bearing Laterites from Halmahera Island (Indonesia)," *Applied Sciences* 12 (2022): 7586.

185 **various energy-intensive processes:** "Smeltering," *Economist*, October 26, 2024.

185 **Indonesia overtook Australia:** Harry Dempsey, "Indonesia Emerges as World's Second-Largest Cobalt Producer," *Financial Times*, May 9, 2023.

185 **clearly many differences:** In recent years, notably, David van Reybrouck, a historian and one of the most elegant contemporary chroniclers of Congo's history, has turned his attention to Indonesia's colonial story. It is perhaps a testament to how little Indonesia appears on the radar of the English-speaking world that van Reybrouck's powerful study of Indonesian history, *Revolusi: Indonesia and the Birth of the Modern World* (2020), was only translated into English after four years. Of course, to those seeking minerals there, the country is front and center of consciousness.

185 **exploited for at least:** Theo M. van Leeuwen and P. E. Pieters, "Mineral Deposits of Sulawesi," paper presented at the Proceedings of the Sulawesi Mineral Resources 2011 Seminar MGEI-IAGI, November 28–29, 2011, 7.

186 **marked by significant brutality:** In 2020, King Willem-Alexander of the Netherlands apologized for "excessive violence" during the colonial epoch, and in 2022, Mark Rutte, the Dutch prime minister, apologized for his country's role in slavery. "Dutch King Apologizes for 'Excessive Violence' in Colonial Indonesia," Reuters, March 10, 2020; and Jon Henley, "Dutch PM Apologises for Netherlands' Role in Slave Trade," *Guardian*, December 19, 2022.

186 **Dutch geologists realized:** E. C. Abendanon, as cited in Han Van Gorsel, "Geological Investigations of Sulawesi (Celebes) Before 1930," *Berita Sedimentologi* [Indonesian Journal of Sedimentary Geology] 48, no. 1 (2022): 91.

186 **the Japanese invaded:** Van Leeuwen and Pieters, "Mineral Deposits of Sulawesi," 8.

186 **Indonesia became independent:** Vincent Bevins, *The Jakarta Method* (PublicAffairs, 2020).

186 **a consortium headed:** "Our History in Indonesia," Vale, vale.com/indonesia/our-history

-in-indonesia. Researchers from the Canadian NGO Mining Watch reported in 2000 that researchers linked to INCO may have been searching for nickel there as early as 1964, in con-
travention of the Sukarno-era laws.

186 **Bahodopi's nickel laterite:** "INCO in Indonesia: A Report for Canadian People," Mining Watch,
January 7, 2000, miningwatch.ca/blog/2000/1/7/inco-indonesia-report-canadian-people.

186 **"a top producer of cathode":** "Materials Business," Sumitomo Metal Mining, smm.co.jp/en
/business/material/.

186 **the company displaced locals:** "INCO in Indonesia"; and Kathryn M. Robinson, *Stepchildren
of Progress: The Political Economy of Development in an Indonesian Mining Town* (State University of
New York Press, 1986), 178.

186 **"People lost their livelihoods":** Richard Kent, interview with the author, November 2024.

186 **As a popular protest:** See, for example, Emily Harwell, *Without Remedy: Human Rights Abuse
and Indonesia's Pulp and Paper Industry* (Human Rights Watch, January 2003), part IV, hrw.org
/report/2003/01/06/without-remedy/human-rights-abuse-and-indonesias-pulp-and-paper
-industry.

187 **perhaps tens of millions:** Numbers of transmigrants are difficult to come by, but census data
from 2010 estimated that there were more than five million transmigrants in Indonesia. See
Statistics Indonesia, 2010, "National Census," Badan Pusat Statistik (BPS), archived at https://
web.archive.org/web/20240911102329/https://sensus.bps.go.id/topik/tabular/sp2010/31/0/0.

Chapter 27: Crossing the River by Feeling for the Stones

189 **"I was born and raised":** Peter Zhou, interview with the author, May 2020.

190 **"has become a rival power":** Ray Dalio, *Principles* (Avid Reader Press, 2021), 13.

190 **hit a major industrial milestone:** Fang Yan and Jason Subler, "China Tops Global Auto Market in 2009, Challenges Ahead," Reuters, January 11, 2010.

190 **In the year after:** Jonathan Watts, "Beijing Keeps Olympic Restrictions on Cars After Air
Quality Improves," *Guardian*, April 9, 2009.

190 **a former auto engineer:** Zeyi Yang, "How Did China Come to Dominate the World of Electric Cars?," *MIT Technology Review*, February 21, 2023.

191 **fewer than five hundred:** Yang, "How Did China Come to Dominate the World of Electric
Cars?"

191 **provide individual subsidies:** Yang, "How Did China Come to Dominate the World of Electric Cars?"

191 **"more affordable batteries built here":** Tad Friend, "Plugged In," *New Yorker*, August 17, 2009.

191 **Obama administration's American Recovery:** Ayesha Rascoe, "Battery Maker A123 Got
U.S. Funds as It Sought Bankruptcy," Reuters, November 16, 2012. Among the companies that
received the largest tranches of funding was A123, which filed for bankruptcy in 2012. The
other leading firm, Johnson Controls, spun off its battery division to a private equity fund in
2019.

192 **there were up to:** Andrew Erickson and Gregory Collins, "Electric Bikes Are China's Real
Electric Vehicle Story," *China SignPost*, November 7, 2011, chinasignpost.com/wp-content/uploads
/2011/11/China-SignPost-49_E-bikes-are-Chinas-real-electric-vehicle-story_20111107.pdf.

192 **e-bikes cost commuters:** Erickson and Collins, "Electric Bikes Are China's Real Electric Vehicle Story."

192 **"For each mile traveled":** J. D. Goodman, "An Electric Boost for Bicyclists," *New York Times*,
January 31, 2010.

193 **"electric two-wheelers are so":** Akshat Rathi, "To Master Electric-Car Manufacturing,
China Started with Bikes," *Quartz*, December 14, 2018, qz.com/1495191/to-master-electric-car
-manufacturing-china-started-with-bikes.

193 **had called "crossing the river":** The phrase is often attributed to Deng Xiaoping, but it in
fact has its roots in a Chinese folk saying, "Crossing the river by feeling the stones—take a
steady step, then take another step," that was adopted by the Chinese revolutionary leader
Chen Yun. "摸着石头河"的" [The Origin of "Crossing the River by Feeling the Stones"], 人
民网 [*People's Daily Online*], April 12, 2018, cpc.people.com.cn/n1/2018/0412/c69113-29921565
.html.

193 **Elon Musk may have scoffed:** "Back in 2011 Tesla CEO Musk Mocked Chinese Competitor
BYD," video, *Bloomberg*, December 27, 2023, bloomberg.com/news/videos/2023-12-27/in-2011
-tesla-ceo-musk-mocked-chinese-competitor-byd-video.

Chapter 28: China's Answer to Dr. Zee

194 **"small upstart Chinese automaker"**: Matt Hardigree, "Detroit Auto Show: World Exclusive Surreal, Illegal Test Drive of Chinese Hybrid Through Cobo Arena," *Jalopnik*, January 15, 2008, jalopnik.com/detroit-auto-show-world-exclusive-surreal-illegal-tes-344806.

194 **BYD had grown**: BYD Company Limited, *Green Tech for Tomorrow, Annual Report 2008* (BYD Inc., 2009), 4.

194 **"green tech for tomorrow"**: BYD Company Limited, *Green Tech for Tomorrow, Annual Report 2008*, 1.

195 **"Dr. Chuanfu is China's"**: Hardigree, "Detroit Auto Show."

195 **Increased subsidies of**: Elizabeth Economy, "China's Round Two on Electric Cars: Will It Work?," Council on Foreign Relations, April 17, 2014, cfr.org/blog/chinas-round-two-electric-cars-will-it-work.

195 **triple the annual**: "Statistical Communiqué of the People's Republic of China on the 2009 National Economic and Social Development," National Bureau of Statistics of China, January 26, 2010, stats.gov.cn/english/NewsEvents/201002/t20100226_26295.html.

195 **government went even further**: "车 优惠政策延长 减免总 达5200亿元——精准施策 助新能源汽车 量提" [The Extension of the Preferential Policy on Vehicle Purchase Tax Is Expected to Reduce and Exempt a Total Of 520 Billion Yuan—Precise Measures to Help New Energy Vehicles Expand And Improve Quality], China Government Network, June 25, 2023 https://www.gov.cn/zhengce/202306/content_6888094.htm#:~:text=%E8%B4%A2%E6%94%BF%E9%83%A8%E3%80%81%E7%A8%8E%E5%8A%A1%E6%80%BB%E5%B1%80%E3%80%81%E5%B7%A5%E4%B8%9A,%E5%8F%91%E5%B1%95%E7%9A%84%E9%87%8D%E8%A6%81%E6%94%BF%E7%AD%96%E5%B7%A5%E5%85%B7%E3%80%82.

195 **China produced only**: "5,579 Electric Cars Sold in China in 2011," REVE, January 16, 2012, evwind.es/2012/01/16/5579-electric-cars-sold-in-china-in-2011/15911.

195 **Among the plan's**: "CHINA: 12th Five-Year Plan (2011–2015) for National Economic and Social Development," Asia Pacific Energy Portal, 2011, policy.asiapacificenergy.org/node/37.

195 **Such a figure**: "New Energy Vehicles Enjoyed a High-Speed Growth," China Association of Automobile Manufacturers, January 20, 2016, archived January 28, 2016, at web.archive.org/web/20160128212046/http://www.caam.org.cn/AutomotivesStatistics/20160120/1305184260.html.

195 **increase by 223 percent**: James Ayre, "China Electric Car Sales Increased 223% in 2015," *EV Obsession*, March 7, 2016, https://evobsession.com/1-4-china-auto-market-2015/.

195 **Three writers at**: Christopher Marquis et al., "China's Quest to Adopt Electric Vehicles," *Stanford Social Innovation Review*, Spring 2013, 53.

196 **to have been "dismal"**: Economy, "China's Round Two on Electric Cars."

196 **home to two million**: "EVs in China Will Account for 40% of Sales, 50 Million Vehicles and 200 TWh Demand in 2030," Global Sustainable Electricity Partnership, globalelectricity.org/success-story/evs-in-china-will-account-for-40-of-sales-50-million-vehicles-and-200-twh-demand-in-2030/.

196 **China was *mandating***: Nancy W. Stauffer, "China's Transition to Electric Vehicles," *MIT News*, April 29, 2021, news.mit.edu/2021/chinas-transition-electric-vehicles-0429.

197 **four times what the market**: Daniel Ren, "Overcapacity in China's EV Battery Industry to Reach Four Times Demand by 2025, Putting Small Players at Risk," *South China Morning Post*, July 20, 2023. A gigawatt-hour is a billion watt-hours. Watt-hours are units that are used to express energy in relation to time. Therefore, in terms of batteries, a gigawatt-hour is the amount of energy that a set of batteries with the combined power of a gigawatt produce in an hour.

197 **Chorzempa blamed this**: Martin Chorzempa, interview with the author, December 2022.

197 **34,637 charging points**: Ross Lydall, "London to See Surge in Ultra-Rapid Charging Points for Electric Cars—As Experts Gather for Standard's First Plug It In Summit," *Evening Standard*, November 24, 2022.

197 **electric-vehicle dream**: "Best EV Charging States in United States," PlugShare, July 29, 2023, plugshare.com/directory/us.

197 **1.26 million electric**: "Now We Know Who Is Really Buying Electric Vehicles in China," *Automotive News China*, January 28, 2019.

197 **in a single month**: Blagojce Krivevski, "China Is First Country to Sell Over a Million Elec-

tric Vehicles in a Month," ElectricCarsReport, September 13, 2024, electriccarsreport.com
/2024/09/china-is-first-country-to-sell-over-a-million-electric-vehicles-in-a-month/.

198 **"If its bet succeeds"**: Akshat Rathi, "Five Things to Know About China's Electric-Car Boom,"
Quartz, January 8, 2019, qz.com/1517557/five-things-to-know-about-chinas-electric-car-boom.

Chapter 29: The Deal of the Century

200 **The president's policy rested**: The French slogan that Kabila used, *"Cinq chantiers,"* technically means "five construction sites," but I have gone with a translation used in various reports
of the time in order to better render the sense of Kabila's words.

200 **"Congo will surprise"**: Colette Braeckman, "Joseph Kabila: 'Le Congo va surprendre, car il
se redressera beaucoup plus vite que prévu,'" *Le Soir*, November 16, 2006.

200 **slow to materialize**: *China and Congo: Friends in Need* (Global Witness, March 2011), 11, archiv
.kongo-kinshasa.de/dokumente/ngo/gw_rep_0311_en.pdf.

200 **"There he discovered that China"**: Jean Mpisi, *La Sicomines* (L'Harmattan, May 2020), 21.

200 **one of the more unpleasant**: In 2011, I worked on a story involving these debts for *The Guardian*.
It was my first time working on issues relating to the Democratic Republic of the Congo. See Greg
Palast et al., "Vulture Funds Await Jersey Decision on Poor Countries' Debts," *Guardian*, Novem-
ber 15, 2011, theguardian.com/global-development/2011/nov/15/vulture-funds-jersey-decision.

201 **$6 billion infrastructure**: Mpisi, *La Sicomines*, 119.

201 **tons of gold**: Marie-France Cros, "RDC: Le contrat chinois est modifié," *La Libre*, November
16, 2009.

201 **"We did not explain"**: Barnabé Kikaya Bin Karubi, interview with the author, June 2025.

201 **Surprisingly for such**: *The Backchannel: State Capture and Bribery in Congo's Deal of the Century*
(Sentry, November 2021), 7, thesentry.org/wp-content/uploads/2021/11/TheBackchannel
-TheSentry-Nov2021.pdf.

202 **"working with babies"**: Simon Clark et al., "China Lets Child Workers Die Digging in Congo
Mines for Copper," *Bloomberg*, July 23, 2008.

203 **company related to Huayou**: "Chen Xuehua: Minor Metal, Big World," Huayou Holding,
March 18, 2019, en.huayouholding.com/news/503.html. (Huayou declined to reply to questions for this book, but I include answers to a few questions put to Huayou during fact-checking
for a piece I wrote for *The New Yorker*.)

203 **working as a minor**: "的华友" [*Advancing Huayou*], Huayou Cobalt, huayou.com/about/culture
/video/10.

203 **the company website explains**: "Chen Xuehua: Minor Metal, Big World."

203 **with $3.19 million**: Si Chen, "The Emerging Role of Chinese Transnational Corporations as
Non-State Actors in Transnational Labour Law: A Case Study of Huayou Cobalt in the Global
Cobalt Supply Chain," *Journal of Asian Sociology* 50, no. 1 (March 2021): 149.

203 **traditional homes connected**: Scott Fagerstrom, "Tongxiang's Humble Silkworm a Symbol
of China's Culture, Trade," *China Daily*, April 17, 2019, chinadaily.com.cn/cndy/2019-04/17
/content_37459165.htm.

204 **"The source of raw"**: "的华友" [*Advancing Huayou*].

204 **a $4 million loan**: *2023 Semi-Annual Report of Huayou Cobalt* (Huayou Cobalt, August 2023), 152–53.

204 **It also gave charitable**: "Huayou ESG Series Report (Part 3): With Great Power Comes Great
Responsibility," Huayou Cobalt, huayou.com/news/corporate-news/127.html.

204 **almost $8.7 billion**: "Steady Progress! Huayou Cobalt Continues Its Third Consecutive Year
on China's Top 500 Private Enterprises List," Huayou Cobalt, October 22, 2024, huayou.com
/en/news/corporate-news/256.

204 **"complete industrial chain"**: "Chen Xuehua: Minor Metal, Big World."

204 **"not develop sustainably"**: "的华友" [*Advancing Huayou*].

204 **"Taking the flag over there"**: *Keep Moving Forward with Your Dreams: Huayou Cobalt 2023
Cultural Case Story Collection* (Zhejiang Huayou, 2023), 67–69, online (Mandarin version) at
yunzhan365.com/basic/51-100/34178235.html.

204 **"To be bestowed"**: "的华友" [*Advancing Huayou*].

205 **always left the office**: *Keep Moving Forward with Your Dreams*, 67–69.

205 **"domestic-garden-style residential areas"**: *Keep Moving Forward with Your Dreams*, 68.

205 **"on the islands of Indonesia"**: *Keep Moving Forward with Your Dreams*, 66.

205 **ideas about strength**: *Keep Moving Forward with Your Dreams*, 60.

205 **"With the dream of lithium"**: *Keep Moving Forward with Your Dreams*, 164.

205 **"a global leader in lithium-ion"**: "的华友" [*Advancing Huayou*].

206 **Xi Jinping's speeches and edicts**: "Returning from Duty! Chairman Chen Xuehua Immediately Shares the 'Good Voice' of the Two Sessions," Huayou Cobalt, March 21, 2024, huayou .com/en/news/corporate-news/173.

206 **"Huayou's success is inseparable"**: "的华友" [*Advancing Huayou*].

206 **Chen understood that**: "Returning from Duty! Chairman Chen Xuehua Immediately Shares the 'Good Voice' of the Two Sessions."

206 **"God rewards the diligent"**: "Chen Xuehua: Minor Metal, Big World."

206 **"The almost unanimous sentiment"**: Mpisi, *La Sicomines*, 91.

206 **"a spanner in the works"**: *Le Monde*, March 26, 2008, as cited in Mpisi, *La Sicomines*, 92.

206 **might be to Congo's advantage**: Stefaan Marysse and Sara Geenen, "Les contrats chinois en RDC: L'impérialisme rouge en marche?," in *L'Afrique des Grands Lacs: Annuaire 2007–2008* (L'Harmattan), 311.

206 **China might saddle Congo**: At the end of the negotiations, $3 billion was to be repaid at 4.4 percent interest, $2.3 billion at 6.1 percent interest, and $1.07 billion at 0 percent interest.

206 **would be difficult to sell**: *China and Congo: Friends in Need*, 20.

206 **KCC was the result**: Rebecca Bream, "A Bid for Front-Line Command in Africa," *Financial Times*, November 7, 2007, ft.com/content/eec69b16-8d6e-11dc-a398-0000779fd2ac.

207 **"price to pay"**: *A State Affair: Privatizing Congo's Copper Sector* (Carter Center, November 2017), 44, cartercenter.org/resources/pdfs/news/peace_publications/democracy/congo-report-carter -center-nov-2017.pdf.

207 **Global Witness described**: *China and Congo: Friends in Need*, 24.

207 **Du represented himself as**: *The Backchannel*, 9.

207 **"shell company at the center"**: *The Backchannel*, 3.

207 **the Lubumbashi jurist**: Marcel Yabili, *Chine-RD Congo*, vol. 2, *Il Manque un Détail!* (Les Impliqués Éditeurs, 2022), 151.

208 **"the power behind the throne"**: Cable "09KINSHASA1084_a," U.S. Embassy Kinshasa, December 15, 2009, wikileaks.org/plusd/cables/09KINSHASA1084_a.html.

208 **"the new universe"**: Augustin Katumba Mwanke, *Ma vérité* (EPI, 2013), 196.

208 **"the President's strategic constructions"**: Katumba Mwanke, *Ma verité*, 198–99.

208 **catapulted out of his seat**: Jonny Hogg, "Death of Kabila Deal-Maker Leaves Void in Congo," Reuters, February 13, 2012.

209 **"Our friend is happy"**: Gertner and Gertner v. Gertler, פסק בוררות בפני כבוד הבורר, השופט (בדימוס) איתן אורנשטיין [Arbitration Award Before the Honorable Arbitrator, Judge (Ret.) Eitan Orenstein], Arb. Eitan Orenstein, April 22, 2024, 1,195.

209 **"I will have to help"**: Plea Agreement, Exhibit 3 at 18, United States v. OZ Africa Management GP, LLC, 16-CR-515 (NGG) (E.D.N.Y. 2016), justice.gov/criminal/criminal-fraud/file /900276/dl?inline.

PART 4: CONSOLIDATION

211 **"You have all probably heard"**: John Hays Hammond, "Good Advice to Mine Investors," *Gilpin Observer*, March 30, 1911.

Chapter 30: No Such Thing as Death

214 **"I developed a network"**: André, interview with the author, March 2022. I am withholding André's full name because of his age.

214 **"There is food"**: I conducted interviews with eight children at Pamoja, a Still I Rise school in Kolwezi, Congo, in March 2022.

215 **"I am a Black Man"**: "Poem," unpublished poem, 2020.

215 **Pamoja's stated aim**: "Still I Rise Academy: Kolwezi," stillirise.org/en/our-schools/emergency -schools/congo/.

217 **a geologist who has covered**: Uwe Naeher, interview with the author, March 2023.

218 **"Until now [Musompo]"**: Many traders remained frustrated that Musompo was not operational by 2025. Norbert Kalenga, text message to the author, April 2025.

220 **released a landmark report**: *"This Is What We Die For": Human Rights Abuses in the Democratic Re-*

public of the Congo Power the Global Trade in Cobalt (Amnesty International, January 2016), amnesty
.org/en/documents/afr62/3183/2016/en/.

220 **the wake of the report:** Todd C. Frankel, "Apple Cracks Down Further on Cobalt Supplier in
Congo as Child Labor Persists," *Washington Post*, March 3, 2017.

Chapter 31: Kasulo

221 **with the two other men:** Yannick Mputu, interview with the author, September 2019. They
had come to Lubumbashi because of the lack of work in Likasi. "When Gécamines closed, we
had to go to Kolwezi," Yannick said.

223 **"There was a lot":** Mputu, interview.

224 **"if it had been bombed":** Michael Kavanagh, "This Is Our Land," *New York Times*, January
26, 2019.

225 **a thousand holes:** "Lualaba: 5 morts dans un incendie à Kasulo," Radio Okapi, September 5,
2019, radiookapi.net/2015/09/05/actualite/societe/lualaba-5-morts-dans-un-incendie-kasulo.

226 **first visited Kolwezi:** "His Excellency Governor Richard Muyej," *Mining and Business*, 2018,
en.miningandbusiness.com/%20interviews/his-excellency-governor-richard-muyej-71.

226 **fifteen thousand *creuseurs*:** Email from Bryce Lee, Huayou CSR, May 19, 2021.

Chapter 32: Bare Branches

230 **"We can say that thirty":** Vidiye Tshimanga Tshipanda, interview with the author, May 2023.

231 **"When there was a problem":** Ady Nawezi, interview with the author, April 2022.

234 **babies being abandoned:** As a 2009 study in the *British Medical Journal* puts it, "Sex selective
abortion accounts for almost all the excess males." See Wei Xing Zhu et al., "China's Excess
Males, Sex Selective Abortion, and One Child Policy: Analysis of Data from 2005 National Intercensus Survey," *BMJ* (Clinical Research Edition) 338, no. 7700 (April 2009): b1211, 923.

234 **Kalala's home province:** Zhu et al., "China's Excess Males, Sex Selective Abortion, and One
Child Policy," 922.

234 **left the poorest men:** The global health scholar Thérèse Hesketh once wrote that, in China,
"when there is a shortage of women in the marriage market, women can 'marry up'; this inev-
itably leaves the least desirable men with no marriage prospects." She continued: "So in many
communities today there are growing numbers of young men in the lower echelons of society
who are marginalised because of lack of family prospects and who have little outlet for sexual
energy." See Thérèse Hesketh, "Too Many Males in China: The Causes and the Consequences,"
Significance 6, no. 1 (2009): 9–13.

234 **"all over Africa":** Deborah Brautigam, interview with the author, November 2022.

235 **Sata won the 2011:** "Zambian President Urged to Protect Workers at Chinese-Owned Mines,"
Associated Press, November 3, 2011.

235 **"a purported network":** Leyland Cecco, "'A Brazen Intrusion': China's Foreign Police Stations
Raise Hackles in Canada," *Guardian*, November 7, 2022.

235 **"uncover the mystery":** "Weixin Official Account," "点左上方，注我," accessed April 11,
2025, https://mp.weixin.qq.com/s/yUC7RWQnFirsNiAl50Dw0g.

235 **official integration policy:** Germain Ngoie Tshibambe, "Sans Chinatown? L'intégration des
migrants chinois à Lubumbashi (DRC)," *Revista da Faculdade de Direito da UFMG* 63 (2013): 266.

235 **Chinese community website has posted:** "Inventory of Recent Violent Incidents Against
Chinese in the Mining Provinces of Haut-Katanga and Lualaba," Weixin, June 20, 2022,
mp.weixin.qq.com/s/yUC7RWQnFirsNiAl50Dw0g.

Chapter 33: Papa Solution

237 ***Creuseurs* from an approved:** Email from Bryce Lee, May 19, 2021. In response to emailed
questions, a Huayou spokesperson said that Congo Dongfang followed international standards
in developing Kasulo.

237 **A consortium of local:** Schadrak Mukad Mway End Naw, interview with the author, October 2019.

239 **Korean conglomerate LG:** DNV GL Business Assurance Ltd., *Audit Report on Congo Dongfang
International Mining sarl* (LG Chem, May 2018), lgchem.com/upload/file/principle/Audit_Report
_CDM_2018.pdf.

241 **"My percentage of the investment"**: Clément Fayol et al., "'I Will Take My Percentage': Congolese Presidential Adviser Caught on Tape Negotiating Corrupt Mineral Deal," OCCRP, September 15, 2022. In December 2022, a U.S. anti-corruption nonprofit called the Sentry published a long report detailing opaque dealings by a Canadian firm named Ivanhoe that had worked with Tshimanga in Congo. Ivanhoe denied wrongdoing. See *Gaming the System: How a Canadian Mining Giant Undermined the Law in the DRC* (Sentry, December 2022), thesentry.org/reports/gaming-the-system/.

241 **"These people just steal"**: Charlotte Cime Jinga, interview with the author, March 2022.

241 *Grands Lacs* **magazine:** "Richard Muyej: Désigné Meilleur Gouverneur 2018 en RDC," *Grands Lacs* 111 (February/March 2018): 47–48.

241 *Forbes* **praised Muyej's:** "Best of Africa," *Forbes*, 2018, forbes.com/custom/2018/10/16/best-of-africa-2018/.

242 **The governor's son:** On LinkedIn, one of his employees describes himself as the site supervisor of the Congo Dongfang mine. See Joel Kamoj, "Joel Kamoj," LinkedIn, linkedin.com/in/joel-kamoj-b8302012b/.

242 **"Kasulo is a village":** Jean-Jacques Kayembe, interview with the author.

243 **A representative for Muyej:** Representative for Muyej, interview with the author, May 2021.

243 **"a good cohabitation":** "Lualaba: La Jeunesse Ruund adresse une lettre ouverte au vice-gouverneur l'appelant à 'revenir au bon sens,'" *Politico.cd*, April 2, 2021, politico.cd/encontinu/2021/04/02/lualaba-la-jeunesse-ruund-adresse-une-lettre-ouverte-au-vice-gouverneur-lappelant-a-revenir-au-bon-sens.html/80729/.

243 **Masuka was listed:** "Extrait du procès verbal de l'assemblée générale du 15 Novembre 2012," Lubumbashi, November 15, 2012.

243 **the cooperative was accused:** Olivier Liffran and Joan Tilouine, "DRC: Kazakh Miner ERG Clashes with Presidential Family in Lualaba," *Africa Intelligence*, July 28, 2023, africaintelligence.com/central-africa/2023/07/28/kazakh-miner-erg-clashes-with-presidential-family-in-lualaba,110008413-art.

243 **Katangese political analyst:** Serge Noël Ngoy Mwanabute, *Toute la vérité sur les guerres de l'est du Congo-Kinshasa: Comment mettre fin définitivement à ces guerres?* (Éditions Universitaires Européennes, 2023), 343.

243 **"That's all Tshisekedi's government":** Martin Fayulu, interview with the author, April 2022.

243 **Banza and Masuka to artisanal:** Olivier Liffran, "Tshisekedi Clan Involved in Former Katanga Province's Mining Wild West," *Africa Intelligence*, February 2, 2024. africaintelligence.com/central-africa/2024/02/05/tshisekedi-clan-involved-in-former-katanga-province-s-mining-wild-west,110155456-ge0.

244 **"She's just a front":** Bradley Barnett, interview with the author, June 2023.

244 **fallen out with Tshisekedi:** "Despite the Ongoing M23 war, Tshisekedi Ousts Key Katanga Figure," *Africa Intelligence*, March 12, 2025, africaintelligence.com/central-africa/2025/03/12/despite-the-ongoing-m23-war-tshisekedi-ousts-key-katanga-figure,110386022-art.

244 **"Banza is at the heart":** Congolese analyst, WhatsApp message to author, March 13, 2025; Pistons Brothers, "Billionaire Capo in Monaco #monaco #billionaire #luxury #girls," TikTok video. Soon after I received this, Pistons Brothers, a TikTok account that focuses on cars and videos of the wealthy entering and exiting the Hôtel de Paris, appears to have removed this video.

Chapter 34: The Steve Jobs of Metals

246 **a complex proposition:** For more on other jurisdictions where nickel is being mined, see Ben Crair's excellent piece about New Caledonia, "The Island Where Environmentalism Implodes," *New Yorker*, November 23, 2024, newyorker.com/news/the-weekend-essay/the-island-where-environmentalism-implodes.

246 **ban was challenged:** In 2022, the World Trade Organization sided with the European Union, saying that Indonesia was flouting international rules. See "WTO Backs EU in Nickel Dispute, Indonesia Plans Appeal," Reuters, November 30, 2022.

246 **"Our trade balance":** Office of Assistant to Deputy Cabinet Secretary for State Documents & Translation, "President Jokowi Pushes for Down Streaming of Mining Products," Cabinet Secretariat of the Republic of Indonesia, November 30, 2022, setkab.go.id/en/president-jokowi-pushes-for-down-streaming-of-mining-products/.

246 **The swings in the nickel:** Alfred Cang et al., "Tycoon Whose Bet Broke the Nickel Market Walks Away a Billionaire," *Bloomberg*, July 6, 2022.

246 **"the Steve Jobs of metals"**: Neil Hume et al., "Xiang Guangda, the Metals 'Visionary' Who Brought the Nickel Market to a Standstill," *Financial Times*, March 11, 2022.

246 **"The loss has been roughly"**: Alfred Cang et al., "Tycoon Whose Bet Broke the Nickel Market Walks Away a Billionaire."

247 **the Huayue nickel plant:** "的华友" [*Advancing Huayou*], Huayou Cobalt, huayou.com/about /culture/video/10.

247 **article praising Xiang:** "Xiang Guangda's 3 Short Stories Will Show You How Qingshan Steel Has Become One of the World's Top 500!," Tsingtuo Group, August 22, 2020, tsingtuo .com/companyNews/280.html.

247 **to dominate Indonesian:** Evi Fitriani, "Indonesia's Wary Embrace of China," Mercator Institute for China Studies (MERICS), merics.org/en/indonesias-wary-embrace-china.

247 **"I do think Jokowi"**: William Yuen Yee, interview with the author, January 2023.

248 **The message that Xiang:** "Xiang Guangda's 3 Short Stories Will Show You How Qingshan Steel Has Become One of the World's Top 500!"

Chapter 35: Dirty Nickel

249 **Primarily for nickel:** Henry Sanderson, "Nickel Drama Highlights Tsingshan's Role in Energy Transition," Dialogue Earth, May 13, 2022, chinadialogue.net/en/business/nickel-drama -highlights-tsingshans-role-in-energy-transition/.

249 **a total of $29 billion:** Erwida Maulia, "Dirty Metals for Clean Cars: Indonesian Nickel Could Be Key to EV Battery Industry," *Nikkei Asia*, October 19, 2022, asia.nikkei.com/Spotlight/The -Big-Story/Dirty-metals-for-clean-cars-Indonesian-nickel-could-be-key-to-EV-battery -industry.

250 **IMIP was burning:** Verda Nano Setiawan, "Didesak Pemegang Saham, Kawasan Industri Morowali Bangun PLTS 150 MW," *Katadata*, September 16, 2021, katadata.co.id/ekonomi-hijau /energi-baru/6141804a08964/didesak-pemegang-saham-kawasan-industri-morowali-bangun -plts-150-mw.

250 **IMIP focused on producing:** Maulia, "Dirty Metals for Clean Cars."

250 **"The practice of nickel"**: "Sufferings of Residents and Environment Behind the Business Transactions of Tesla and Chinese Companies in Indonesia," JATAM, August 10, 2022, https://www .jatam.org/en/sufferings-of-residents-and-environment-behind-the-business-transactions -of-tesla-and-chinese-companies-in-indonesia.

251 **three environmental groups:** "Press Release: Aksi Ekologi dan Emansipasi Rakyat (AEER), Jatam Sulawesi Tengah and Yayasan Tanah Merdeka," AEER, archived April 14, 2023, at web .archive.org/web/20230414055416/https://aeer.info/en/hua-pioneers-steps-to-cancel-request -for-permit-to-dispose-of-tailings-in-the-morowali-sea-should-be-the-standard-for-all -companies/.

252 **brings much-needed jobs:** I spoke to an IMIP rep in Morowali, but he refused to comment for the record.

252 **the plant has created:** Alfian al-Ayubby, "The Workers Paying the Price for Indonesia's Nickel Boom," *New Mandala*, June 14, 2024, newmandala.org/the-workers-paying-the-price -for-indonesias-nickel-boom/.

253 **"productivity of the factory up"**: Sahlun Sahidi, interview with the author, December 2022.

253 **The strikes went ahead:** Muhammad Rushdi et al., *Fast and Furious for Future: The Dark Side of Electric Vehicle Battery Components and Their Social and Ecological Impacts in Indonesia* (Rosa-Luxemburg -Stiftung, 2021), 87–88, https://www.rosalux.de/fileadmin/images/publikationen/Studien /Fast_and_Furious_for_Future.pdf.

Chapter 36: A Young Continent

254 **in advising Glencore:** "Glencore Reaps $7 Billion from Las Bambas Mine Sale to China's MMG," Reuters, August 1, 2014, reuters.com/article/uk-glencore-mmg-lasbambas-idUKKBN0 G12XN20140801.

254 **the Australian firm that had:** Anvil owned a 70 percent controlling stake in AMCK, a subsidiary that Moïse Katumbi Chapwe set up to control the mine. See "Annual Information Form for Financial Year Ended December 31, 2004," Anvil Mining Limited, 2004, 4.

254 **MMG stopped using:** Stanis Bujakera Tshiamala, "RDC: Kinsevere, la mine de la discorde," *Jeune Afrique*, April 24, 2020.

255 **Makola told us:** Yannick Makola Kasonde, MMG Mining, visit to MMG Kinsevere Mine, Haut-Katanga, Democratic Republic of the Congo, 2019.

256 **Huayou Cobalt deemed:** *2020 Environmental, Social & Governance Report* (Huayou Cobalt, 2021), 4, huayou.com/Public/Uploads/uploadfile2/files/20230824/2020EnvironmentalSocial GovernanceReportHuayueNickleCobaltIndonesiaCo.,Ltd.pdf.

257 **parts of Shanxi:** "The New Generation of Yaodong Cave Dwellings, Loess Plateau," World Habitat Awards 2006, world-habitat.org/world-habitat-awards/winners-and-finalists/the-new -generation-of-yaodong-cave-dwellings-loess-plateau/.

257 **conditions there were abysmal:** The mines in Shanxi were unsafe for much of Zhou's youth. Official figures show that seventy-four miners were killed in an underground explosion in 2009 and more than a hundred in similar circumstances in 2007. At the time, the BBC reported that "China has the world's deadliest mining industry." See "Toll Climbs in China Mine Blast," BBC, February 22, 2009, news.bbc.co.uk/2/hi/asia-pacific/7904122.stm.

257 **province's brick kilns:** Howard W. French, "Reports of Forced Labor Unsettle China," *New York Times*, June 16, 2007.

257 **"They eat food for pigs":** "Urban Channel Reporters and Parents of Abducted Children Tell Their Harrowing Stories of Survival at Illegal Brick Kilns," "都市 道 者和被拐孩子的家长 述 黑窑 险" [*Dahe*], June 8, 2007, archived June 15, 2007, at web.archive.org/web/20070615154328 /http://www.dahe.cn/xwzx/zt/sh/zehr/xgbd/t20070608_995613.htm.

258 **When the slave labor:** "Convictions in China Slave Trial," BBC, July 17, 2007, news.bbc .co.uk/1/hi/world/asia-pacific/6902459.stm.

258 **Chinese artist in exile:** Chinese artist, conversation with the author, February 2023.

Chapter 37: Tokyo Drift

259 **All Sony's production:** Jon Topham, "Sony to Move Lithium Battery Assembly Abroad: Report," Reuters, January 21, 2012.

259 **Sony's battery division had mainly:** Kana Inagaki, "Sony Sells Battery Business to Murata," *Financial Times*, July 28, 2016.

259 **During the relocation:** Topham, "Sony to Move Lithium Battery Assembly Abroad."

259 **"Severe price competition":** Emi Emoto and Tim Kelly, "Exclusive: Banks Offer to Help Sony Offload Battery Unit—Sources," Reuters, November 27, 2012.

260 **China's industrial heartland:** Tatsuya Terazawa, "How Japan Solved Its Rare Earth Minerals Dependency Issue," World Economic Forum, October 13, 2023, weforum.org/agenda /2023/10/japan-rare-earth-minerals/#:~:text=On%207%20September%202010%2C %20a,rare%20earth%20minerals%20to%20Japan.

260 **Prices for rare:** Yen Nee Lee, "A Massive, 'Semi-Infinite' Trove of Rare-Earth Metals Has Been Found in Japan," CNBC, April 12, 2018, cnbc.com/2018/04/12/japan-rare-earths-huge -deposit-of-metals-found-in-pacific.html.

260 **"It is critically important":** "Japan Loosens China's Grip on Rare Earths Supplies," Reuters, September 4, 2014.

260 **dependency on Chinese rare:** Yen, "A Massive, 'Semi-Infinite' Trove of Rare-Earth Metals Has Been Found in Japan."

260 **"rare metals war":** Guillaume Pitron, *La guerre des métaux rares*, nouvelle ed. (Les Liens qui Libèrent, 2023).

260 **"Batteries were never":** Pitron, *La guerre des métaux rares*.

260 **focus on gaming:** Leo Lewis et al., "Sony and Honda Plan Electric Vehicle Tie-Up to Take on Tesla," *Financial Times*, March 4, 2022. The *Financial Times*' Kana Inagaki mused that by forming the joint venture, Sony also hoped to win over Tesla to adopt its image sensors. See Kana Inagaki, "Why Sony Wants to Win Over Tesla Despite Honda Tie-Up," *Financial Times*, October 25, 2022.

261 **This executive told me:** Robin Harding would later argue the same point in the *Financial Times* a few months later. Battery technology has "advanced at a slow, linear pace. Batteries are a matter of chemistry. You cannot just make them smaller, like a transistor." See Robin Harding, "Beware the Great Battery Fallacy," *Financial Times*, February 2, 2023.

261 **leapfrogged the goal:** Kevin Clemens, "CATL Condensed Battery Could Make Electric Aviation Possible," EEPower, April 27, 2023, eepower.com/market-insights/catl-condensed-battery -could-make-electric-aviation-possible/#.

Chapter 38: No Guts, No Glory

263 **Humans have mined:** A feasibility study from 2007 says that, on the TFM site, "archaeological evidence indicates that mining occurred over two hundred thousand years ago and that smelting of copper was practiced on the concession several thousand years before present." *Environmental and Social Impact Assessment: Executive Summary, Submitted to Tenke Fungurume Mining S.A.R.L. (TFM), Democratic Republic of the Congo* (Golder Associates, March 2007), 2.

263 **"boast cobalt concentrations":** Ryan C. Rosenfels and Bjorn P. von der Heyden, "A Critical Comparison Between the Fungurume 8 and 88 Cu-Co Deposits, Central African Copperbelt," *Ore Geology Reviews* 140 (January 2022): 1. The study in which this remark is made also notes that there are probably many more undiscovered supplies within the bend in the Lufilian Arc upon which Fungurume sits. The authors suggest "vast exploration potential for similar deepwater sub-basins which may host significant, and yet-undiscovered metal (notably Co) resources."

263 **Union Minière knew:** "Environmental and Social Impact Assessment: Executive Summary, Submitted to Tenke Fungurume Mining S.A.R.L. (TFM), Democratic Republic of the Congo," Golder Associates, March 2007, 2.

263 **Tempelsman hired Larry:** Jeff Gerth, "Former Intelligence Aides Profiting from Old Ties," *New York Times*, December 6, 1981.

263 **spent $250 million:** Robert Eriksson, *Adolf H. Lundin: No Guts No Glory* (Ehrenblad Editions, 2003), 216–17.

263 **The truth was:** *A State Affair: Privatizing Congo's Copper Sector* (Carter Center, November 2017), 9, cartercenter.org/resources/pdfs/news/peace_publications/democracy/congo-report-carter-center-nov-2017.pdf.

264 **told the Swede to:** Eriksson, *No Guts No Glory*, 227.

264 **"A lot of analysts":** Petter Bolme, interview with the author, May 2024.

264 **"with a big ego":** Eriksson, *No Guts No Glory*, 9.

264 **The hard-charging attitude:** The trial was ongoing as I wrote this, and the Lundin executives denied responsibility for the killings. See Anne-Françoise Hivert, "Swedish Oil Executives Face Landmark Trial for Alleged Complicity in Sudanese War Crimes," *Le Monde*, September 5, 2023.

264 **Lundin would recall:** Lundin, who died of leukemia in 2006, might have been telling a lie, but it may also have been true that Mobutu, drawn into the vortex of civil war, never did get around to asking Lundin for money. "Unlike some other mining companies, Lundin Mining still does not have any general prohibition against contributions to political parties, election campaigns or candidates," the authors of a Swedish NGO report wrote. See Raf Custers and Sara Nordbrand, *Risky Business: The Lundin Group's Involvement in the Tenke Fungurume Mining Project in the Democratic Republic of Congo* (International Peace Information Service, Swedwatch, and Diakonia, February 2008), 3.

264 **Gécamines and Congolese officials:** Custers and Nordbrand, *Risky Business*, 17.

265 **"President Kabila, because of us":** Barnabé Kikaya Bin Karubi, interview with the author, June 2025.

266 **a "backroom deal":** "AXS TV, Dan Rather Reports, 'All Mine' Full Episode," posted September 23, 2008, by Dan Rather Reports, YouTube, youtube.com/watch?v=PzkmX2ROL0k.

266 **the Arizona-based minerals:** Steve James, "Phelps Dodge to Proceed with Congo Copper Mine," Reuters, August 9, 2007.

266 **acquired Phelps Dodge:** James, "Freeport Acquires Phelps Dodge, Launches Offering."

266 **"the common perception":** City of New York, Office of the Comptroller, "27th February 2003," letter to the U.S. Securities and Exchange Commission, as cited in *Paying for Protection: The Freeport Mine and the Indonesian Security Forces* (Global Witness, 2005).

266 **Allegations abounded that:** Jones, Walker, Waechter, Poitevent, Carrère & Denègre LLP, letter to the U.S. Securities and Exchange Commission, 3 March 2003, as cited in *Paying for Protection: The Freeport Mine and the Indonesian Security Forces* (Global Witness, 2005). (At the time, lawyers for Freeport said that such claims were "irrelevant and false.")

266 **"unsuitable for aquatic life":** Jane Perlez and Raymond Bonner, "Below a Mountain of Wealth, a River of Waste," *New York Times*, December 27, 2005.

267 **Norway's sovereign wealth:** Christopher Eaton, "James 'Jim Bob' Moffett, Who Helped Build Freeport-McMoRan, Dies at 82," *Wall Street Journal*, January 11, 2021.

267 **Rather framed the displacement:** "AXS TV, Dan Rather Reports, 'All Mine' Full Episode."
267 **"This land is not fertile":** Custers and Nordbrand, *Risky Business*, 43.
267 **built ninety-one wells:** "An Investment in the Future of the Democratic Republic of the Congo," Tenke Fungurume Mining, 2013.
268 **They were prosecuted:** Joe Bavier, "DRC Holds Freeport Staff in Alleged Visa Scam," Reuters, August 15, 2009.
268 **"The stinkin' government":** Melissa Sanderson, interview with the author, August 23, 2023.
268 **Freeport agreed to pay:** Joe Bavier, "Freeport to Pay $16 Million in DRC Visa Settlement," Reuters, August 17, 2009.
268 **"action by a foreign official":** Foreign Corrupt Practices Act of 1977, as amended, 15 U.S.C. §§ 78dd-1, et seq. ("FCPA").
268 **"huge pallets of cash":** Mining official, interview with the author, March 2022.
269 **uneasy about the risks:** Source close to Freeport-McMoRan's board, interview with the author.
269 **dig as they might:** Eaton, "James 'Jim Bob' Moffett."
269 **examine what assets:** James Wilson, "BHP to Expand Copper and Oil Exploration Despite Weak Prices," *Financial Times*, May 10, 2016.
269 **found a buyer in China:** Dionne Searcey et al., "A Power Struggle Over Cobalt Rattles the Clean Energy Revolution," *New York Times*, November 20, 2021.
269 **"It breaks my heart":** Wilson, "BHP to Expand Copper and Oil Exploration Despite Weak Prices."
269 **Congolese politicians tried:** Searcey et al., "A Power Struggle Over Cobalt Rattles the Clean Energy Revolution."
269 **China Molybdenum was certainly:** "The Backchannel: State Capture and Bribery in Congo's Deal of the Century," The Sentry, November 2021, https://thesentry.org/wp-content/uploads/2021/11/TheBackchannel-TheSentry-Nov2021.pdf.
270 **took 51 percent of the Grasberg:** Wilda Asmarini and Bernadette Christina Munthe, "Freeport, Rio Sell Majority Stake in Grasberg Mine to Indonesia," Reuters, September 27, 2018, reuters.com/article/indonesia-freeport-rio-tinto/update-3-freeport-rio-sell-majority-stake-in-grasberg-mine-to-indonesia-idUSL4N1WD1L8.

Chapter 39: Tying Up the Supply

273 **Glencore doesn't publish:** "Glencore Congo Copper Mine Hit with $894 Million Royalty Row," *Bloomberg*, September 27, 2024.
273 **a whopping 68 percent:** *The Road to Ruin?: Electric Vehicles and Workers' Rights Abuses at DR Congo's Industrial Cobalt Mines* (Rights and Accountability in Development, November 2021), 26, raid-uk.org/post-library/the-road-to-ruin-electric-vehicles-and-workers-rights-abuses-at-dr-congos-industrial-cobalt-mines/.
273 **KCC maintenance engineers:** Hereward Holland, "Glencore's KCC Mine in Congo Had Acid Spill on March 16," Reuters, April 6, 2021.
274 **"make a contribution":** *Beneath the Green* (Rights and Accountability in Development, March 2024), 77, raid-uk.org/post-library/report-beneath-the-green/.
274 **"failed to ensure adequate management":** Federal Prosecutor's Office of Switzerland, "Office of the Attorney General Closes Its Criminal Investigation Against Glencore International AG with a Summary Penalty Order and an Abandonment Order," News Service Bund, August 4, 2024, https://www.news.admin.ch/en/nsb?id=101995.
274 **Another criticism leveled:** *Out of Africa: British Offshore Secrecy and Congo's Missing $1.5 Billion* (Global Witness, May 2016), globalwitness.org/en/campaigns/corruption-and-money-laundering/out-of-africa/.
275 **"Glencore has made loans":** Glencore, "Response to Global Witness," May 2, 2012, media.business-humanrights.org/media/documents/files/documents/Glencore_response_to_Global_witness_2012.pdf.
275 **"the mining giant gets":** *Glencore and the Gatekeeper* (Global Witness, May 2014), 10, globalwitness.org/en/campaigns/corruption-and-money-laundering/glencore-and-the-gatekeeper/.
275 **Former Glencore officials with whom:** Glencore official, interview with the author, March 2022; Glencore official, interview with the author, June 2022; and Glencore official, interview with the author, July 2019.
275 **for $922 million:** "Glencore Purchases Stakes in Mutanda and Katanga," Glencore, February

13, 2017, glencore.com/media-and-insights/news/glencore-purchases-stakes-in-mutanda-and -katanga.

276 **"worth of opaque"**: U.S. Department of the Treasury, "United States Sanctions Human Rights Abusers and Corrupt Actors Across the Globe," press release, December 21, 2017, home.treasury .gov/news/press-releases/sm0243.

276 **$28 million in 2018**: Barbara Lesi, "Glencore Settles with Gertler over Congo Royalties," Reuters, June 15, 2018.

276 **almost $3 billion**: Lesi, "Glencore Settles with Gertler Over Congo Royalties"; and "Settlement of Dispute with Ventora and Africa Horizons," Glencore, June 15, 2018, archived May 24, 2024, at web.archive.org/web/20240524183659/https://www.glencore.com/media-and-insights /news/Settlement-of-dispute-with-Ventora-and-Africa-horizons.

277 **"pay the relevant royalties"**: "Settlement of Dispute with Ventora and Africa Horizons."

277 **"U.S. authorities must hold Glencore"**: Global Witness, "Glencore Must Not Pay Millions to Sanctioned Individual," press release, June 15, 2018, archived January 17, 2025, at web.archive .org/web/20250117191409/https://www.globalwitness.org/en/press-releases/glencore-must-not -pay-millions-sanctioned-individual/.

277 **"Bribery is a highly corrosive"**: Daniel Thomas, "DR Congo: Miner Glencore Pays $180m in Latest Corruption Case," *BBC News*, December 5, 2022, bbc.com/news/business-63858295.

278 **"no longer necessary"**: Consent Motion to Modify Conditions of Probation, United States v. Glencore International A.G., No. 22-cr-580440 (S.D.N.Y. filed Mar. 20, 2025), 1, https://ppl-ai-file -upload.s3.amazonaws.com/web/direct-files/31530623/ab4d25ca-d0dd-4860-88c5 -1d9661eda303/gov.uscourts.nysd.580440.63.0.pdf.

278 **"This kind of corruption"**: Bridget Nsimenta, "Kinshasa Govt Lacks Resources and Capacity to Control Eastern DRC—US Congressman," *Nile Post*, March 26, 2025, nilepost.co.ug/news /250232/kinshasa-govt-lacks-resources-and-capacity-to-control-eastern-drc---us -congressman.

279 **"concerns and discussions"**: Andrew Stevenson, interview with the author, July 2023.

279 **"Communist Party is firmly"**: Jack Nicas, "He Warned Apple About the Risks in China. Then They Became Reality," *New York Times*, June 17, 2021.

280 **"not going to export batteries"**: Ivan Glasenberg, keynote speech, Financial Times Future of the Car Summit, May 12, 2021.

Chapter 40: Green-Tinted Glasses

281 **But other mines seem**: George Arthur Forrest, *Un siècle de rêves: Ensemble, bâtissons l'avenir* (Le Cherche Midi, 2022), 105.

281 **Yu also held**: "#296 Yu Yong," Bloomberg Billionaires Index, *Bloomberg*, 2024, bloomberg.com /billionaires/profiles/yong-yu/.

282 **Because of its acquisitions**: Mark Burton, "Glencore Set to Lose Crown as Top Cobalt Miner to China's CMOC," *Bloomberg*, March 21, 2023, bloomberg.com/news/articles/2023-03-21 /glencore-set-to-lose-crown-as-top-cobalt-miner-to-china-s-cmoc.

282 **create powerful alloys**: Thomas J. Vogl et al., *Diagnostic and Interventional Radiology* (Springer, 2016), 593.

282 **founders of BHR Partners**: The firm's name was Bohai Harvest RST (Shanghai) Equity Investment Fund Management Co., Ltd., and in 2023 changed to BHR Partners (Shanghai) Equity Investment Fund Management Co., Ltd.

283 **"Oligarchs used to"**: Devon Archer, interview with the author, January 2022.

283 **China Moly was interested**: "Lundin Mining Announces Agreement to Sell Interest in TF Holdings for $1.136 Billion," Lundin Mining, November 15, 2016, lundinmining.com/news /lundin-mining-announces-agreement-to-sell-interest-122523/.

283 **Gécamines did protest**: Thomas Wilson, "Congo Said to Get $100 Million to Clear China Moly Mine Purchase," *Bloomberg*, February 22, 2017, bloomberg.com/news/articles/2017-02 -22/congo-said-to-get-100-million-to-clear-china-moly-mine-purchase.

284 **greenfield cobalt site**: "CMOC Announces Acquisition of Kisanfu Copper-Cobalt Deposit in DRC," CMOC, December 13, 2020, en.cmoc.com/html/2020/News_1213/39.html.

284 **Kisanfu has an estimated**: Tom Daly, "CATL Takes Stake in China Moly Cobalt Mine for $137.5 Million," Reuters, April 11, 2021.

284 **most important cobalt deposits**: Daly, "CATL Takes Stake in China Moly Cobalt Mine for $137.5 Million."

284 **web of connections:** "的华友" [*Advancing Huayou*], Huayou Cobalt, huayou.com/about/culture/video/10.

284 **Other Chinese battery:** "China's Gotion Follows CATL Playbook with Nickel Investment," September 20, 2023, source.benchmarkminerals.com/article/chinas-gotion-follows-catl-playbook-with-nickel-investment.

Chapter 41: Seizing the Space

286 **China Moly's subsidiary:** "Democratic Republic of the Congo: Artisanal Miners at Risk as the Army Moves In," Amnesty International, June 28, 2019, https://www.amnesty.org/en/documents/afr62/0625/2019/en/.

286 **Eight hundred men:** "Treasury Sanctions Two Individuals for Threatening the Stability of and Undermining Democratic Processes in the Democratic Republic of the Congo," United States Justice Department, September 26, 2016, https://home.treasury.gov/news/press-releases/jl0560.

286 **force the miners out:** John Numbi, interview with the author, March 2023. "I asked the gov' to remove people from the mines of Katanga," Numbi told me. "I was the only one who went to chase people from the mines who were stealing." (Numbi went into hiding somewhere in Southern Africa after falling out with the government, and he denied killing the human-rights activist.) See also Amnesty International to Excellency Mister Félix Tshisekedi, "Artisanal Miners at Risk as the Army Moves In," letter, June 27, 2019, amnesty.org/en/documents/afr62/0625/2019/en/.

287 **"Imagine yourself on":** "Lualaba: Le Gouverneur Muyej condamne l'érection de la cité 'Kafwaya,'" Radio Okapi, January 19, 2018, radiookapi.net/2018/01/19/actualite/en-bref/lualaba-le-gouverneur-muyej-condamne-lerection-de-la-cite-kafwaya.

287 **"If the investors complain":** Aaron Ross, "Send in the Troops: Congo Raises the Stakes on Illegal Mining," Reuters, July 17, 2019. Parentheses appear in the original.

289 **obvious in Congo:** It was not just China Moly that had enacted such policies. Such an arrangement had been here since before the Chinese, since the days when the mine was owned by Freeport-McMoRan.

290 **The GOP pointed:** Louis Jacobsen, "Hunter Biden and China: Sorting Through a Murky Business Deal," PolitiFact, May 22, 2020, politifact.com/article/2020/may/22/hunter-biden-and-china-sorting-through-murky-busin/.

PART 5: WAKING UP

293 **"High speed is":** Ivan Illich, *Energy and Equity* (Harper & Row, 1974), 12.

Chapter 42: God Is Smiling on Congo

295 *God is smiling:* Amos Hochstein, interview with the author, May 2022.

296 **a Congolese newscaster:** "RDC-USA: Amos Hochstein, Émissaire de Joe Biden à Kinshasa," posted on September 12, 2022, by b-one TV Congo, YouTube, youtube.com/watch?v=NvMFuAcMsDo.

296 **its critical minerals and where:** *Securing Defense-Critical Supply Chains: An Action Plan Developed in Response to President Biden's Executive Order 14017* (Department of Defense, February 2022), 10, media.defense.gov/2022/Feb/24/2002944158/-1/-1/1/DOD-EO-14017-REPORT-SECURING-DEFENSE-CRITICAL-SUPPLY-CHAINS.PDF.

296 **supply-chain review by:** *Securing Defense-Critical Supply Chains*, 19.

297 **a seven-way memorandum:** "Memorandum of Understanding (MOU) on Working Arrangements Between the Government of the United States of America, the European Commission, the Government of the Republic of Zambia, the Government of the Republic of Angola, the Government of the Democratic Republic of the Congo, the African Development Bank, and Africa Finance Corporation Relating to the Development of the Lobito Corridor and the Zambia–Lobito Rail Line," U.S. Embassy Zambia, 2024, zm.usembassy.gov/wp-content/uploads/sites/44/2024/05/Lobito_Corridor_MOU.pdf.

297 **no U.S. businesses in Congo:** U.S. State Department official, conversation with the author, November 2023.

297 **A Washington lobbyist told:** U.S. lobbyist, conversation with the author, March 2024.

298 **famous in the U.S. business:** Zeitlin had resigned from the firm that owns the brands Kate Spade and Coach in July 2020, after it emerged that he was involved in what he called an "inappropriate relationship" with a woman who later complained about it. See Jesse Eisinger, "The Bizarre Fall of the CEO of Coach and Kate Spade's Parent Company," *ProPublica*, July 22, 2020, propublica.org/article/the-bizarre-fall-of-the-ceo-of-coach-and-kate-spades-parent-company.

298 **"laser focus" on Congo's:** Eric Lipton and Dionne Searcey, "Who Are Congo's Cobalt Entrepreneurs?," *New York Times*, December 7, 2017, www.nytimes.com/interactive/2021/12/07/world/congo-cobalt-investors.html.

298 **"his investment plan could":** Dionne Searcey et al., "On the Banks of the Furious Congo River, a 5-Star Emporium of Ambition," *New York Times*, December 7, 2021.

298 **even showed the paper:** Letter to President Joseph R. Biden from Senator Christopher A. Coons et al., United States Senate, July 13, 2021.

298 **Glencore certainly hadn't:** Glencore representative, interview with the author, June 2022.

298 **After more geological:** "Probe into Sodimico Deal with US Rapper Akon's Firm," *Africa Intelligence*, December 12, 2022, africaintelligence.com/central-africa/2022/12/08/probe-into-sodimico-deal-with-us-rapper-akon-s-firm,109872137-art.

298 **report by the Cobalt Institute:** *Cobalt Market Report 2023* (Cobalt Institute, May 2024), 46.

298 **twenty Chinese workers had died:** "Feature: Chinese-Built Benguela Railway Revitalizes Angola's Economy," *Xinhua*, October 3, 2019, xinhuanet.com/english/2019-10/03/c_138446624_2.htm.

299 **Angola's huge offshore:** "Angola to Leverage Chinese Investment to Stimulate Oil Production," Energy Capital & Power, September 19, 2024, energycapitalpower.com/angola-to-leverage-chinese-investment-to-stimulate-oil-production/.

299 **saw a man crushed:** Anna Hollmer, "Unveiling the True Nature of the Artisanal Cobalt Mines in the Democratic Republic of Congo: Systemic Conditions in the Mines and Implications for Global Supply Chains and Social Justice" (thesis, Anglo-American University in Prague, 2024), 1.

299 **Congolese on the Cobalt:** "Ivanhoe Mines' Exports Commence from Kamoa-Kakula Copper Complex Along Lobito Atlantic Rail Corridor," Ivanhoe Mines, January 2, 2024, ivanhoemines.com/news-stories/news-release/ivanhoe-mines-exports-commence-from-kamoa-kakula-copper-complex-along-lobito-atlantic-rail-corridor/.

300 **heralded the project:** Andres Schipani, "The US Railway That Could Set Off a Copper War," *Financial Times*, August 21, 2024.

300 **constantly asked them for money:** Message to author from State Department consultant, December 4, 2024.

300 **Tshisekedi hailed the project:** "Remarks by President Biden Participating in the Lobito Corridor Trans-Africa Summit," White House, December 4, 2024, whitehouse.gov/briefing-room/speeches-remarks/2024/12/04/remarks-by-president-biden-participating-in-the-lobito-corridor-trans-africa-summit-benguela-angola/.

300 **a few printed pages:** E. D. Wala Chabala, "Lobito Corridor—A Reality Check," African Policy Research Institute, February 2, 2024, afripoli.org/lobito-corridor-a-reality-check.

300 **"fostering U.S. private sector investment":** "Trump Envoy Says US-Congo Mining and Security Deal Moving Ahead," *Bloomberg*, April 3, 2025.

301 **"I mean, the Lobito Corridor":** Businessman connected to Congolese mining, conversation with the author, February 2025.

301 **After an agreement to end:** William Clowes, Archie Hunter, and Michael J Kavanagh, "Special Forces Veterans Lead US Bid to Buy Congo Cobalt Miner," *Bloomberg*, July 17, 2025, https://www.bloomberg.com/news/articles/2025-07-17/special-forces-veterans-lead-us-bid-to-buy-congo-cobalt-miner.

301 **James Story, the U.S. ambassador:** Miguel Gomes, "US Committed to Funding Angola's Lobito Rail Corridor Despite Spending Cuts, Diplomat Says," Reuters, April 3, 2025.

Chapter 43: The Spice of Life

303 **has been nicknamed:** "Aminatou Haidar, the 'Gandhi of Western Sahara,' Wins Right Livelihood Award," Democracy Now, September 25, 2019, democracynow.org/2019/9/25/aminatou_haidar_the_gandhi_of_western.

303 **"Life can multiply":** Isaac Asimov, *Asimov on Chemistry* (Anchor Books, 1975), 164, 170.

304 **No cobalt or nickel:** Arumugam Manthiram and J. B. Goodenough, "Lithium Insertion into Fe2(MO4)3 Frameworks: Comparison of M = W with M = M," *Journal of Solid State Chemistry* 71, no. 2 (1987): 349.

304 **The two scientists' attention:** Manthiram and Goodenough, "Lithium Insertion into Fe2(MO4)3 Frameworks," 349.

304 **"a cathode of good capacity":** A. K. Padhi et al., "Phospho-Olivines as Positive-Electrode Materials for Rechargeable Lithium Batteries," *Journal of the Electrochemical Society* 144, no. 4 (1997): 1191.

304 **"shows this material to be":** Padhi et al., "Phospho-Olivines as Positive-Electrode Materials for Rechargeable Lithium Batteries," 1188.

305 **LFP cathodes represented:** *Batteries and Secure Energy Transitions* (International Energy Agency, 2024), iea.blob.core.windows.net/assets/cb39c1bf-d2b3-446d-8c35-aae6b1f3a4a0/Batteriesand SecureEnergyTransitions.pdf.

305 **low cost and high safety:** LFMP, another olivine cathode that is structurally similar to LFP but contains manganese and has 15 to 20 percent more energy density than the original cathodes, is also rising in popularity. According to battery manufacturers, LMFP cathodes can be produced at much the same price as the other phosphate batteries. See Phate Zhang, "CATL Said to Mass Produce LMFP Batteries Within This Year," CnEVPost, July 12, 2022, cnevpost .com/2022/07/12/catl-said-to-mass-produce-lithium-manganese-iron-phosphate-batteries -within-this-year/.

305 **acquire in the 1990s:** As Goodenough and two other scientists in his lab wrote, "The availability and cost of the transition metals used in these compounds are unfavourable as the Wh/$ [watt-hours per dollar] is a more important figure of merit than Wh/g [watt-hours per gram] in the case of large batteries to be used in an electric vehicle or a load-levelling system." They noted: "These consideration [*sic*] have motivated the investigation of iron-based oxides." See Padhi et al., "Phospho-Olivines as Positive-Electrode Materials for Rechargeable Lithium Batteries," 1189.

306 **those of Morocco:** Directorate of Intelligence, "Spanish Sahara: Phosphates and Sovereignty," Intelligence Memorandum, September 1970, Central Intelligence Agency, www.cia.gov/reading room/document/cia-rdp85t00875r001600030125-0. The memo was declassified, in part, on October 31, 2011.

306 **For every dirham of profit:** Fouad Abdelmoumni (Transparency International), interview with the author, June 2018.

307 **prices on the phosphate market:** Dan Egan, *The Devil's Element: Phosphorus and a World Out of Balance* (W. W. Norton, 2023), 59.

307 **"the spice of life":** "Morocco and the Spice of Life: The Phosphates," *Morocco World News*, December 6, 2015, moroccoworldnews.com/2015/12/174397/morocco-and-the-spice-of-life-the -phosphates.

Chapter 44: Lithium Is Sexy

309 **"Mr. Baker's proposals":** Mohammed VI of Morocco to George W. Bush, letter, April 6, 2004, box 322–328, James A. Baker III Collection, Mudd Manuscript Library, Princeton Library.

310 **"We're definitely not going":** Claire Bushey and Oliver Roeder, "US Electric Vehicle Batteries Poised for New Lithium Iron Age," *Financial Times*, March 4, 2023.

311 **"If all batteries today":** *Global EV Outlook 2023: Catching Up with Climate Ambitions* (International Energy Agency, 2023), iea.blob.core.windows.net/assets/dacf14d2-eabc-498a-8263-9f97fd5dc 327/GEVO2023.pdf.

311 **to electrify 60 percent:** "Beijing's Green New Deal Pledges Factories in Africa," *Africa Confidential* 65, no. 18 (September 6, 2024), africa-confidential.com/home/issue/id/1342.

311 **"LFP battery chemistries are":** Email from Undersecretary Jose W. Fernandez to the author, June 25, 2024.

Chapter 45: Next Energy

313 **employed a staggering:** Phate Zhang, "BYD's Workforce Exceeds 900,000," CnEVPost, September 13, 2024, cnevpost.com/2024/09/13/byd-workforce-exceeds-900000/.

314 **didn't emit smoke:** "BYD's Revolutionary Blade Battery: All You Need to Know," BYD, January 22, 2023, byd.com/eu/blog/BYDs-revolutionary-Blade-Battery-all-you-need-to-know.html.

314 **"could not do what"**: Yang Jie, "Building Apple Products Has Become a Side Hustle for China's Biggest EV Maker," *Wall Street Journal*, November 30, 2024, wsj.com/tech/building-apple -products-has-become-a-side-hustle-for-chinas-biggest-ev-maker-f26e251c.

314 **Elon Musk admitted**: Keith Bradsher, "How China Built BYD, Its Tesla Killer," *New York Times*, February 12, 2024.

314 **Wang was worth almost**: "#110 Wang Chuan-Fu," Bloomberg Billionaires Index, *Bloomberg*, 2024, bloomberg.com/billionaires/profiles/chuanfu-wang/.

314 **The latest Blade**: Damian Smy, "BYD to Introduce New Second-Generation 1000km Battery Technology," Drive, April 11, 2024, drive.com.au/news/byd-to-introduce-new-second-gen -1000km-battery-technology/.

314 *Bloomberg* **mused that**: Danny Lee, "Why BYD's Wang Chuanfu Could Be China's Version of Henry Ford," *Bloomberg*, June 24, 2024, bloomberg.com/news/articles/2024-06-24/byd-s-wang -chuanfu-china-s-henry-ford.

315 **"That SUV traveled"**: "Gemini: 608 Miles on a Single Charge," Our Next Energy (ONE), accessed July 12, 2024, one.ai/products/gemini.

316 **U.S. company's battery blueprints**: Gabrielle Coppola, "America's Long, Tortured Journey to Build EV Batteries," *Bloomberg Businessweek*, June 8, 2023.

316 **"much more stable supply chain"**: Shirley Meng, interview with the author, May 2020.

316 **cobalt's demand growth**: *Cobalt Market Report* (Cobalt Institute, May 2024), 13–14.

317 **the exploration permits**: AVZ Minerals Limited, "AVZ Minerals Completes Acquisition of Additional 5% of Manono Lithium and Tin Project," Australian Securities Exchange, April 18, 2019.

317 **blocked them in a series**: *A New Rush for Lithium in Africa Risks Fuelling Corruption and Failing Citizens* (Global Witness, November 2023), globalwitness.org/en/campaigns/transition-minerals /a-rush-for-lithium-in-africa-risks-fuelling-corruption-and-failing-citizens/.

317 **was suffering economically**: Pierre Mukamba Kaseya, interview with the author, April 2022.

317 **"We are facing a concerted"**: Nicolas Niarchos, "Power Metals," *Granta*, June 5, 2024, granta .com/power-metals/.

317 **Someone captured photographs**: Jens Emil Soderlund (@emil_jens), "I have advised Jennifer Allison Fendrick of #KoBold Metals, who even visited $AVZ Manono site under false pretenses arranged by @cominiereSA @CelestinKibeya @AlainMONGA8 . . . ," X, April 10, 2025, https://x.com/emil_jens/status/1851798275051098489.

317 **"It is an amazing deposit"**: Jennifer Fendrick, interview with the author, March 2025.

318 **"We are now seeing"**: "Undersecretary Jose Fernandez's Answers for Journalist Nicolas Niarchos," author email with Isabel Malowany, senior adviser to Jose W. Fernandez, undersecretary for economic growth, energy, and the environment, June 25, 2024.

318 **"grants for domestic production"**: H.R. 5376–117th Congress: Inflation Reduction Act of 2022, 2044.

318 **"foreign entity of concern"**: H.R. 5376–117th Congress: Inflation Reduction Act of 2022, 1957.

318 **Administration officials including Amos Hochstein**: Eric Lipton, "Seeking Access to Congo's Metals, White House Aims to Ease Sanctions," *New York Times*, May 16, 2024.

318 **"People think that"**: Lawyer involved in advising on critical metals policy, conversation with the author, November 2023.

319 **China controls the market**: Matthew Blackwood and Catherine DeFilippo, "Germanium and Gallium: U.S. Trade and Chinese Export Controls," U.S. International Trade Commission, Executive Briefings on Trade, March 2024.

319 **"China is leveraging"**: Emily Benson and Thibault Denamiel, "China's New Graphite Restrictions," CSIS, October 23, 2023, csis.org/analysis/chinas-new-graphite-restrictions.

319 **By the time of Trump's**: Amanda Chu et al., "Delays Hit 40% of Biden's Major IRA Manufacturing Projects," *Financial Times*, August 12, 2024.

320 **ONE was three**: Garrett Hering, "IRA at 1: US Climate Law Cues $63B Spending Spree on Battery Factories," S&P Global, August 9, 2023, spglobal.com/marketintelligence/en/news -insights/latest-news-headlines/ira-at-1-us-climate-law-cues-63b-spending-spree-on -battery-factories-76839524.

320 **building a factory in Michigan**: "This Just In: Team Michigan Continues to Deliver as Our Next Energy Rolls Out First Cell Production," Michigan Economic Development Corporation, November 1, 2023, michiganbusiness.org/press-releases/2023/11/our-next-energy-cel l-production/.

Chapter 46: Red Seas

322 **"They ran this cash cow"**: Adrian Vickers, *A History of Modern Indonesia* (Cambridge University Press, 2013), 190–91.

323 **a distant second**: *Cobalt Market Report* (Cobalt Institute, May 2024), 23–24.

323 **dollars in his homeland**: "Chinese Investment Project to Inject More than US$200 Million to Boost National Lithium Strategy," Gob.cl, October 16, 2023, gob.cl/en/news/chinese-investment-project-to-inject-more-than-us200-million-to-boost-national-lithium-strategy/.

323 **contract for Indonesian nickel**: "Indonesia Says Tesla Strikes $5 Billion Deal to Buy Nickel Products," Reuters, August 8, 2022.

323 **BYD would follow suit**: Stefanno Sulaiman, "China's BYD to Complete $1 Billion Indonesia Plant by End 2025, Executive Says," Reuters, January 20, 2025, reuters.com/business/autos-transportation/chinas-byd-complete-1-billion-indonesia-plant-by-end-2025-executive-says-2025-01-20/. For BYD being coy, see: "Semua Mobil Listrik BYD di RI Pakai LFP, Janji Pelajari Gunakan Nikel," CNN Indonesia, January 22, 2024, cnnindonesia.com/otomotif/2024 0122143320-603-1052861/semua-mobil-listrik-byd-di-ri-pakai-lfp-janji-pelajari-gunakan-nikel.

323 **abundant in Indonesia**: Indonesia, which has a population of 274 million, has around 125 million scooters and motorcycles. Only 32,000 of these are electric. (Scooters are also big business: In Indonesia, two-wheelers are a $6.03-billion-dollar-a-year industry.) See David Waterworth, "Electric Two-Wheelers in Indonesia," Clean Technica, July 30, 2023, cleantechnica.com/2023/07/30/electric-two-wheelers-in-indonesia/.

325 **the isle had received**: "Obi HPAL Nickel-Cobalt Project," NS Energy, May 19, 2021, nsenergy business.com/projects/obi-hpal-nickel-cobalt-project/?cf-view.

326 **destroyed their organs**: Muhammad Aris and T. Tamrin, "Heavy Metal (Ni, Fe) Concentration in Water and Histopathological of Marine Fish in the Obi Island, Indonesia," *Jurnal Ilmiah Platax* 8, no. 2 (July–December 2020): 222.

327 **pressured residents to sell**: Shelley Marshall et al., "Access to Justice for Communities Affected by the PT Weda Bay Nickel Mine—Interim Report," Non-Judicial Human Rights Redress Mechanisms Project, September 2013.

327 **"epoch-making impact"**: "Chen Xuehua, Chairman of the Board of Directors of the Group, Goes to the Site of Huafei Project in Indonesia for Inspection and Guidance," Huayou Cobalt, April 17, 2024, huayou.com/en/news/corporate-news/186.

328 **its owner a billionaire**: Netty Ismail, "Hidden Billionaire in Indonesia Reaps Gains from Palm Oil," *Bloomberg*, November 4, 2012.

328 **"Mum was wrong"**: Wan Phing Lim, "On Rushing in a Finite World," Wan Phing Lim (personal website), June 11, 2012, archived June 8, 2023, at web.archive.org/web/20230608175701 /https://wanphing.com/2012/06/11/on-rushing-in-a-finite-world/.

328 **refinement of nickel ore**: Robby Irfany, "Harita Group Takes Chinese Investor to Build Smelter," *Tempo.Co* (English version), June 15, 2015, en.tempo.co/read/675204/harita-group-takes-chinese-investor-to-build-smelter.

328 **Financed by Harita**: Tom Daly et al., "China's Lygend Starts Milestone Nickel Project in Indonesia," Reuters, May 19, 2021.

328 **GEM and Easpring**: "Easpring to Buy Nickel, Cobalt Chemical from Lygend Project in Indonesia," Asian Metal, February 5, 2021, asianmetal.com/news/1635315/Easpring-to-buy-nickel,-cobalt-chemicals-from-Lygend-project-in-Indonesia.

328 **"complete supporting facilities"**: Lygend Resources, "The First Batch of MHP Products Are Rolling off the Production Line from Lygend HPAL Project in OBI Island Indonesia," Lygend Media Center, May 26, 2021, lygend.com/media/125.html.

329 **waste into the deep sea**: Rabul Sawal, "Red Seas and No Fish: Nickel Mining Takes Its Toll on Indonesia's Spice Islands," trans. Basten Gokkon, *Mongabay*, February 16, 2022, news.mongabay.com/2022/02/red-seas-and-no-fish-nickel-mining-takes-its-toll-on-indonesias-spice-islands/.

329 **the Indonesian magazine**: Bagja Hidayat and Dini Pramita, "Kawasi Muddied by Nickel," *Tempo*, February 7, 2022, accessed at pulitzercenter.org/stories/kawasi-muddied-nickel.

329 **"Harita Nickel is fully"**: Anie Rahmi (Harita Nickel), correspondence with the author, February 3, 2023. Harita, Rahmi said, contributes to local prosperity. Rahmi noted that more than 85 percent of Harita's workforce is Indonesian, and that the company upholds strict environmental standards. "The presence of Harita Nickel has been contributing to the national and regional development since 2010," Rahmi wrote, "and economic growth in North Maluku Province has increased since the existence of the smelter industry."

329 **"conservation-dependent" species:** Mei Lin Neo and Peter A. Todd, "Conservation Status Reassessment of Giant Clams (Mollusca: Bivalvia: Tridacninae) in Singapore," *Nature in Singapore* 6 (2013): 125.

329 **despite the pressure:** Aris and Tamrin, "Heavy Metal (Ni, Fe) Concentration in Water and Histopathological of Marine Fish in the Obi Island, Indonesia," 214–20; and T. Tamrin and Muhammad Aris, "Health Condition of *Tridacna* sp. in the Waters of Obi Island, Indonesia," *Jurnal Ilmiah Platax* 8, no. 2 (2020): 242–50.

330 **chemical hexavalent chromium:** Febriana Firdaus and Tom Levitt, "'We Are Afraid': Erin Brockovich Pollutant Linked to Global Electric Car Boom," *Guardian*, February 19, 2022.

Chapter 47: Buried Underground

331 **money that Ilunga:** Françoise Ilunga, interview with the author, October 2019.

334 **Glencore shares briefly plummeted:** "Glencore Says 19 People Were Killed in Congo Mine Collapse," *Bloomberg*, June 29, 2019, bloomberg.com/news/articles/2019-06-27/glencore-says-19 -illegal-miners-killed-in-congo-mine-collapse.

334 **The European Union has:** Chris Stretton et al., "EU Battery Passport Regulation Requirements," Circularise, March 20, 2025, https://www.circularise.com/blogs/eu-battery-passport -regulation-requirements.

335 **very dirty indeed:** Christoph N. Vogel, *Conflict Minerals, Inc.: War, Profit and White Saviourism in Eastern Congo* (Hurst Publishers, 2022).

335 **poorer member states:** Ilka von Dalwigk (European Battery Alliance), interview with the author, Stuttgart, May 2023.

335 **"put the cat back in":** Charles Carron Brown, interview with the author, July 2024.

335 **ended in March 2020:** Ostensibly this was for COVID-related reasons, but the economics of the project also did not add up.

336 **"Since the end of formalization":** Dorothée Baumann-Pauly, "Cobalt Mining in the Democratic Republic of the Congo: Addressing Root Causes of Human Rights Abuses," white paper (NYU Stern Center for Business and Human Rights and Geneva Center for Business & Human Rights, February 2023), 7, bhr.stern.nyu.edu/publication/cobalt-mining-in-the-democratic -republic-of-the-congo-addressing-root-causes-of-human-rights-abuses/.

337 **"I am trying to fight":** Odilon Kajumba Kilanga, author interview between Jeef Kazadi Kamwanga and Kajumba, May 2024.

Chapter 48: Detained

338 **a court in Likasi:** "DR Congo: Militia Leader Guilty in Landmark Trial," Human Rights Watch, March 10, 2009, hrw.org/news/2009/03/10/dr-congo-militia-leader-guilty-landmark -trial.

338 **escaped from prison:** Moïse Katumbi Chapwe, interview with the author, May 2019. "The day he got out from the prison is the day the army replaced the police at the facility. He escaped when the army was there. It was easier for him to run away when the army was there than when the police was there," Katumbi, who was the governor of Katanga at the time, told me.

338 **fled into the hills:** A BBC journalist, Maud Jullien, also heard this when she reported on the Bakata Katanga in 2013. See Maud Jullien, "Katanga: Fighting for DR Congo's Cash Cow to Secede," *BBC News*, August 12, 2013, bbc.com/news/world-africa-23422038.

338 **went on the run again:** When I spoke with Thierry Mukelekele in May 2019, he remembered the conversation he had about disarmament with Gédéon, who was suspicious of what might happen to him. "The first stage, you should abandon armed struggle," Mukelekele had told him. Gédéon decided to follow Mukelekele's advice but then headed back into the bush under Tshisekedi, whom he did not trust. See also Lewis Mudge, "Convicted Congolese Warlord Escapes. Again," Human Rights Watch, April 7, 2020, hrw.org/news/2020/04/07/convicted-congolese -warlord-escapes-again.

339 **the lam for two years:** Mudge, "Convicted Congolese Warlord Escapes. Again."

343 **journalists were tortured:** "Three Journalists Tortured by DRC Intelligence Agency," May 23, 2022, Reporters Without Borders (RSF), rsf.org/en/three-journalists-tortured-drc-intelligence -agency/.

343 **newly built by the Chinese:** On a bus in Addis Ababa, I once met an Israeli investor who told me about the new terminal at N'djili. "I have just been in Kinshasa, where they have a new air-

port," he said. "It was built by the Chinese. A whole terminal without an air conditioner. Think about that. The Congo—the heat is tropical, and people get angry."

344 **previously healthy Beya:** Pascal Mulegwa, "RDC: François Beya obtient une libération provisoire pour raisons de santé," Radio France Internationale, August 16, 2022, rfi.fr/fr/afrique /20220816-rdc-françois-beya-obtient-une-libération-provisoire-pour-raisons-de-santé.

345 **"a man of the":** Kamanda Wa Kamanda, "RDC: Du changement à la tête de l'Agence nationale du renseignement," Radio France Internationale, December 17, 2021, rfi.fr/fr/afrique /20211217-rdc-du-changement-à-la-tête-de-l-agence-nationale-du-renseignement.

346 **detention of Beya:** Romain Gras, "RDC: Dix choses à savoir sur Jean-Hervé Mbelu Biosha, le patron de l'ANR," *Jeune Afrique*, March 9, 2022, jeuneafrique.com/1314124/politique/rdc -dix-choses-a-savoir-sur-jean-herve-mbelu-biosha-le-patron-de-l-anr/.

347 **a "mafia network":** Stanis Bujakera and Hereward Holland, "Congo Virus Funds Embezzled by 'Mafia Network,' Says Deputy Minister," Reuters, July 8, 2020, reuters.com/article/world /congo-virus-funds-embezzled-by-mafia-network-says-deputy-minister-idUSKBN249224/.

Chapter 49: An African Electric Car

349 **Uganda's chameleonic president:** Conrad Comrade, "President Museveni Calls for Self-Sufficiency Among African Countries," Record FM, October 28, 2021, archived at web.archive .org/web/20211209020037/https://recordradio.co.ug/president-museveni-calls-for-self -sufficiency-among-african-countries/.

350 **overwhelmingly young population:** "President to Launch Kiira Motors Electric Trike at National Science Week 2023," *Independent*, September 28, 2023, independent.co.ug/president-to -launch-kiira-motors-electric-trike-at-national-science-week-2023/.

350 **Uganda should focus:** "Uganda Economic Update: Improving Public Spending on Health to Build Human Capital," World Bank, June 27, 2024, worldbank.org/en/country/uganda/publication /uganda-afe-economic-update-improving-public-spending-on-health-to-build-human-capital.

351 **"is a renaissance":** "Museveni Pledges Better Pay for Scientists," *New Vision*, November 24, 2011, www.newvision.co.ug/news/1003744/museveni-pledges-pay-scientists.

351 **progress has been slow:** "RDC: Félix Tshisekedi veut mettre en place une industrie de fabrication des batteries toute suite," Desk Eco, November 25, 2021, deskeco.com/index.php /2021/11/25/rdc-felix-tshisekedi-veut-mettre-en-place-une-industrie-de-fabrication-des -batteries-toute-suite.

351 **"Faced with the challenge":** Jonas Gerding and Carole Assignon, "Des batteries électriques bientôt produites en RDC?," Deutsche Welle (DW), October 6, 2023, p.dw.com/p/4X9Ta.

352 **are exported unprocessed:** "Blue Lines," *Africa Confidential* 65, no. 18 (September 6, 2024), africa-confidential.com/home/issue/id/1342.

352 **The report's authors urged:** Abdurrehman Naveed and Cina Vazir, "Value Amidst Transition: Evaluating Strategic Opportunities for Value Addition in the Democratic Republic of Congo," M-RCBG Associate Working Paper No. 204 (Mossavar-Rahmani Center for Business and Government, Harvard Kennedy School, June 2023), 60, hks.harvard.edu/centers/mrcbg /publications/awp/awp204.

353 **only around twenty thousand:** "Beijing's Green New Deal Pledges Factories in Africa," *Africa Confidential* 65, no. 18 (September 6, 2024), africa-confidential.com/home/issue/id/1342.

Chapter 50: Uncle Bunker

355 **launch the Minerals Security Partnership:** "Minerals Security Partnership: Media Note," U.S. Department of State, June 14, 2022, archived September 25, 2023, at web.archive.org/web /20230925075850/https://www.state.gov/minerals-security-partnership-june-14-2022/.

357 **dumped toxic waste:** Reuse and the Benefit to Community: Bunker Hill Mining and Metallurgical Complex Superfund Site (U.S. Environmental Protection Agency, December 2017), semspub.epa.gov/work/HQ/100001209.pdf.

357 **chain-smoking owner:** "Silver Valley," *Living on Earth*, August 8, 2003, loe.org/shows/segments .html?programID=03-P13-00032&segmentID=9.

357 **owned Bunker Hill:** "Hecla Mining Company to Pay $263 Million in Settlement to Resolve Idaho Superfund Site Litigation and Foster Cooperation," U.S. Department of Justice, June 13, 2011, justice.gov/opa/pr/hecla-mining-company-pay-263-million-settlement-resolve-idaho -superfund-site-litigation-and.

358 **"building an ecosystem"**: "The Regeneration Vision," Bunker Hill Mining Corporation, accessed July 16, 2024, bunkerhillmining.com/esg/the-regeneration-vision/.

358 **"drop in the bucket"**: Michael Holtz, "Idaho Is Sitting on One of the Most Important Elements on Earth," *Atlantic*, January 24, 2022, theatlantic.com/science/archive/2022/01/cobalt-clean-energy-climate-change-idaho/621321/.

358 **"arsenic is associated"**: Matthew Sletten et al., *Idaho Cobalt Operations: Form 43-101F1 Technical Report Feasibility Study, Idaho, USA* (Jervois Global, November 2020), 286, jervoisglobal.com/wp-content/uploads/2021/06/190348_Idaho_Cobalt_13112020_NI_43_101_Technical_Report-FILED-r1.pdf.

359 **"We're a long way"**: Senior U.S. official, interview with the author, March 2025.

361 **invested $40 million**: "Teck Invests $40M to Support Historic Bunker Hill Mine Restart in Idaho," Mining.com, March 6, 2025, www.mining.com/teck-invests-us40m-to-support-bunker-hill-mine-restart-in-idaho/.

Afterword: Power Dreams

366 **Leopold was desperate for a colony**: The king, ambitious and greedy for land overseas and the profit that might come of it, had tried, and failed, to buy a colony in the Philippines and in Fiji, as well as in Argentina and in Uruguay. See Adam Hochschild, *King Leopold's Ghost: A Story of Greed, Terror, and Heroism in Colonial Africa* (Houghton Mifflin, 1999), 38–41.

366 **the "flowering progenitor"**: Charles d'Ydewalle, *L'Union minière du Haut Katanga: De l'âge colonial à l'indépendance* (Librairie Plon, 1960), 39.

366 **Leopold used philanthropic**: The 1884 conference in Berlin was convened by Otto von Bismarck, the first German chancellor. Its object was ostensibly "furthering the moral and material well-being of the native populations," but it was, in fact, the forum at which Europe's "great powers" cemented their colonial claims over West and Central Africa. See "Preamble, General Act of the Berlin Conference," February 26, 1885, in Edward Hertslet, ed., *The Map of Africa by Treaty*, vol. 2, *Great Britain and France to Zanzibar* (Harrison and Sons, 1909), 468.

367 **they insulated the king**: This was the Compagnie du Congo pour le Commerce et l'Industrie. See Georges Defauwes, *Albert Thys, de Dalhem au Congo* (Collection Comté de Dalhem, 1995), 14.

367 **but Congo's forests**: Henry Morton Stanley, the progenitor of Congo's colonization, observed that "if every warrior living on the immediate banks of the Congo and its navigable affluents were to pick about a third of a pound in rubber each day throughout the year . . . and convey it to the trader for sale, five million pounds [over $665 million in 2025 U.S. dollars] worth of vegetable produce could be obtained without exhaustion of the wild forest productions." (Traveling some seven thousand miles between 1876 and 1877, Stanley made his way from the Congo River to the sea in a well-publicized expedition that brought him immense fame. Leopold used him as an agent to colonize the Congo Free State.) See Henry M. Stanley, *The Founding of the Congo Free State* (London: Sampson Low, Marston, Searle & Rivington, 1885), 355.

367 **If they refused to work**: Hochschild, *King Leopold's Ghost*, 125.

367 **"Leopold had not given any"**: Judy Pollard Smith, *Don't Call Me Lady: The Journey of Lady Alice Seeley Harris* (Abbott Press, 2014), 55.

368 **Thys certainly didn't**: Thys appears to have been convinced that he was some kind of charity worker. On his first visit, in 1887, he wrote to his wife that what the Belgians were doing in Congo was "an absolutely new colonial conception, and, truth be told, it is not a colony, which is something that dispossesses the indigenous man of his land, and considers the indigenous as a conquered race. In fact, here, the indigenous people are the citizens of the new State and the whites sent to Congo will be the provisional tutors to the Black population who will only be called to run public affairs when their education has been sufficiently carried out." See Albert Thys, "Lettre du 6 décembre 1887," as cited in Defauwes, *Albert Thys*, 3.

368 **Belgian-French materials technology**: "Sustainability," Umicore, accessed March 28, 2024, umicore.com/en/sustainability/.

368 **"would have been English"**: "Un Colonial par Semaine," *Journal du Congo*, 1928; and Defauwes, *Albert Thys*, 18.

369 **Tanganyika Concessions was**: For more on Tanganyika Concessions, see Kwame Nkrumah, *Neo-Colonialism: The Last Stage of Imperialism* (Thomas Nelson & Sons, 1965), 197.

369 **"the uncrowned queen of Belgium"**: Pierre Loppe, "L'histoire s'arrête pour la Société générale de Belgique," *La Libre*, October 29, 2003, lalibre.be/economie/entreprises-startup/2003

/10/29/lhistoire-sarrete-pour-la-societe-generale-de-belgique-JFPYEZQK2NAY7EAYN432 NOFKB4/.

369 **the current ownership structures:** Just look at the ownership of a copper-and-cobalt mine near the city of Kolwezi called the Congolaise des Mines et du Développement (COMIDE). On February 13, 2024, the state-owned mining firm had put what it called a "firm proposal" on the table to buy assets owned by the Eurasian Resources Group (ERG), a Kazakh company, which owned the COMIDE mine. ERG was partly owned through a corporate structure based in Holland, but it wasn't a Dutch company. Headquartered in the European microstate of Luxembourg, the company was 40 percent owned by the Kazakh government, and it had been founded by three of Central Asia's most powerful oligarchs. See Felix Njini and Veronica Brown, "Congo's Gecamines Offers to Buy Some of Kazakh Miner ERG's Copper Assets," Reuters, February 13, 2024, reuters.com/markets/commodities/congos-gecamines-offers-buy-some-khazakh-miner-ergs-copper-assets-2024-02-13/.

369 **"modest engineering success":** Hochschild, *King Leopold's Ghost*, 171.

370 **the Qing court:** David Shinn and Joshua Eisenman, *China and Africa: A Century of Engagement* (University of Pennsylvania Press, 2012), 26.

371 **the firm's general manager announced:** Peter Johnson, "This Chinese Brand Is Launching a Semi-Solid-State Battery EV in 2025, and It Won't Be Expensive," Electrek, December 16, 2024, https://electrek.co/2024/12/16/new-semi-solid-state-battery-ev-launching-2025/.

371 **counted thirty-six Chinese companies:** "Firms Are Exploring Sodium Batteries as an Alternative to Lithium," *Economist*, October 25, 2023.

372 **Scientists at research institutions:** Prabhat Ranjan Mishra, "Sodium-Ion Batteries Hit 458 Wh/kg: Breakthrough Material Closes Gap with Lithium," Interesting Engineering, December 22, 2024, https://interestingengineering.com/energy/sodium-batteries-breakthrough-material.

373 **money made at TFM:** Jean Luc Kayoko, interview with the author, March 2022.

373 **The firm had to pay:** *Voluntary Announcement: Announcement on the TFM Copper-Cobalt Mine in the DRC* (CMOC Group Limited, July 18, 2023), 1–2, https://www1.hkexnews.hk/listedco/listconews/sehk/2023/0718/2023071800667.pdf.

375 **"External actors have frequently":** Kevin C. Dunn, *Imagining the Congo: The International Relations of Identity* (Palgrave Macmillan, 2003), 9.

375 **"own self-interested objectives":** Howard W. French, "A History of Denial," *New York Review of Books*, April 19, 2018, nybooks.com/articles/2010/04/19/africa-history-of-denial/.

375 **"even subhuman nature":** French, "A History of Denial."

376 **"If investment doesn't happen":** Brian Menell, interview with the author, November 2023.

SELECTED BIBLIOGRAPHY

Afoaku, Osita G. "The U.S. and Mobutu Sese Seko: Waiting on Disaster." *Journal of Third World Studies* 14, no. 1 (Spring 1997): 65–90.

Africa Watch. "Zaire: Inciting Hatred: Violence Against Kasaiens in Shaba." *News from Africa Watch* 5, no. 10 (June 1993).

Amnesty International. "Democratic Republic of the Congo: Artisanal Miners at Risk as the Army Moves In." June 28, 2019. amnesty.org/en/documents/afr62/0625/2019/en/.

————. *"This Is What We Die For": Human Rights Abuses in the Democratic Republic of the Congo Power the Global Trade in Cobalt.* Amnesty International, 2016. amnesty.org/en/documents/afr62/3183/2016/en/.

Anderson, Jon Lee. *Che Guevara: A Revolutionary Life.* Grove Press, 1997.

Anvil Mining Limited. "Annual Information Form for Financial Year Ended December 31, 2004," 2004.

Aris, Muhammad, and T. Tamrin. "Heavy Metal (Ni, Fe) Concentration in Water and Histopathological of Marine Fish in the Obi Island, Indonesia." *Jurnal Ilmiah Platax* 8, no. 2 (July–December 2020): 221–33.

Asimov, Isaac. *Asimov on Chemistry.* Doubleday, 1975.

Askin, Steve, and Carole Collins. "External Collusion with Kleptocracy: Can Zaire Recapture Its Stolen Wealth?" *Review of African Political Economy*, no. 57 (1993): 72–85.

Assemblée Générale du Haut-Katanga. "Extrait du Procès Verbal de l'Assemblée Générale du 15 Novembre 2012." Lubumbashi, November 15, 2012.

AVZ Minerals Limited. "AVZ Minerals Completes Acquisition of Additional 5% of Manono Lithium and Tin Project." Australian Securities Exchange, April 18, 2019.

Bakajika Banjikila, Thomas. *Épuration ethnique en Afrique: Les "Kasaïens" (Katanga 1961–Shaba 1992).* L'Harmattan, 1997.

Baptista, P. J., and Amaro José. "Across Africa from Angola to Tette on the Zambeze." In *The Lands of Cazembe: Lacerda's Journey to Cazembe in 1798,* translated and annotated by Captain R. F. Burton. John Murray, 1873.

Baumann-Pauly, Dorothée. *Cobalt Mining in the Democratic Republic of the Congo: Addressing Root Causes of Human Rights Abuses.* NYU Stern Center for Business and Human Rights and Geneva Center for Business & Human Rights, February 2023. gcbhr.org/insights/2023/02/cobalt-mining-in-the-democratic-republic-of-the-congo-addressing-root-causes-of-human-rights-abuses.

Bazano. "Response to the Article Published by PREMICONGO on the Subject Mining Exploitation,

Environmental Management and Social Responsibility of Mining Companies in Katanga."
January 2013.

Benson, Emily, and Thibault Denamiel. "China's New Graphite Restrictions." CSIS, October 23, 2023. csis.org/analysis/chinas-new-graphite-restrictions.

Berkeley, Bill. "Zaire: An African Horror Story." *Atlantic*, August 1993.

Bertieaux, Raymond. "The Economy of the Belgian Congo in 1958–59." *Civilisations* 9, no. 3 (1959): 385–90.

Bevins, Vincent. *The Jakarta Method*. PublicAffairs, 2020.

Bigwood, E. J. *Du problème alimentaire des travailleurs de l'U.M.H.K. et de leurs familles*. 1965, 20–21. Box 2, folder 718. Union Minière Collection: Archives sur l'Exploitation des Mines de Cuivre Katangaises, Premier Série, Quatrième Partie. Archives de l'État en Belgique, Dépôt Joseph Cuvelier, Brussels, Belgium.

Blackwood, Matthew, and Catherine DeFilippo. "Germanium and Gallium: U.S. Trade and Chinese Export Controls." U.S. International Trade Commission, Executive Briefings on Trade, March 2024. usitc.gov/publications/332/executive_briefings/ebot_germanium_and_gallium.pdf.

Blas, Javier, and Jack Farchy. *The World for Sale: Money, Power, and the Traders Who Barter the Earth's Resources*. Oxford University Press, 2021.

Bujakera Tshiamala, Stanis. "RDC: Kinsevere, la mine de la discorde." *Jeune Afrique*, April 24, 2020. jeuneafrique.com/933847/economie-entreprises/rdc-kinsevere-la-mine-de-la-discorde/.

BYD. *2002 Annual Report*. BYD Company Ltd., 2003.

———. "BYD's Revolutionary Blade Battery: All You Need to Know." January 22, 2023. byd.com /eu/blog/BYDs-revolutionary-Blade-Battery-all-you-need-to-know.

———. *Green Tech for Tomorrow, Annual Report 2008*. BYD Inc., 2009.

———. *Interim Report*. BYD Inc., June 30, 2003.

Caijing. "Tsingshan's Indonesia Morowali Industrial Park: Build, and They Will Come." *Caijing* 30 (2019).

Carter Center. *A State Affair: Privatizing Congo's Copper Sector*. November 3, 2017. cartercenter.org /news/pr/drc-110317.html.

Chen, Si. "The Emerging Role of Chinese Transnational Corporations as Non-State Actors in Transnational Labour Law: A Case Study of Huayou Cobalt in the Global Cobalt Supply Chain." *Journal of Asian Sociology* 50, no. 1 (March 2021).

Choi, Yuri, Insung Lee, and Inkyeong Moon. "Geochemical and Mineralogical Characteristics of Garnierite from the Morowali Ni-Laterite Deposit in Sulawesi, Indonesia." *Frontiers in Earth Science* 9 (November 2021).

Cobalt Institute. *Cobalt Market Report 2023*. May 2024. cobaltinstitute.org/resource/cobalt-market -report-2023/.

Conrad, Joseph. "Geography and Some Explorers." In *Last Essays*. Doubleday, Page & Company, 1926.

Consulate General of the People's Republic of China in New York. "High Tech Research and Development (863) Programme." October 21, 2003.

Coons, Christopher A., et al. Letter to President Joseph R. Biden from Senator Christopher A. Coons et al. United States Senate, July 13, 2021.

Copetas, A. Craig. *Metal Men: Marc Rich and the 10-Billion-Dollar Scam*. Harper & Row, 1985.

Covert, Bryce. "The Toxic Culture at Tesla." *Nation*, April 9, 2024.

Crair, Ben. "The Island Where Environmentalism Implodes." *New Yorker*, November 23, 2024.

Craven, Matthew. "Between Law and History: The Berlin Conference of 1884–1885 and the Logic of Free Trade." *London Review of International Law* 3, no. 1 (2015): 31–59.

Custers, Raf, and Sara Nordbrand. *Risky Business: The Lundin Group's Involvement in the Tenke Fungurume Mining Project in the Democratic Republic of Congo*. International Peace Information Service, Swedwatch, and Diakonia, February 2008.

Dalio, Ray. *Principles*. Avid Reader Press, 2021.

Dayal, Rajeshwar. *Report on Recent Developments in Northern Katanga from the Special Representative of the Secretary-General*. United Nations Security Council, February 1961. digitallibrary.un.org/record /630681?v=pdf.

Deberdt, Raphael. *Baseline Study of Artisanal and Small-Scale Cobalt Mining in the Democratic Republic of the Congo*. Responsible Sourcing Network and UBC Anthropology, July 2021. securityhuman rightshub.org/media/pdf/resources/Baseline+Study+Cobalt+ASM+Mining+-+7.28.2021.pdf.

Defauwes, G. *Albert Thys, de Dalhem au Congo*. Collection Comté de Dalhem, 1995.

Demarais, Agathe. "How the U.S.-Chinese Technology War Is Changing the World." *Foreign Policy*, November 19, 2022.

Democratic Republic of the Congo. "Jugement RAC 122." *Journal Officiel de la République Démocratique du Congo*, première partie, no. 15, August 1, 2020.

———. "Loi n° 007/2002 du 11 juillet 2002 portant code minier." *Journal Officiel de la République Démocratique du Congo*, no. spécial du 15 juillet 2002 (Kinshasa, 2002).

Denys Perier, Gaston. *Moukanda*. 2nd ed. Office de Publicité, J. Lebègue & Cie., 1924.

Desmond, Kevin. *Gustave Trouvé: French Electrical Genius (1839–1902)*. McFarland & Co., 2015.

———. *Innovators in Battery Technology: Profiles of 95 Influential Electrochemists*. McFarland, 2016.

Dietrich, Christian. "Have African-Based Diamond Monopolies Been Effective?" *Central Africa Minerals and Arms Research Bulletin*, International Peace Information Service (IPIS), June 28, 2001.

Dikötter, Frank. *China After Mao: The Rise of a Superpower*. Bloomsbury, 2022.

Dunn, Kevin C. *Imagining the Congo: The International Relations of Identity*. Palgrave Macmillan, 2003.

Dusausoy, Jean. *Kolwezi 1977: Un Technicien Belge dans les Mines du Katanga*. Lucpire Éditions, 2018.

Eastern District of New York. United States v. OZ Africa Management GP, LLC, 2016.

Economy, Elizabeth. "China's Round Two on Electric Cars: Will It Work?" Council on Foreign Relations, April 17, 2014.

Edsall, Larry. *Chevrolet Volt: Charging into the Future*. Motorbooks, 2010.

Egan, Dan. *The Devil's Element: Phosphorus and a World Out of Balance*. W. W. Norton, 2023.

Eriksson, Rolf. *Adolf H. Lundin: No Guts No Glory*. Ehrenblad Editions, 2003.

Esser, Christian, Manka Heise, and Tina Kaiser. "Vorfall in Teslas Gigafactory: Wie eine illegale Tankstelle unter einem Partyzelt verschwand." *Stern*, September 20, 2023.

Fage, J. D., and Roland Anthony Oliver, eds. *The Cambridge History of Africa*. 8 vols. Cambridge University Press. 1975–86.

Fayol, Clément, Antoine Harari, and Pete Jones. "'I Will Take My Percentage': Congolese Presidential Adviser Caught on Tape Negotiating Corrupt Mineral Deal." OCCRP, September 15, 2022.

Fletcher, Seth. *Bottled Lightning: Superbatteries, Electric Cars, and the New Lithium Economy*. Hill and Wang, 2011.

Forrest, George Arthur. *Un siècle de rêves: Ensemble, bâtissons l'avenir*. Le Cherche Midi, 2022.

French, Howard W. "A History of Denial." *New York Review of Books*, April 19, 2018.

———. *China's Second Continent: How a Million Migrants Are Building a New Empire in Africa*. Knopf, 2014.

Friend, Tad. "Plugged In." *New Yorker*, August 17, 2009.

Gately, Dermot. "Lessons from the 1986 Oil Price Collapse." *Brookings Papers on Economic Activity* 1986, no. 2 (1986): 237–84. www.brookings.edu/wp-content/uploads/1986/06/1986b_bpea_gately _adelman_griffin.pdf.

La Générale de Carrières et de Mines. "Contrat conclu entre GECAMINES et SAMREF CONGO Sprl portant création de la société MUTANDA YA MUKONKOTA MINING (MUMI)," No. 474/10300/SG/GC/2001, Annexe I au procès-verbal synthétique de la réunion extraordinaire du CA, December 3, 2008. Kinshasa, Democratic Republico of the Congo.

———. "Contrat de création de société entre la Générale de Carrières et de Mines et SAMREF-Congo pour l'exploitation du gisement de Mutanda ya Mukonkota," No. 474/10300/SG/GC /2001, May 2001. Kinshasa, Democratic Republic of the Congo.

Gilliam, Eva. "DRC: Mai Mai Leader Gedeon of Manono Territory—Known 'Good Guy,' Accused Cannibal." MONUC, April 14, 2004.

Global Witness. *China and Congo: Friends in Need*. March 2011. archiv.kongo-kinshasa.de/dokumente /ngo/gw_rep_0311_en.pdf.

———. *Paying for Protection: The Freeport Mine and the Indonesian Security Forces*. July 2005.

———. *Glencore and the Gatekeeper*. May 2014. globalwitness.org/en/campaigns/corruption-and -money-laundering/glencore-and-the-gatekeeper/.

———. *Out of Africa*. May 2016. globalwitness.org/en/campaigns/corruption-and-money-laundering /out-of-africa/.

———. *A Rush for Lithium in Africa Risks Fuelling Corruption and Failing Citizens*. November 2023. globalwitness.org/en/campaigns/transition-minerals/a-rush-for-lithium-in-africa-risks -fuelling-corruption-and-failing-citizens/.

Golder Associates. *Environmental and Social Impact Assessment: Executive Summary, Submitted to Tenke Fungurume Mining S.A.R.L. (TFM), Democratic Republic of the Congo*. March 2007.

Goodenough, John B. *Witness to Grace*. PublishAmerica, 2008.

Gras, Romain. "RDC: Dix choses à savoir sur Jean-Hervé Mbelu Biosha, le patron de l'ANR." *Jeune Afrique*, March 9, 2022.

Guevara, Che. *Congo Diary: Episodes of the Revolutionary War in the Congo.* Seven Stories Press, 2011.

Gulley, Andrew L. "China, the Democratic Republic of the Congo, and Artisanal Cobalt Mining from 2000 Through 2020." *Proceedings of the National Academy of Sciences* 120, no. 26 (June 2023): art. e2212037120.

———. "One Hundred Years of Cobalt Production in the Democratic Republic of the Congo." *Resources Policy* 79 (2022): art. 103007.

Hammond, John Hays. "Good Advice to Mine Investors." *Gilpin Observer*, March 30, 1911.

Harrison, Mark. "The Soviet Economy, 1917–1991: Its Life and Afterlife." *Independent Review* 22, no. 2 (2017): 199–206.

Herbert, Eugenia W. *Red Gold of Africa: Copper in Precolonial History and Culture.* University of Wisconsin Press, 1984.

Hertslet, Edward, ed. *The Map of Africa by Treaty.* Vol. 2. *Great Britain and France to Zanzibar.* Harrison and Sons, 1909.

Hesketh, Thérèse. "Too Many Males in China: The Causes and the Consequences." *Significance* 6, no. 1 (2009): 9–13.

Hidayat, Bagja, and Dini Pramita. "Kawasi Muddied by Nickel." *Tempo*, February 7, 2022.

Hochschild, Adam. *King Leopold's Ghost: A Story of Greed, Terror, and Heroism in Colonial Africa.* Houghton Mifflin, 1998.

Hollmer, Anna. "Unveiling the True Nature of the Artisanal Cobalt Mines in the Democratic Republic of Congo: Systemic Conditions in the Mines and Implications for Global Supply Chains and Social Justice." Thesis, Anglo-American University in Prague, 2024.

Holtz, Michael. "Idaho Is Sitting on One of the Most Important Elements on Earth." *Atlantic*, January 24, 2022.

Huayou Cobalt. *2020 Environmental, Social & Governance Report.* 2021.

———. *2023 Semi-Annual Report of Huayou Cobalt.* August 2023.

———. *Keep Moving Forward with Your Dreams: Huayou Cobalt 2023 Cultural Case Story Collection.* 2023.

Huber, Isabel. "Indonesia's Nickel Industrial Strategy." Center for Strategic and International Studies, December 8, 2021.

Huckman, Robert S., and Alan D. MacCormack. "BYD Company, Ltd." Harvard Business School Case No. 606-139. HBS Case Collection, April 2006, revised September 15, 2009.

Human Rights Watch. *Without Remedy: Human Rights Abuse and Indonesia's Pulp and Paper Industry.* January 2003. hrw.org/report/2003/01/06/without-remedy/human-rights-abuse-and-indonesias-pulp-and-paper-industry.

Iguma Wakenge, Claude. *Eating the Congo: Unveiling State Governance of Copper and Cobalt Mining in Former Katanga.* Lambert Academic Publishing, 2019.

Illich, Ivan. *Energy and Equity.* Harper & Row, 1974.

International Crisis Group. *How Kabila Lost His Way: The Performance of Laurent-Désiré Kabila's Government.* ICG Democratic Republic of Congo Report No. 3. May 1999.

———. *Katanga: The Congo's Forgotten Crisis.* ICG Africa Report No. 103. January 2006. crisisgroup.org/africa/central-africa/democratic-republic-congo/katanga-congo-s-forgotten-crisis.

———. *Storm Clouds Over Sun City: The Urgent Need to Recast the Congolese Peace Process.* ICG Africa Report No. 44. May 2002. crisisgroup.org/africa/central-africa/democratic-republic-congo/storm-clouds-over-sun-city-urgent-need-recast-congolese-peace-process.

International Energy Agency. *Global EV Outlook 2023: Catching Up with Climate Ambitions.* April 2023. iea.org/reports/global-ev-outlook-2023/trends-in-batteries.

Isaacson, Walter. *Elon Musk.* Simon & Schuster, 2023.

Ishioka, Munehiro, Toshihiko Okada, Kenji Matsubara, Michio Inagaki, and Yoshihiro Hishiyama. "Electrical Resistivity, Magnetoresistance, and Morphology of Vapor-Grown Carbon Fibers Prepared in a Mixture of Benzene and Linz–Donawitz Converter Gas by Floating Catalyst Method." *Journal of Materials Research* 8, no. 8 (August 1993): 1866-74.

Itoh, Masanao, Ryoji Koike, and Masato Shizume. "Bank of Japan's Monetary Policy in the 1980s: A View Perceived from Archived and Other Materials." *Monetary and Economic Studies* 33 (November 2015): 97–200.

Ivanhoe Mines. "Ivanhoe Mines' Exports Commence from Kamoa-Kakula Copper Complex Along Lobito Atlantic Rail Corridor." January 2, 2024.

Jaringan Advokasi Tambang. "Sufferings of Residents and Environment Behind the Business Transactions of Tesla and Chinese Companies in Indonesia." JATAM, August 10, 2022, archived May 12,

2023. https://web.archive.org/web/20230512120915/jatam.org/en/sufferings-of-residents-and-environment-behind-the-business-transactions-of-tesla-and-chinese-companies-in-indonesia/.

Jervois Global. *Idaho Cobalt Operations: Form 43-101F1 Technical Report Feasibility Study, Idaho, USA.* November 2020. jervoisglobal.com/wp-content/uploads/2021/06/190348_Idaho_Cobalt_13112020_NI_43_101_Technical_Report-FILED-r1.pdf.

Johnson, Chalmers. *MITI and the Japanese Miracle: The Growth of Industrial Policy, 1925–1975.* Stanford University Press, 1982.

Kalonji Ditunga, Albert. *Congo 1960: La sécession du Sud-Kasaï.* L'Harmattan, 2005.

Kara, Siddharth. *Cobalt Red: How the Blood of the Congo Powers Our Lives.* St. Martin's Press, 2023.

Katanga Mining Limited. *Kamoto Copper Company Technical Report.* November 2019.

Katumba Mwanke, Augustin. *Ma verité.* EPI, 2013.

Kennes, Erik. "Le secteur minier au Congo: 'Déconnexion' et descente aux enfers." In *L'Afrique des Grands Lacs: Annuaire 1999–2000.* Edited by Filip Reyntjens and Stefaan Marysse. L'Harmattan, 2000.

Kennes, Erik, and Miles Larmer. *The Katangese Gendarmes and War in Central Africa: Fighting Their Way Home.* Indiana University Press, 2016.

el-Khawas, Mohamed A. "China's Changing Policies in Africa." *Issue: A Journal of Opinion* 3, no. 1 (1973): 24–28.

Kisangani, Emizet François. *The Belgian Congo as a Developmental State: Revisiting Colonialism.* Routledge, 2023.

Konopka, Gustaw et al. "Ni-Co Bearing Laterites from Halmahera Island (Indonesia)." *Applied Sciences* 12, no. 15 (2022).

Kummer, Joseph T., and Neill Weber. "A Sodium-Sulfur Secondary Battery." *SAE Transactions* 76 (1968): 1003–7, 1023–28.

Lefever, Ernest. *Crisis in the Congo.* Brookings Institution, 1965.

LeVine, Steve. *The Powerhouse: Inside the Invention of a Battery to Save the World.* Viking, 2015.

Li, Daqian. *The Creative Wisdom of Wang Chuanfu.* Translated by Denis Mair. China Intercontinental Press, 2013.

Lin Neo, Mei, and Peter A. Todd. "Conservation Status Reassessment of Giant Clams (Mollusca: Bivalvia: Tridacninae) in Singapore." *Nature in Singapore* 6 (2013): 125–33.

Lumumba, Patrice. *La pensée politique de Patrice Lumumba.* Edited by Jan Van Lierde. Présence Africaine, 2003.

Lynn, Leonard. "Japanese Technology: Successes and Strategies." *Current History* 82, no. 487 (1983): 366.

Macola, Giacomo, and James Hogan. "Guerrilla Warfare in Katanga: The Sanga Rebellion of the 1890s and Its Suppression." *Small Wars and Insurgencies* 30, no. 4–5 (2019).

Mallick, Sourav, et al. "Low-Cobalt Active Cathode Materials for High-Performance Lithium-Ion Batteries: Synthesis and Performance Enhancement Methods." *Journal of Materials Chemistry A* 11, no. 8 (2023): 3789–821.

Manthiram, Arumugam, and John B. Goodenough. "Lithium Insertion into Fe2(MO4)3 Frameworks: Comparison of M = W with M = M." *Journal of Solid State Chemistry* 71, no. 2 (1987).

Marquis, Christopher, Hongyu Zhang, and Lixuan Zhou. "China's Quest to Adopt Electric Vehicles." *Stanford Social Innovation Review* (Spring 2013): 52–57.

Marshall, Shelley, Samantha Balaton-Chrimes, and Omar Pidani. *Access to Justice for Communities Affected by the PT Weda Bay Nickel Mine—Interim Report.* Non-Judicial Human Rights Redress Mechanisms Project, September 2013.

Marysse, Stefaan, and Sara Geenen. "Les contrats chinois en RDC: L'impérialisme rouge en marche?" In *L'Afrique des Grands Lacs: Annuaire 2007–2008.* L'Harmattan, 2008.

Mathieu, P. L. "Résumé de l'étude 'African Development,'" 1960. Box A.4, Folder 843, Union Minière Collection: Archives sur l'Exploitation des Mines de Cuivre Katangaises, Premier Série, Quatrième Partie. Archives de l'État en Belgique, Dépôt Joseph Cuvelier, Brussels, Belgium.

McNish, Jacquie. *The Big Score: Robert Friedland, INCO, and the Voisey's Bay Hustle.* Doubleday Canada, 1998.

Meerts, Pierre, and Michel Hasson. *Arbres et arbustes du Haut Katanga.* Jardin Botanique Meise, 2016.

Melosi, Martin V. "Energy and Environment in the United States: The Era of Fossil Fuels." *Environmental Review* 11, no. 3 (1987).

Miatto, Alessio, Barbara K. Reck, James West, and Thomas E. Graedel. "The Rise and Fall of American Lithium." *Resources, Conservation and Recycling* 162 (November 2020).

Michigan Economic Development Corporation. "This Just In: Team Michigan Continues to Deliver as Our Next Energy Rolls Out First Cell Production." November 1, 2023.

Mining Watch. "INCO in Indonesia: A Report for Canadian People." January 7, 2000.

Mohammed VI of Morocco. Letter to George W. Bush, 2004. James A. Baker III Collection. Mudd Manuscript Library, Princeton Library.

Montrie, Chad. "'To Have, Hold, Develop, and Defend': Natural Rights and the Movement to Abolish Strip Mining in Eastern Kentucky." *Journal of Appalachian Studies* 11, no. ½ (2005): 64–82.

Mpisi, Jean. *La Sicomines (Sino-congolaise des mines).* L'Harmattan, 2020.

Mukenge, Tshilemalema. *Culture and Customs of the Congo.* Bloomsbury Academic, 2002.

Murray, Charles. *Long Hard Road: The Lithium-Ion Battery and the Electric Car.* Purdue University Press, 2022.

Nagle, John C. "The Earth Day Pioneer Nobody Remembers." *Scientific American,* April 22, 2016.

Naipaul, V. S. "A New King for the Congo." *New York Review of Books,* June 26, 1975.

National Bureau of Statistics of China. "Statistical Communiqué of the People's Republic of China on the 2009 National Economic and Social Development." January 26, 2010.

Naveed, Abdurrehman, and Cina Vazir. "Value Amidst Transition: Evaluating Strategic Opportunities for Value Addition in the Democratic Republic of Congo." M-RCBG Associate Working Paper No. 204. Mossavar-Rahmani Center for Business and Government, Harvard Kennedy School, June 2023.

Ngoie Tshibambe, Germain. "Sans Chinatown? L'intégration des migrants chinois à Lubumbashi (DRC)." *Revista da Faculdade de Direito da UFMG* 63 (2013).

Ngoy Mwanabute, Serge-Noël. *Toute la vérité sur les guerres de l'est du Congo-Kinshasa: Comment mettre fin définitivement à ces guerres?* Éditions Universitaires Européennes, 2023.

Niarchos, Nicolas. "Beijing Calls Washington's Bluff on Strategic Metals." *Nation,* January 2, 2025.

———. "Buried Dreams." *New Yorker,* May 24, 2021.

———. "Dirty Nickel, Clean Power: Making the Ocean Bleed Red." *Nation,* March 6/13, 2023.

———. "In Congo's Cobalt Mines." *New York Review of Books,* December 7, 2023.

———. "Power Metals." *Granta,* June 5, 2024.

———. "What Happened to Patrick Masengo Kalasa?" *Nation,* October 1, 2024.

Nkrumah, Kwame. *Neo-Colonialism: The Last Stage of Imperialism.* Thomas Nelson & Sons, 1965.

Ober, Joyce. "Lithium." In *Minerals Yearbook: Metals and Minerals.* Edited by the U.S. Geological Survey and the U.S. Department of the Interior. U.S. Government Printing Office, 1998.

Office of Assistant to Deputy Cabinet Secretary for State Documents & Translation. "President Jokowi Pushes for Down Streaming of Mining Products." Cabinet Secretariat of the Republic of Indonesia. November 30, 2022.

Olivier, Mathieu, and Romain Gras. "Exclusif—Dan Gertler: 'Tous étaient effrayés par le Congo, mais pas moi.'" *Jeune Afrique,* August 1, 2022.

———. "RDC: Dan Gertler, l'irrésistible ascension du businessman de Kabila." *Jeune Afrique,* August 2, 2002.

Padhi, A. K., K. S. Nanjundaswamy, and John B. Goodenough. "Phospho-Olivines as Positive-Electrode Materials for Rechargeable Lithium Batteries." *Journal of the Electrochemical Society* 144, no. 4 (1997).

Pakenham, Thomas. *The Scramble for Africa, 1876–1912.* Random House, 1991.

People's Liberation Army of China. *The Politics of the Chinese Red Army.* Edited by James Chester Cheng. Hoover Institution, 1966.

Pereira, Nathalie, Glenn G. Amatucci, M. Stanley Whittingham, and Robert Hamlen. "Lithium–Titanium Disulfide Rechargeable Cell Performance After 35 Years of Storage." *Journal of Power Sources* 280, no. 15 (April 2015): 18–22.

Peyer, Chantal, and François Mercier. *Glencore in the Democratic Republic of Congo: Profit Before Human Rights and the Environment.* Swiss Catholic Lenten Fund, 2012.

Pilling, David. *Bending Adversity.* Penguin Books, 2014.

Pitron, Guillaume. *La Guerre des Métaux Rares.* Nouvelle Édition. Les Liens qui Libèrent, 2023.

Pollard Smith, Judy. *Don't Call Me Lady: The Journey of Lady Alice Seeley Harris.* Abbott Press, 2014.

Pratt, J. A. "Exxon and the Control of Oil." *Journal of American History* 99, no. 1 (2012).

Prunier, Gérard. *Africa's World War: Congo, the Rwandan Genocide, and the Making of a Continental Catastrophe.* Oxford University Press, 2009.

Radmann, Wolf. "The Nationalization of Zaire's Copper: From Union Minière to Gecamines." *Africa Today* 25, no. 4 (1978).

Reagan, Ronald. "Our Environment Crisis." *Nation's Business*, February 1970.

Reefe, Thomas Q. *The Rainbow and the Kings: A History of the Luba Empire to 1891.* University of California Press, 1981.

Reeves, Richard. *President Nixon: Alone in the White House.* Simon & Schuster, 2001.

Reid, Stuart A. *The Lumumba Plot: The Secret History of the CIA and a Cold War Assassination.* Alfred A. Knopf, 2023.

Rights and Accountability in Development. *Anvil Mining Limited and the Kilwa Incident: Unanswered Questions.* October 2005. raid-uk.org/wp-content/uploads/2023/04/qq-anvil.pdf.

———. *Beneath the Green.* March 2024. raid-uk.org/post-library/report-beneath-the-green/.

———. "Military Court Delivers a Not Guilty Verdict in Kilwa Trial." July 2, 2007. https://raid-uk.org/post-library/pr-kilwa-verdict/.

———. *The Road to Ruin?: Electric Vehicles and Workers' Rights Abuses at DR Congo's Industrial Cobalt Mine.* November 2021. raid-uk.org/post-library/the-road-to-ruin-electric-vehicles-and-workers-rights-abuses-at-dr-congos-industrial-cobalt-mines/.

Rosenfels, Ryan C., and Bjorn P. von der Heyden. "A Critical Comparison Between the Fungurume 8 and 88 Cu-Co Deposits, Central African Copperbelt." *Ore Geology Reviews* 140 (January 2022).

Ross, John, and David de Vries. "Mufulira Smelter Upgrade Project." Mopani Copper Mines PLC, Glencore Technologies, 2005. glencoretechnology.com/.rest/api/v1/documents/3f912f e463bebc7f0de40789ad95b319/XTpaper_Mopani_MSUPyromet05.pdf.

Rotberg, Robert I. "Plymouth Brethren and the Occupation of Katanga, 1886–1907." *Journal of African History* 5, no. 2 (1964).

Rubbens, Antoine. "Political Awakening in the Belgian Congo." *Civilisations* 10, no. 1 (1960).

Rubbers, Benjamin. "La dislocation du secteur minier au Katanga (RDC)." *Politique Africaine* 1, no. 93 (2004).

Sanderson, Henry. *Volt Rush: The Winners and Losers in the Race to Go Green.* Oneworld Publications, 2022.

Sawai, Tomoki. "The Invention of Rechargeable Batteries: An Interview with Dr. Akira Yoshino, 2019 Nobel Laureate." *WIPO Magazine*, September 2020.

Tamrin, T., and Muhammad Aris. "Health Condition of *Tridacna* sp. in the Waters of Obi Island, Indonesia." *Jurnal Ilmiah Platax* 8, no. 2 (2020): 234–41.

Tenke Fungurume Mining. "An Investment in the Future of the Democratic Republic of the Congo." 2013. Archived at https://int.nyt.com/data/documenttools/2013-freeport-tfm-tenke-fast-facts-english/09789d8b8b64c1fe/full.pdf.

Schouten, Pieter. *Roadblock Politics: The Origins of Violence in Central Africa.* Cambridge University Press, 2022.

ScienceInsider. "How Many Have Died Due to Congo's Fighting? Scientists Battle Over How to Estimate War-Related Deaths." *Science*, January 21, 2010.

Scrosati, Bruno. *Fast Ion Transport in Solids.* Edited by S. Pizzini, B. Scrosati, A. Magistris, C. M. Mari, and G. Mariotto. Springer Netherlands, 1993.

The Sentry. *The Backchannel: State Capture and Bribery in Congo's Deal of the Century.* November 2021. thesentry.org/reports/backchannel/.

Shedd, K. B. "Cobalt." In *Metal Prices in the United States Through 2010.* Scientific Investigations Report 2012–5188. Edited by the U.S. Geological Survey National Minerals Information Center Staff. U.S. Department of the Interior and U.S. Geological Survey, 2013.

Shinn, David Hamilton, and Joshua Eisenman. *China and Africa: A Century of Engagement.* University of Pennsylvania Press, 2012.

Shnayerson, Michael. *The Car That Could: The Inside Story of GM's Revolutionary Electric Vehicle.* Random House, 1996.

Stanley, Henry M. *The Founding of the Congo Free State.* Sampson Low, Marston, Searle & Rivington, 1885.

Stearns, Jason. *Dancing in the Glory of Monsters: The Collapse of Congo and the Great War of Africa.* PublicAffairs, 2011.

Stockwell, John. *In Search of Enemies: A CIA Story.* Norton, 1978.

Takeuchi, Hiroki. "Political Economy of Trade Protection: China in the 1990s." *International Relations of the Asia-Pacific* 13, no. 7 (2012).

Tarnoff, Ben. "Ultra Hardcore." *New York Review of Books*, January 18, 2024.

Terazawa, Tatsuya. "How Japan Solved Its Rare Earth Minerals Dependency Issue." World Economic Forum, October 13, 2023.

Thackeray, Michael M. *Running with Lithium—Empowering the Earth: A Personal Journey.* Archway Publishing, 2019.

Thomson, Joseph B. *Joseph Thomson, African Explorer.* London, 1896.

Tibbetts, Gary G. "Vapor-Grown Carbon Fibers." In *Carbon Fibers Filaments and Composites.* Edited by J. L. Figueiredo, C. A. Bernardo, R. T. K. Baker, and K. J. Hüttinger. NATO ASI Series, vol. 177. Springer, 1990.

Trapido, Joe. "Africa's Leaky Giant." *New Left Review* 92 (Mar/Apr 2015).

UNESCAP. "CHINA: 12th Five-Year Plan (2011–2015) for National Economic and Social Development." 2011. policy.asiapacificenergy.org/node/37.

Union Minière du Haut-Katanga. *Économie katangaise et économie congolaise à la veille de l'indépendance,* SD No. 61/5, 1 July 1961. Box 4, folder 854. Union Minière Collection: Archives sur l'Exploitation des Mines de Cuivre Katangaises, Premier Série, Quatrième Partie. Archives de l'État en Belgique, Dépôt Joseph Cuvelier, Brussels, Belgium.

———. "Les causes de la situation économique et financière désastreuse au Congo," 1962. Box A4, Folder 870. Union Minière Collection: Archives sur l'Exploitation des Mines de Cuivre Katangaises, Premier Série, Quatrième Partie. Archives de l'État en Belgique, Dépôt Joseph Cuvelier, Brussels, Belgium.

———. "NOTE concernant les contrats de raffinage Union Minière / Hoboken," UMHK, 23 February 1960. Box C4, folder 1002. Union Minière Collection: Archives sur l'Exploitation des Mines de Cuivre Katangaises, Premier série, Quatrième partie. Archives de l'État en Belgique, Dépôt Joseph Cuvelier, Brussels, Belgium.

———. "Note sur les Grandes entreprises, seul soutien actuel de l'économie congolaise," 4 June 1962. Box A4, folder 869. Union Minière Collection: Archives sur l'Exploitation des Mines de Cuivre Katangaises, Premier Série, Quatrième Partie. Archives de l'État en Belgique, Dépôt Joseph Cuvelier, Brussels, Belgium.

United Nations. *Addendum to the Report of the Panel of Experts on the Illegal Exploitation of Natural Resources and Other Forms of Wealth of DR Congo.* United Nations Document No. S/2001/1072. November 10, 2001. securitycouncilreport.org/un-documents/document/drc-s-2001-1072.php.

———. *Report of the Panel of Experts on the Illegal Exploitation of Natural Resources and Other Forms of Wealth of DR Congo.* United Nations Document No. S/2001/357. April 12, 2001. securitycouncilreport .org/atf/cf/%7B65BFCF9B-6D27-4E9C-8CD3-CF6E4FF96FF9%7D/DRC%20S%202001 %20357.pdf.

———. *Sixteenth Report of the Secretary-General on the United Nations Organization Mission in the Democratic Republic of the Congo.* United Nations Document No. S/2004/1034. December 2004. digital library.un.org/record/538578?ln=en&v=pdf.

U.S. Central Intelligence Agency. "Angola: UNITA vs. the Benguela." National Foreign Assessment Center Memorandum. November 30, 1978, declassified December 2, 2004. cia.gov/readingroom /docs/CIA-RDP80T00634A000500010013-7.pdf.

———. *China's Role in Africa.* Special Report Weekly Review. February 25, 1972, declassified August 24, 2012. cia.gov/readingroom/docs/CIA-RDP08S02113R000100080001-0.pdf.

———. "Spanish Sahara: Phosphates and Sovereignty." Intelligence Memorandum. September 1970, partially declassified October 31, 2011. cia.gov/readingroom/docs/CIA-RDP85T00875R 001600030125-0.pdf.

———. *Zaire: Mobutu and the Military.* August 1982, declassified July 30, 2008. cia.gov/readingroom /docs/CIA-RDP83S00855R000100080001-6.pdf.

U.S. Congress. *House Hearing: The Future of U.S.-China Relations and the Possible Accession of China into the World Trade Organization.* 105th Cong., 106 (1995) (testimony of Robert R. Aronson).

———. *U.S.-China Trade Relations and Renewal of China's Most-Favored-Nation Status.* 104th Cong. 132 (1995) (testimony of Robert R. Aronson).

U.S. Congress, Senate. *Strip Mining and Its Impact: Hearings Before the Senate Committee on Interior and Insular Affairs.* 90th Cong. 87-95 (1968) (testimony of Harry M. Caudill).

U.S. Department of Defense. *Securing Defense-Critical Supply Chains: An Action Plan Developed in Response to President Biden's Executive Order 14017.* February 2022. media.defense.gov/2022/Feb/24 /2002944158/-1/-1/1/DOD-EO-14017-REPORT-SECURING-DEFENSE-CRITICAL -SUPPLY-CHAINS.PDF.

U.S. Department of Justice. "Hecla Mining Company to Pay $263 Million in Settlement to Resolve Idaho Superfund Site Litigation and Foster Cooperation." June 13, 2011. justice.gov/archives /opa/pr/hecla-mining-company-pay-263-million-settlement-resolve-idaho-superfund-site -litigation-and.

U.S. Department of State. "Minerals Security Partnership: Media Note." June 14, 2022.

U.S. Department of the Treasury. "Treasury Sanctions Two Individuals for Threatening the Stability of and Undermining Democratic Processes in the Democratic Republic of the Congo." September 26, 2016. treasury.gov/news/press-releases/jl0560.

———. "United States Sanctions Human Rights Abusers and Corrupt Actors Across the Globe." December 21, 2017. treasury.gov/news/press-releases/sm0243.

U.S. Embassy Zambia. Memorandum of Understanding (MOU) on Working Arrangements Between the Government of the United States of America, the European Commission, the Government of the Republic of Zambia, the Government of the Republic of Angola, the Government of the Democratic Republic of the Congo, the African Development Bank, and Africa Finance Corporation Relating to the Development of the Lobito Corridor and the Zambia–Lobito Rail Line." May 2024. zm.usembassy.gov/wp-content/uploads/sites/44/2024/05/Lobito_Corridor _MOU.pdf.

U.S. Environmental Protection Agency. *Reuse and the Benefit to Community Bunker Hill Mining and Metallurgical Complex Superfund Site.* December 2017. semspub.epa.gov/work/HQ/100001209 .pdf.

Van Brusselen, Daan, et al. "Metal Mining and Birth Defects: A Case-Control Study in Lubumbashi, Democratic Republic of the Congo." *Lancet: Planetary Health* 4, no. 4 (2020).

Van Gorsel, H. "Geological Investigations of Sulawesi (Celebes) Before 1930." *Berita Sedimentologi* [Indonesian Journal of Sedimentary Geology] 48, no. 1 (2022).

Van Leeuwen, Theo, and P. E. Pieters. "Mineral Deposits of Sulawesi." Paper presented at the Proceedings of the Sulawesi Mineral Resources 2011 Seminar MGEI-IAGI, November 28–29, 2011.

Van Reybrouck, David. *Congo: The Epic History of a People.* Translated by Sam Garrett. HarperCollins, 2014.

———. *Revolusi: Indonesia and the Birth of the Modern World.* W. W. Norton & Company, 2024.

Vansina, Jan. *Kingdoms of the Savanna.* University of Wisconsin Press, 1966.

Vickers, Adrian. *A History of Modern Indonesia.* Cambridge University Press, 2013.

Vinckel, Sandrine. "Violence and Everyday Interactions Between Katangese and Kasaians: Memory and Elections in Two Katanga Cities." *Africa: Journal of the International African Institute* 85, no. 1 (2015).

Vogel, Christoph N. *Conflict Minerals, Inc.: War, Profit and White Saviourism in Eastern Congo.* Hurst Publishers, 2022.

Vogl, Thomas J., Wolfgang Reith, and Ernst J. Rummeny. *Diagnostic and Interventional Radiology.* Springer, 2016.

Wala Chabala, E. D. "Lobito Corridor—A Reality Check." African Policy Research Institute, February 2, 2024.

Whitaker, John C. *Striking a Balance: Environment and Natural Resources Policy in the Nixon-Ford Years.* AEI-Hoover Policy Studies, 1976.

Whittingham, M. Stanley. *Intercalation Chemistry.* Edited by M. Stanley Whittingham and Allan J. Jacobson. Academic Press, 1982.

———. "Lithium Titanium Disulfide Cathodes." *Nature Energy,* February 19, 2021.

Whittingham, M. Stanley, and Robert A. Huggins. "Beta Alumina—Prelude to a Revolution in Solid State Electrochemistry." In *Solid State Chemistry—Proceedings of the 5th Materials Research Symposium Sponsored by the Institute for Materials Research, National Bureau of Standards, October 18–21, 1971, Held at Gaithersburg, Maryland.* Edited by Robert S. Roth and Samuel J. Schneider Jr. U.S. Department of Commerce and National Bureau of Standards, 1972.

Williams, Stephen. *Spies in the Congo.* PublicAffairs, 2016.

Wood, Christopher. *The Bubble Economy.* Solstice, 2006.

Wrong, Michela. *In the Footsteps of Mr. Kurtz: Living on the Brink of Disaster in Mobutu's Congo.* Fourth Estate, 2000.

Yabili, Marcel. *Chine—RD Congo.* Vol. 1. *Chronique d'une colonisation choisie.* L'Harmattan, 2020.

———. *Chine—RD Congo.* Vol. 2. *Il manque un détail!* L'Harmattan, 2022.

Yang, Zeyi. "How Did China Come to Dominate the World of Electric Cars?" *MIT Technology Review,* February 21, 2023.

Yao, Yung-Fang Yu, and J. T. Kummer. "Ion Exchange Properties of and Rates of Ionic Diffusion in Beta-Alumina." *Journal of Inorganic and Nuclear Chemistry* 29, no. 9 (1967): 2453–75.

Yav, André. *Vocabulaire de ville de Elisabethville: A History of Elisabethville from Its Beginnings to 1965.*

Translated and edited by Johannes Fabian with Kalundi Mango. *Archives of Popular Swahili* 4, no. 2 (2001): lpca.socsci.uva.nl/aps/vol4/facsimile/toc.html.

d'Ydewalle, Charles. *L'Union minière du Haut Katanga: De l'âge colonial à l'indépendance.* Librairie Plon, 1960.

Zhang, Hua, Simeng Zhang, and Zheda Liu. "Evolution and Influencing Factors of China's Rural Population Distribution Patterns Since 1990." *PLOS One* 15, no. 5 (May 2020).

Zhu, Wei Xing, Li Lu, and Thérèse Hesketh. "China's Excess Males, Sex Selective Abortion, and One Child Policy: Analysis of Data from 2005 National Intercensus Survey." *BMJ* (Clinical Research Edition) 338, no. 7700 (April 2009): 920–23.

IMAGE CREDITS

INDEX

Italicized page numbers indicate material in photographs.